An Introduction to
The Botany of Tropical Crops

An Introduction to
The Botany of Tropical Crops

Second edition

Leslie S. Cobley

revised by **W. M. Steele**

Longman
London and New York

Longman Group Limited London

*Associated companies, branches and representatives
throughout the world*

*Published in the United States of America
by Longman Inc., New York*

© Longman Group Limited 1976

All rights reserved. No part of this publication may be
reproduced, stored in a retrieval system, or transmitted
in any form or by any means, electronic, mechanical,
photocopying, recording, or otherwise, without the
prior permission of the Copyright owner.

*First published 1956
Third edition 1977*

Library of Congress Cataloging in Publication Data

Cobley, Leslie, S.
    An introduction to the botany of tropical crops.

    Includes bibliographical references and indexes.
    1. Tropical crops.   2. Botany — Tropics.
    I. Steele, W. M., joint author.   II. Title.
SB111.C62 1976   581.6'1'0913   76—7447
ISBN 0 582 44153 6

Set in IBM Journal 10 on 11pt
and printed in Great Britain by
Lowe & Brydone Printers Limited,
Thetford, Norfolk

Preface to the first edition

A surprising diversity of plants is grown throughout the tropics for economic purposes and the object of this book is to survey the botanical features of the more important ones. The tropical crops range from those which have been cultivated for thousands of years and which, during that time, have evolved into a wide range of varieties differing markedly from their wild ancestors to those plants whose economic possibilities have been explored only recently and whose potentialities as crop plants have hardly begun to be exploited. They range from the crops which grow happily in the hot arid parts of the great land masses lying within the tropics to those crops which originated in the warm humid tropical forests and which demand a similar environment for their cultivation. They range from the major cash crops of the tropics which usually require elaborate and expensive processing to the simple plants grown in every peasant's garden or small holding under conditions only slightly better than those provided by the natural environment. When one considers the vastly differing environments included between the tropics of Cancer and Capricorn, the ranges of temperature encountered, the rainfall differences, the different altitudes at which cultivation is carried on, the diversity of soil types, the occurrence of large land masses and the hundreds of small islands, this large assemblage of economic plants seems less surprising.

With increasing ease of communications and transport, and increasing demands of a growing world population focusing more and more attention on the tropics as a potential source of food and raw materials, the time seems appropriate for a review of the present day tropical crops. With such a wide range of plants to cover, treatment has, of necessity, been superficial and little attention has been given in the text to those plants which offer possibilities for future development as crop plants. Except for some of the major crops such as cotton, sugar cane, cacao, tea, rice, rubber and the oil-palm, many of the plants grown for economic purposes in the tropics have not yet been developed to their full potentiality as crop plants, and in the case of many of the more recently developed crops the selection or breeding of the ideal genotype for any given environment can hardly be said to have been attempted.

The present text is based on 14 years' experience in various parts of the wet and dry tropics, the last 9 of which have been spent in lecturing in tropical crop botany to students of the Faculty of Agriculture of the University College of Khartoum. It has been designed mainly for use as a text by the increasing numbers of agricultural students, both in the tropics and elsewhere, and as an aid for all those who deal with tropical crops and their products. The various crops have been classified under the headings of cereals, fibres, oil-seeds, pulses, root crops, spices, beverages and drugs, fruits, vegetables, rubber and the essential oils. Such a method of classification has much to commend it when surveying this large and widely differing range of plants. From the purely botanical point of view it might have been more appropriate to adopt the natural system of classification of the Angiosperms as a general framework, but the fact that the examples are retricted to tropical crops means that the student gains little appreciation of the inter-relationships of the different families, nor of the evolutionary trends exhibited by such a system. The artificial system of classification on the basis of crop products still allows for the major families to be discussed in detail, and experience has shown that it provides a much more effective framework for a botanical survey of the tropical crops.

Although all the most important and many of the minor crops have been dealt with in this work there are several omissions, especially among those crops which are as yet of little but local importance. The tropical gums and resins have been omitted as being more in the nature of forest products: this group of plants has recently been very adequately dealt with in Howes' monograph, 'Vegetable Gums and Resins'. The omission of any reference to tropical forage and fodder crops is perhaps more serious; this is such a wide subject that a work of equal size would be required to provide an adequate survey. Limitations of space have also been responsible for the omission, except where necessary, of any description of the internal anatomy of the plants, nor has any mention been made of actual or potential yields. Perhaps such omissions are justified in a book intended as an introduction to the range of tropical crops and their main botanical features — as such it is offered to all those who are, or intend to be, concerned with tropical agriculture.

A large number of people have been of assistance in the preparation of this book. I would like to acknowledge my indebtedness to Mr J. L. Read, Senior Lecturer in Agriculture at the University College of Khartoum, for his unfailing help, encouragement and criticism during the writing of the text; to Dr E. E. Cheesman of the Agricultural Research Council for his comments and criticisms on the chapter on 'Fruits' and of the section on cacao; to Dr F. N. Howes of the Royal Botanic Gardens, Kew, for his advice on the chapters relating to fibre crops, vegetables, legumes, beverages and drugs; to Professor T. P. Hilditch of the University of Liverpool for his help with the chapter on 'Oil-seeds'; to Professor G. E. Blackman of the Department of Agriculture, Oxford University, for his comments on the chapter on 'Cereals'; to Dr H. H. Mann of the Woburn Experimental Station for his criticisms and comments on the section on millets; to Dr E. D. C. Baptist of the Rubber Research Institute of Malaya for his comments on the chapter devoted to 'Rubber'; to Dr T. Eden of the Tea Research Institute of East Africa for his help with the section on tea; to Mr E. Brown of the Colonial Products Laboratory for his advice on the chapters relating to 'Spices' and 'Essential oils', and to Miss R. M. Johnson for her comments on the chapter on 'Sugar cane'.

Finally my sincere thanks are due to Mrs M. F. Hubbard who typed the manuscript, to Mrs R. Brain for her help with the completion of the typescript, and to my wife for all her help with the proofs.

Preface to the second edition

The 18 years which have passed since the first edition of Mr Cobley's book was published have been a time of unprecedented activity in the study of tropical crops, and of development in tropical agriculture. A very large amount of new knowledge, and a vast new literature about tropical crop botany has accumulated since the mid-1950s as a result of the work of agricultural scientists in every specialized discipline. The need to revise this book in keeping with these developments has become urgent in recent years.

Perhaps the chief criticism of the first edition was its scant reference to botany in the context of practical farming. To correct this omission some colleagues urged me to make major changes in the revised edition, and in particular to decrease the number of crops covered in order to concentrate on important principles by reference to fewer crops in each of the major groups which constitute the chapters. I have tended to agree with others who believe that the first edition of *An Introduction to the Botany of Tropical Crops* has played a restricted, but very useful, role in undergraduate teaching, and that it should be revised in the same format to meet the same continuing need. However, I have added a new first chapter which includes a section about the tropical environment and about farming systems which I hope will go some way towards meeting the demand for a practical approach to crop botany in the book. Under the individual crops references are made to agronomy, physiology, genetics and breeding on the assumption that students will be able to discuss them with their teachers, because the limitations of space do not allow more than brief mention of these topics. It has been my experience that some discussion of these important aspects of agricultural botany, though not perhaps an essential part of an introductory course, does help to hold the students' interest while they learn what to many are the dull facts of crop morphology. It is apparent that students in the laboratory find Mr Cobley's simple line drawings easy to understand and useful in practical classes, and so their style has been retained, though they have been redrafted by the publisher's artists.

The principal reference work for students of tropical crop botany is Purseglove's *Tropical Crops* published by Longman, and it is important that it should be available to them. Of the many journals and books relevant to the subject it is difficult to select a few which might be recommended for an introductory course. *Economic Botany* must be one, and students will find the review articles in *Field Crop Abstracts*, books in the Longman 'Tropical Agriculture' series and the Leonard Hill 'World Crops' series very useful.

Throughout this revised edition, and in the Appendix, crop production data are taken from the FAO 'Production Yearbook, 1973'. I am indebted to the FAO as well for permission to reproduce their photographs in many new figures; to the Cotton Research Corporation for Figs 10.1 and 10.6; to the Cocoa Research Institute of Nigeria for Fig. 8.8 from their publication No. 59, *Kola Tree*; and to the Director, Institute for Agricultural Research, Ahmadu Bello University, Zaria, Nigeria, for Fig. 2.19.

Finally, I would like to thank my colleagues in the Department of Agricultural Botany at Reading, and Professor A. H. Bunting of Reading, for their advice and encouragement, and Dr W. Clayton of the Royal Botanic Garden, Kew, for his advice on the taxonomy of the *Gramineae*.

W.M.S.

August 1975 *Renhold, Bedford*

Contents

PREFACE TO THE FIRST EDITION   v
PREFACE TO THE SECOND EDITION   viii
LIST OF PHOTOGRAPHS   xiv

1   Introduction   1

A — The tropical environment   1
    Solar radiation   2
    Temperature   2
    Day-length   3
    Rainfall   4
    The root and tuber farming system   6
    The cereal farming system   6
B — The origin of crop plants   9
Further reading   14

2   The Cereal Crops   16

The family *Gramineae*   16
    Rice: *Oryza sativa*   26
    Maize, corn: *Zea mays*   33
    Grain sorghum: *Sorghum bicolor*   43
The millets   49
    Pearl or bulrush millet: *Pennisetum americanum*   50
    Foxtail millet: *Setaria italica*   54
    Common millet; proso: *Panicum miliaceum*   55
    Japanese barnyard millet: *Echinochloa frumentacea*   56
    Finger millet: *Eleusine coracana*   57
    Hungry rice: *Digitaria exilis*   60
    Teff, Ingera: *Eragrostis tef*   61
Further reading   61

3   Sugar Cane: *Saccharum* spp.   64

Further reading   70

4   The Legumes   71

The sub-family *Papilionoideae*   73
Root nodules   76

The grain legumes	79
✗Groundnut, peanut, monkey nut: *Arachis hypogaea*	80
✗Soyabean: *Glycine max*	85
Phaseolus species	87
Common bean: *Phaseolus vulgaris*	88
Lima bean: *Phaseolus lunatus*	89
Tepary bean: *Phaseolus acutifolius*	90
Mat bean: *Phaseolus aconitifolius* (now *Vigna aconitifolia*)	91
Adzuki bean: *Phaseolus angularis* (now *Vigna angularis*)	91
Vigna species	91
➤ Cowpea: *Vigna unguiculata*	92
Black gram: *Vigna mungo*	95
Green gram: *Vigna radiata*	95
Hyacinth bean: *Lablab purpureus*	96
Chick pea: *Cicer arietinum*	98
Lentil: *Lens esculenta*	98
Pigeon pea: *Cajanus cajan*	99
Bambarra groundnut: *Voandzeia subterranea*	101
Sword bean: *Canavalia* spp.	102
Cluster bean: *Cyamopsis tetragonolobus*	104
Winged bean: *Psophocarpus tetragonolobus*	105
✗Forage and pasture legumes, cover and shade crops	105
Other legumes	108
Yam bean: *Pachyrrhizus* spp.	108
Derris: *Derris elliptica*	108
Further reading	109

5   Root and Tuber Crops   111

✗Sweet potato: *Ipomoea batatas*	111
✗Cassava: *Manihot esculenta*	116
⋎Yams: *Dioscorea* spp.	119
Dasheen: *Colocasia esculenta*	123
Tannia: *Xanthosoma sagittifolium*	126
Arrowroot: *Maranta arundinacea*	126
Sago: *Metroxylon* spp.	128
Further reading	129

6   Vegetable Crops   131

The *Cucurbitaceae*	133
Squashes, pumpkins, marrows: *Cucurbita* spp.	134
Sweet melons and cucumbers: *Cucumis* spp.	138
Water melon: *Citrullus lanatus*	139
The *Solanaceae*	142
Egg-plant, aubergine: *Solanum melongena*	142
Tomato: *Lycopersicon esculentum*	143
The *Cruciferae*	145
Purslane: *Portulaca oleracea*	145
Jew's mallow: *Corchorus olitorius*	146
Okra, lady's finger: *Hibiscus esculentus*	146
Onion: *Allium cepa*	148
Further reading	151

7   The Cultivated Tropical Fruits (and Nuts)   153

⋎Bananas: *Musa* spp.	153
Citrus fruits: *Citrus* spp.	159

xii Contents

    Date palm: *Phoenix dactylifera*   168
    Mango: *Mangifera indica*   174
    Pineapple: *Ananas comosus*   178
    Pawpaw: *Carica papaya*   181
    Guava: *Psidium guajava*   185
    Mangosteen: *Garcinia mangostana*   187
    Breadfruit: *Artocarpus altilis*   188
    Avocado pear: *Persea americana*   190
    Nuts — Cashew: *Anacardium occidentale*   192
    Further reading   193

8 Beverage, Masticatory and Drug Plants   195

    The beverages   196
        Tea: *Camellia sinensis*   196
        Coffee: *Coffea* spp.   202
        Cocoa: *Theobroma cacao*   207
        Maté: *Ilex paraguariensis*   214
    The masticatories   214
        Kola: *Cola* spp.   214
        Betel-palm and betel-pepper: *Areca catechu* and *Piper betle*   217
    Drugs and fumitories   218
        Tobacco: *Nicotiana tabacum*   218
        Bhang, ganja, marijuana, hashish: *Cannabis sativa*   222
        Quinine: *Cinchona* spp.   222
        Opium: *Papaver somniferum*   225
    Other tropical drug plants   226
    Further reading   227

9 The Spices   230

    Ginger: *Zingiber officinale*   230
    Turmeric: *Curcuma domestica*   232
    Cardamom: *Elettaria cardamomum*   233
    Cinnamon: *Cinnamomum zeylanicum*   235
    Pepper: *Piper nigrum*   236
    Vanilla: *Vanilla fragrans*   238
    Clove: *Eugenia caryophyllus*   240
    Allspice: *Pimenta dioica*   243
    Nutmeg and mace: *Myristica fragrans*   244
    Chilli peppers: *Capsicum* spp.   245
    The umbelliferous spices   248
    Further reading   250

10 Vegetable Fibres   252

    The surface hairs   253
        Cotton: *Gossypium* spp.   253
        The cultivated cottons   258
    The bast fibres   268
        Jute: *Corchorus capsularis* and *C. olitorius*   269
    Other bast fibres   272
        Kenaf: *Hibiscus cannabinus*   273
        Sunn hemp: *Crotalaria juncea*   276
        Ramie: *Boehmeria nivea*   278
        Aramina: *Urena lobata*   279

Contents xiii

The structural fibres 280
  Sisal: *Agave sisalana* 280
  Henequen: *Agave fourcroydes* 283
  Abaca: *Musa textilis* 284
  Other structural fibres 286
  Further reading 287

11 The Vegetable Oils and Fats 289

Drying oils 291
  Safflower: *Carthamus tinctorius* 291
  Niger seed: *Guizotia abyssinica* 294
  Tung: *Aleurites montana* and *A. fordii* 294
Other drying oils 297
Semi-drying oils 298
  Sesame: *Sesamum indicum* 299
Non-drying oils and fats 301
  Castor oil: *Ricinus communis* 302
  Coconut oil: *Cocos nucifera* 306
  Oil-palm: *Elaeis guineensis* 312
  Cocoa butter: *Theobroma cacao* 320
  Shea butter: *Butryospermum paradoxum* 320
Further reading 321

12 The Essential Oil Crops 323

Citrus oils: *Citrus* spp. 325
Geranium oil: *Pelargonium* spp. 325
Camphor and camphor oil: *Cinnamomum camphora* 326
Ylang-ylang and cananga oils: *Cananga odorata* 328
Patchouli oil: *Pogostemon cablin* 329
Bay oil: *Pimenta racemosa* 329
Basil oil: *Ocimum basilicum* 330
Cymbopogon spp. 331
  Lemon grass oil: *Cymbopogon citratus* 331
  Malabar lemon grass oil: *Cymbopogon flexuosus* 331
  Citronella oil: *Cymbopogon nardus* 333
  Palmarosa oil, gingergrass oil, rosha oil: *Cymbopogon martinii* 334
  Vetiver oil: *Vetiveria zizanioides* 334
Further reading 335

13 Rubber 336

Para rubber: *Hevea brasiliensis* 338
Further reading 344

APPENDIX: CROP PRODUCTION DATA 345

INDEX OF SCIENTIFIC NAMES 357

GENERAL INDEX 362

List of photographs

2.7 *Oryza sativa*: Rice. Harvesting an experiment conducted in Indonesia by the International Rice Research Institute. 27
2.8 *Oryza sativa*: Rice. Recently transplanted paddy in an experiment at the International Institute of Tropical Agriculture, Ibadan, Nigeria. 29
2.9 *Oryza sativa*: Rice. Mature paddy in an experiment at Ibadan, Nigeria. 30
2.11 *Oryza sativa*: Rice. A bunch of mature panicles harvested in Indonesia. 32
2.12 *Zea mays*: Maize. A standing crop in flower in the Yemen Arab Republic. 35
2.14 *Zea mays*: Maize. The male inflorescence or 'tassel'. 37
2.17 *Zea mays*: Maize. Young female inflorescences, the 'cobs' or 'ears' with the styles or 'silks' exserted. 41
2.19 *Sorghum bicolor*: Sorghum. Indigenous tall sorghum ('Guineacorn') and recently bred dwarf sorghum in north Nigeria. 44
2.20 *Sorghum bicolor*: Sorghum. Mature panicles of the Nigerian 'Guineacorn' cultivar *'Farafara'*. 46
2.22 *Pennisetum americanum*: Pearl millet. Condensed panicles with mature grain. In the cultivars illustrated the bristles which subtend spikelet groups are short and not visible. 51
2.24 *Pennisetum americanum*: Pearl millet. Panicles from three cultivars in which the bristles subtending spikelet groups are long. 53
2.25 *Eleusine coracana*: Finger millet. The digitate inflorescence of spikes of two cultivars. 58
3.1 *Saccharum officinarum*: Sugar cane. Mature canes in Queensland, Australia. 65
4.5 *Arachis hypogaea*: Groundnut. A young crop growing on ridges made with a tractor-drawn implement in north Nigeria. 81
4.7 *Arachis hypogaea*: Groundnut. A plant lifted to show immature fruits, the 'nuts'. 84
4.8 *Arachis hypogaea*: Groundnut. A mature crop which has

	been lifted and gathered into stacks to dry before picking the 'nuts' from the haulms. North Nigeria.	85
4.10	*Phaseolus lunatus*: Lima beans. Lima beans in experiments at the International Institute of Tropical Agriculture at Ibadan, Nigeria.	90
4.11	*Vigna unguiculata*: Cowpea. An upright cowpea cultivar with pods held erect on long peduncles.	93
4.15	*Voandzeia subterranea*: Bambarra groundnut. A yield trial in north Nigeria.	101
4.17	Leguminous forage and cover crops.	106
5.1	*Ipomoea batatas*: Sweet potato. Variation in the leaf form of sweet potatoes.	113
5.3	*Ipomoea batatas*: Sweet potato. Carrying sweet potatoes to market in Indonesia.	115
5.4	*Manihot esculenta*: Cassava. Recently harvested cassava tubers and cassava plants in Nigeria.	117
5.5	*Dioscorea rotundata*: White guinea yam. A white guinea yam twining up a stake in Nigeria.	122
5.6	*Colocasia esculenta*: Cocoyam. A cocoyam growing at Ibadan, Nigeria.	124
6.1	*Cucurbita maxima*: Pumpkin or winter squash. A young pumpkin plant.	135
6.4	*Citrullus lanatus*: Water melon. Water melons in Saudi Arabia.	140
6.6	*Hibiscus esculentus*: Okra. A crop of okra with flowers and young fruits.	147
6.7	*Allium cepa*: Onion. Onion cultivation near Khartoum, Sudan.	149
7.1	*Musa* spp.: Banana. A banana plantation in Jamaica.	155
7.3	*Musa* spp.: Banana. A good commercial bunch of banana fruits.	158
7.4	*Citrus sinensis*: Sweet orange. An orchard of orange trees in north Nigeria.	160
7.7	*Citrus paradisi*: Grapefruit. Mature grapefruits ready for picking.	165
7.8	*Citrus sinensis*: Sweet orange. An orange tree bearing fruit in Indonesia.	166
7.9	*Phoenix dactylifera*: Date palm. A date palm in the northern Sudan on which the basal offshoots have been allowed to develop naturally.	170
7.11	*Phoenix dactylifera*: Date palm.	173
7.12	*Mangifera indica*: Mango. Mango trees in Tanzania.	175
7.14	*Mangifera indica*: Mango. 'Julie' mangoes in Trinidad.	177
7.15	*Ananas comosus*: Pineapple. A pineapple plantation.	179
7.16	*Carica papaya*: Pawpaw. A pawpaw tree in full fruit.	182
7.18	*Carica papaya*: Pawpaw.	184

7.19	*Artocarpus altilis*: Breadfruits.	189
7.20	*Persea americana*: Avocado. The foliage of the avocado tree.	190
8.1	*Camellia sinensis*: Tea. Tea planted on the contour at the Tea Research Institute, Sri Lanka.	196
8.2	*Camellia sinensis*: Tea.	199
8.3	*Camellia sinensis*: Tea. A tea bush grown to a convenient plucking height.	200
8.4	*Camellia sinensis*: Tea. Pruning the tea bush.	201
8.6	*Coffea arabica*: Coffee.	206
8.8	*Cola* sp. A pendent cluster of mature kola follicles.	215
9.1	*Zingiber officinale*: Ginger. Weeding ginger in Liberia.	231
9.2	*Piper nigrum*: Pepper. A spike of mature drupes.	237
9.4	*Capsicum annuum*: Chilli pepper. A chilli pepper bearing a heavy crop of immature fruits.	246
10.1	*Gossypium hirsutum*: Upland cotton. A mature crop of upland cotton, Nigerian Allen, cultivar Samaru 26J in Nigeria.	259
10.6	*Gossypium* spp.: Cotton. Lint hairs combed out to illustrate variation in 'staple length'.	267
10.9	*Hibiscus cannabinus*: Kenaf. Flowers.	273
11.3	*Aleurites montana*: Tung. Flowers.	296
11.6	*Ricinus communis*: Castor. Inflorescences and immature fruits on dwarf castor in Brazil.	305
11.7	*Cocos nucifera*: Coconut. Coconut palms growing in Jamaica.	307
11.9	*Cocos nucifera*: Coconut.	310
11.10	*Cocos nucifera*: Coconut. Sun drying copra in Kenya.	311
11.11	*Elaeis guineensis*: Oil-palm. A naturalized community of oil-palms near Ibadan, Nigeria.	313
11.12	*Elaeis guineensis*: Oil-palm. A young specimen tree grown as an ornamental in Ibadan, Nigeria.	314
11.15	*Elaeis guineensis*: Oil-palm. A bunch of ripe fruits (a single female inflorescence) gathered together after harvest in Nigeria.	319
12.1	*Cymbopogon* sp.: Lemon grass.	332
13.1	*Hevea braziliensis*: Rubber. A rubber plantation in Malaysia	339
13.2	*Hevea braziliensis*: Rubber.	341
13.3	*Hevea braziliensis*: Rubber. Tapping a rubber tree.	343

Chapter 1

Introduction

A feature of agriculture in the tropics is the diversity of its crops, and the large numbers of species, genera and families to which they belong. On the other hand a relatively small number of species in few families account for most crop production in temperate agriculture, and it is apparent that we must study tropical crops if we are to have an unrestricted view of the botany of economic plants, and of their role in the affairs of man. In temperate agriculture the principal crops over large areas are the cereals wheat and barley, forage and pasture grasses and legumes, few grain legumes, the Brassica crops of the cabbage family *Cruciferae*, root crops of the genus *Beta* (family *Chenopodiaceae*), and the potato (*Solanum tuberosum*, family *Solanaceae*). In contrast large numbers of cereals, grain legumes, roots, tubers and fruits are grown for food in tropical agricultures, and tropical crops from many families provide commodities such as vegetable fibres, oils and fats, beverages, rubber, drugs, spices and essential oils which are important in world trade.

The first section of this introductory chapter describes the environment in which all these tropical crops grow. The second section deals briefly with the origin and dispersal of crop plants.

A – The tropical environment

To the physical geographer the 'tropics' are those parts of the world lying within the limits of the tropics of Cancer and Capricorn latitude 23½° N. and S. of the equator; but the environments which are suitable for the growth of tropical crops are not restricted within such well-defined geographical limits. Like the environment for plant growth everywhere, they are determined mainly by the amount and distribution of annual rainfall, and of solar radiation, which in turn determines temperature. Furthermore, in undeveloped tropical agricultures peasant and subsistence farmers have very few resources; so the environment for the growth of their crops is often influenced to a much greater extent than in advanced temperate agriculture by the capacity of the soil to provide nutrients and to hold water, and by the incidence of insect pests, diseases and weeds.

It is convenient to discuss the tropical environment in terms of the main features of its climate, and to discuss each of these separately, though they interact in a complex way to influence the growth, development and distribution of crop plants.

Solar radiation

Of all environmental factors solar radiation is most important in agriculture because it is the source of energy used by plants in photosynthesis. The amount of solar energy received at the earth's surface (insolation) varies with latitude, and with season of the year, and on a world scale it is the chief determinant of dry matter production in plants. This is not only because photosynthesis depends upon insolation, but also because major variations of temperature over the world, and seasonal variation at any place, depend upon variation in the amount of solar energy received. Furthermore, the reproductive development of photosensitive crops depends upon changing day-length which is consequently one of a complex of environmental factors which affect their distribution.

The amount of solar energy received at the earth's surface each day depends upon the intensity of the radiation, which varies with the sun's elevation and the amount of cloud cover, and upon the length of day which varies with latitude and with the seasons of the year. The amount is least at extreme high latitudes near the poles where there is no solar radiation at all for part of the winter, and it is so cold all year that there is no agriculture. At around latitude $50°-60°$ N. of the equator, in a temperate climate such as that in Great Britain, it varies widely from around 50 cal/cm^2/day in short winter days to as much as 450 cal/cm^2/day on a long, clear day in mid-summer. In the relatively long, cloud-free summer days of the sub-tropics (for example, in the Mediterranean region) the amount of solar energy received may be as great as 750 cal/cm^2/day. Within the tropics the amount received is greatest in semi-arid regions during the winter dry season, when there is no agriculture without irrigation; but during the summer rains in these areas the amount received varies, depending upon cloud cover, from around 200 cal/cm^2/day to as much as 450 cal/cm^2/day. Insolation is least variable in the wet, equatorial tropics where the amount received is 400–450 cal/cm^2/day throughout the year.

Although the amount of solar energy received by crops is beyond the control of farmers (except of course that they can provide shade), the efficiency with which crops utilize it to produce dry matter, and especially the proportion of this dry matter which goes into economic yield, varies between the cultivars of a species, and may be influenced by aspects of crop husbandry such as sowing date and plant density.

Temperature

Variation of temperature between places, and seasonal variation at any place, tend to follow variation of insolation, and the tropical environment is hot compared with temperate climates. Indeed, a major distinction between the environment for crop growth in high-latitude temperate regions and the tropics is that the duration of the growing season is limited by winter cold in much of temperate agriculture, but by winter drought (often accompanied by very hot days) in much of the tropics. In the lowland equatorial tropics there is little seasonal or diurnal (day/night) variation from a mean temperature of around $28°-30°C$, but in the regions of summer rainfall to the north and south of the equator both seasonal and diurnal variations of temperature are relatively large. In these areas the coolest temperatures of $10°C$ or less are at night in the early to mid-dry season, and the hottest ($40°C$ or hotter) are soon after mid-day in the weeks before the summer rains begin, when the sun is high and there is little or no cloud cover. In these seasonally dry regions there is least diurnal variation in temperature during the growing season when the mean commonly varies around $25°C$.

Temperatures become cooler everywhere with increasing altitude, and so on high plateaux and mountains in the tropics, especially near the equator as in East Africa, cool temperatures are associated with tropical day-length and solar energy receipt. These unusual circumstances make it possible to grow a combination of temperate and tropical crops such as tea (at least where there is enough rain and a suitably deep, acid soil), coffee, pyrethrum, wheat, maize, common beans, bananas, sweet potatoes and 'Irish' potatoes. Though some temperate crops can be grown in the hot tropical lowlands, others cannot be grown anywhere in the tropics, not even in the cool highlands, because they flower and produce fruit only after they have been 'vernalized' by a period of exposure to cold temperature, and after they have experienced long days of 14 hours or longer. On the other hand, crops which have evolved in the tropics are killed by frost (tea is in one respect an exception because, in Russia and China, it survives frosts in a dormant condition). Relatively small variations of a few degrees of temperature, as well as their often subtle effects on the vegetative growth of plants, may also influence their development and so affect their distribution. For example, in some cultivars of cowpeas and soyabeans flowering is markedly delayed by cool nights. Tea and arabica coffee are most productive in the cool tropical highlands; though they grow vigorously in the hotter lowlands they do not give good economic yields there.

Day-length

The length of day (roughly from sunrise to sunset) varies with the seasons and with latitude. Seasonal variation is greatest at high latitudes and least at the equator, where the day-length is constant all year. In the northern hemisphere the longest day is 21 June and the shortest is 22 December (the summer and winter solstices) when the sun is furthest north and south of the equator. The opposite seasonal variation of day-length occurs in the southern hemisphere. On the vernal and autumnal equinoxes (20 March and 22 or 23 September) the sun is overhead at the equator in its passage north and south, and everywhere the days are 12 hours long. Thus at extreme high latitudes near the poles the length of day varies from 24 hours in mid-summer to 0 hours in mid-winter, while at the equator it is 12 hours all year round. The greatest seasonal variation at the highest latitudes within the tropics of Cancer and Capricorn is from winter days of about 9 hours to summer days of about 15 hours.

Many temperate crops are reproductively long-day plants, but tropical crops are either apparently insensitive to day-length (for example, groundnuts, annual cottons, cocoa, some varieties of rice and some grain legumes), while others flower or produce tubers only in response to short or shortening days (for example, some sorghum and pearl millet cultivars, some rice, sweet potatoes and some grain legumes). Here it must be borne in mind that, though we refer to the day-length requirements of plants for flowering, it is the length of the night which is physiologically important in photoperiodism.

Cultivars of photosensitive crops, whether they are temperate or tropical, have day-length requirements for inflorescence initiation and flowering which tend to ensure that they flower and produce fruits close to the end of the growing season in the place where they evolved. They are said to be adapted or acclimatized to their local environment. By this means the fruits and seeds of tropical short-day crops mature free from moulds in the dry weather which follows the end of the rainy season, but while the soil holds stored water. Local adaptation of this kind limits the distribution of photosensitive cultivars, but it must be emphasized that it does not confine them only to the latitude and day-length regime of the place where they evolved. The most important feature of the local environment in terms of their adaptation is the relation in time between changing day-length and rainfall distribution. The way this restricts the distribution of crops can be illustrated by reference to a short-day sorghum cultivar which is locally adapted to the length of the growing season at latitude 12° N. in West Africa. There it flowers close to the average date when the rains end in September. Provided that it is sown no later than the end of July at latitude 12° N., such a cultivar will always flower at about the same time in September in response to shortening days; and it will do this whether it is grown at a place where the rains end so early that grain production fails from drought, or where they end so late that the grains are spoiled by moulds.

Rainfall

Tropical rainfall depends upon global pressure and wind systems. A distinct belt of low atmospheric pressure called the 'Horse Latitudes' or 'Doldrums' occurs around, or slightly to, the north of the equator. Though they are deflected by the earth's rotation, winds blow towards this zone from areas of high atmospheric pressure centred on land masses around latitude 20°–30° N. and S. of the equator where the world's great deserts occur. The student is referred to the books listed at the end of this chapter for detailed discussions of the way these features of the earth's atmosphere influence tropical rainfall, and of the way it is variously affected in different regions of the world by the position of continents in relation to the equator, and by the proximity of oceans, lakes and mountains. Here only the end result of the interaction of all these factors is described in terms of major variations of rainfall which are the most important environmental determinants of the distribution of farming systems, crop species and cultivars in the tropics.

In the low-pressure belt around the equator large amounts of rainfall (2,000 mm or more annually) are more or less evenly distributed throughout the year, and there is no time when plant growth is restricted by lack of water. These wet, lowland tropics are hot and humid with little seasonal variation of rainfall, insolation, temperature or day-length, and the wild vegetation is dense evergreen rain forest. This climax vegetation is typified by the jungles of the East Indies, Indonesia, Malaysia, Zaire, parts of West Africa, Brazil and parts of Central America. Moving towards the poles from these hot, wet climates there is a general tendency for the amount of annual rainfall to decrease, and for it to be unevenly distributed with one or two distinct periods each year when little or no rain falls. Correlated with this change there is a gradual transition from rain forest, through woodland and increasingly sparse savanna vegetation, to the most arid sahel and eventually desert regions. In the wettest of the seasonally dry climates rain falls throughout the year, but it is unevenly distributed in a bimodal pattern with two peaks of rainfall intensity each year, separated by intervals when there may be insufficient rain for crop growth, though perennial crops do survive without irrigation on deep soils which hold water. On the other hand, in the most arid seasonally dry climates there are 500 mm or less rain each year falling in a 3–4-month period during summer, and no rain, and so no crop growth without irrigation for the rest of the year.

The duration of summer rainfall in the seasonally dry tropics is broadly determined by periodic, annual north–south movements of a zone of climatic instability called the Intertropical Convergence Zone (the ITCZ; sometimes called the Intertropical Front or Equatorial Trough), where winds blowing towards the equator converge. It separates dry air moving from regions of high pressure towards the equator from moist air near the equator; as it follows the sun towards the poles it brings summer (high sun) rain to the seasonally dry regions. This seasonal pattern of rainfall distribution is seen in its least disrupted form north of the equator in West and Central Africa where variation in rainfall, and consequently in wild vegetation, is roughly correlated with latitude. Elsewhere the seasonal pattern is more or less disrupted by the geographical location of oceans, continents and mountains in relation to the equator.

These variations of rainfall result in a wide range of tropical environments for plant growth, and so in a diversity of wild vegetation, crop plants and farming systems. None the less, it is possible to account for most crop production in the tropics by reference to only two major environments and farming systems, the root and tuber farming system of the wet, equatorial tropics on the one hand, and the cereal farming system of the seasonally dry, summer rainfall tropics on the other.

The root and tuber farming system

In the lowland equatorial rain forests the main source of food energy comes from vegetatively propagated roots and tubers such as sweet potatoes, yams, cassava, dasheens, eddoes and tania, or from bananas and plantains and in some restricted maritime regions from coconuts. There are no cereal crops and very few grain legumes. Indeed, very few crops of any kind are grown for their seeds because there is no dry weather in which they can mature free from moulds. Human diets tend to be unbalanced and deficient in protein because the starchy staple foods contain so little, and because animal protein from fish, pigs and poultry is scarce. Cattle are rare, for though forage grasses can be cultivated successfully little natural grazing is available in the rain forests, and cattle bear a heavy load of disease, especially of trypanosomiasis which is spread by tsetse flies.

As well as the relatively poor quality of root and tuber diets, food is often scarce in the rain forests because cultivated soils there tend to be infertile, so crop yields tend to be small. In the undisturbed rain forest the soil is protected from the impact of heavy rainfall and from erosion by the tree leaf canopy and by a deep litter of leaves, and it is bound by the tree roots. Nutrients taken from deep in the soil by the trees are eventually returned to it when the tree leaves are shed in a more or less closed nutrient cycle. Arable farming, even on a small scale, requires forest clearing which commonly results in a rapid loss of nutrients from the soil by leaching, and of the soil itself through sheet and then gully erosion. In an attempt to maintain crop yields on these infertile soils many peasant farmers practice a system of shifting cultivation in which the land is left fallow for periods as long as 20 years after short intervals of a few years of cropping. Food shortages which result from small crop yields are aggravated by the generally poor keeping qualities of roots and tubers in store.

The cereal farming system

In marked contrast to the rain forests, cereals are the chief source of food energy in the seasonally dry tropics. Rice predominates in the monsoon regions of South-east Asia where there is enough rain each year for the cultivation of perennial crops; sorghum and the millets are the staple foods in the more arid seasonally dry regions where all the major crops are annuals propagated from seed, and except where irrigation water is plentiful, there are no cultivated perennial crops. Over most of these areas the staple cereal is one which has evolved locally, though maize introduced from the New World within the past 300 years has become very important in the least arid seasonally dry regions of Africa and Asia.

Cereal diets in general are nutritionally better than root and tuber diets, largely because the cereal grains contain around 10 per cent or more of protein, and grain legumes are eaten regularly, but also because cattle, sheep and goats are numerous over large parts of the semi-arid tropics where they are relatively free from diseases and parasites compared with rain forest zones. Though increased meat consumption would improve the nutritive value of most cereal diets, more meat is already available in them than in root and tuber diets.

The tropical cereal farming system has evolved with different associations of crops in different parts of the world. For example, maize and *Phaseolus* beans are associated in the ancient cereal farming system of the New World, whereas sorghum, pearl millet and cowpeas have evolved together in Africa and parts of India, and various grain legumes are grown in rotation with rice in Asia.

Though the predominant feature of the tropical environment where cereal crops are grown is an annual period of little or no rain, seasonally dry climates are also characterized by the unreliability of their rainfall. Thus, even though the success of the farming system which has evolved around rice in South-east Asia is indicated by the very dense human population which it supports, drought or floods occasionally result in food shortages or even famine in these monsoon regions. In the driest semi-arid regions the time when the summer rains begin and the duration and amount of annual rainfall are always uncertain. Furthermore, in such areas it is important to distinguish between the length of the 'rainy season' on the one hand, and the length of the 'growing season' when there is enough water in the soil for plant growth on the other. The growing season can be defined as the period each year when the amount of precipitation ('P', essentially rainfall) exceeds the amount of water lost from the soil by evaporation and transpiration ('E_t', evapotranspiration). Evapotranspiration is greatest in hot, dry air and is increased by wind; so the amount of water which may be lost from the soil as a result of it is potentially greatest in the most extreme semi-arid climates where least rain falls in the shortest period.

Within the seasonally dry tropics, wherever subsistence or peasant farmers depend entirely on hand labour and the hoe to cultivate their land, the farming calendar is determined by the distribution of summer rainfall. The cereal farming system of the West African savanna zone provides a good illustration of this, and of the way the ancient farming systems in such areas have evolved to minimize the adverse effects of uncertain and limited rainfall on the production of staple cereal crops in only one short growing season each year. It is also a good example of the very complex way in which many factors interact to influence the distribution of crops and cultivars.

The cereal farming system in West Africa

The staple sorghum and pearl millet crops are sown in West Africa as soon as the summer rains begin (even before P is greater than E_t) because their grain yields decrease markedly as sowing is delayed. Not only are late-sown crops subject to increased damage from populations of insect pests which have multiplied on earlier sown crops, and to increased competition from weeds, but they do not benefit from a flush of nitrogen which occurs in the soils of semi-arid regions when they are first wetted by rain after a long, hot dry season. Nitrogen from this source is soon lost to crops by leaching once the soil is saturated with rain water, and on these characteristically nitrogen-deficient soils the crops are unable to make good the loss from other sources (unless fertilizers are used), which tends to decrease grain yields. A few weeks after the cereals have been sown they are weeded and cowpeas are interplanted with them. Weeding the cereals requires the greatest labour input of the year on the peasant farms. Though the interplanted cowpeas and cereals decrease each other's yields in experiments where fertilizers are used and insect pests are controlled with insecticides, there is no evidence that they do so on peasant farms. The spreading growth of the legume smothers weeds and makes further weeding unnecessary, it protects the soil from the impact of heavy rain and so decreases rainfall run-off and soil erosion. Late in the season, when there is least nitrogen in the soil because of cumulative losses due to leaching and uptake by crops, the cereals may obtain some from the cowpea root nodules, and nitrogen from this source may increase cereal grain yields.

The cereal sowing dates vary from year to year by as much as 6 weeks in areas with least annual rainfall and the shortest growing season because the time when the summer rains begin is so variable; but whether they are sown with early or late rains both the cereals and the cowpeas interplanted with them flower close to the time when the rains end, and at more or less the same time each year because they are locally adapted by their response to photoperiod. This then is the significance of local adaptation in this farming system. To yield well the staple cereals must be sown with the first rains, and they must flower close to the time when the rains end so that their grains mature free from moulds in the early weeks of the dry season, but while there is water in the soil. Non-photosensitive cultivars which flower in a more or less fixed period of time after sowing, and photosensitive ones adapted to a different rainfall regime would not fulfil these requirements.

The subsistence or peasant farmer in this cereal farming system is committed to provide enough food for his dependants, and so gives priority in land, labour and time to establish his staple food crop, often at the expense of good yields from other crops like groundnuts and cotton which are sown too late to yield well. New, higher yielding cereal cultivars would relieve the peasant farmer of some of his burden of land cultivation because enough food could be obtained from them on less land. But so long as he must cultivate the land by hand such new cultivars must be as well adapted to the local environment as those they replace. Furthermore, they must also yield grain which is accepted as food as readily as the grain of those they replace. Larger grain yields, and other desirable improvements in new cultivars such as a nutritionally better balance of amino-acids in the cereal grains, cannot substitute for the important attributes of local adaptation and acceptability. Of course, the peasant farmer's burden of land cultivation in such a farming system would be greatly relieved if his labour could be replaced by other sources of power like tractors or draught animals, and if fertilizers were used to remedy nutrient deficiencies in the soil. Unfortunately solutions of this kind to the problem of increasing food production in this cereal farming system raise other problems such as the need to find alternative employment for the labour made idle by adopting them.

B – The origin of crop plants

The study of the origin and history of crop plants is not only a subject of great academic interest, but it is important also for a very practical reason. As we work to increase crop production, and to extend the range of new environments in which important crops can be grown, our knowledge of their places of origin, of their centres of genetic diversity, and of their wild and weedy relatives is invaluable in agricultural research. In particular it assists plant breeders in their search for useful new sources of genetic variation, whether they are concerned with breeding improved cultivars to meet the ever changing demands of advanced agriculture, or with increasing production and improving the quality of crops in undeveloped tropical agricultures.

All of our crops were domesticated from wild ancestors, and have since evolved in cultivation while subject to selection by man. Many of the most important ones have been evolving in this way for several thousands of years, while a few (such as coffee and rubber) are much younger, and have a different and better known history in cultivation. As cultivated populations of newly domesticated ancient crops became increasingly large, and were dispersed into new environments with migrating people, genetic diversity accumulated in them as a result of mutation and through the hybridization of the cultivars with their wild relatives. Most mutants were probably inferior types and failed to survive in these cultivated populations; but a few were favoured by man, or were unconsciously selected by him, and though they would not have survived in wild populations they became established in his crops. For example, when neolithic human communities had developed a primitive agriculture, and saved the seeds of their cereals to sow crops instead of gathering food grains from wild plants, cereals whose inflorescence had a non-brittle rachis were unconsciously selected because, unlike brittle rachis wild types, their grains were not readily shed until the crop had been harvested and was threshed. Natural hybridization between cultivars and their wild and weedy relatives has played an important role in the evolution of most of our crops, and it continues to do so in those which are still able to hybridize with their wild relatives, and so to exchange genes with them. On the other hand, when all or part of a crop population became geographically isolated from its wild ancestors during dispersal, or when it had accumulated so many gene and chromosome mutations that it no longer produced fertile hybrids in crosses with them, gene exchange with wild plants became impossible, and the crop evolved independently of them. Indeed, the wild ancestors of many crops cannot be identified because the crops have diverged so much from them since they were domesticated. A third major feature in the evolution of many crops has been polyploidy, which is well illustrated by bread wheat, cotton, sugar cane, finger millet and banana.

In recent times a new phase in the evolution of crop plants began with the application of modern scientific principles to plant breeding, though of course man has practised selection on his crop plants since time immemorial. Improved cultivars produced by selection and breeding have replaced ancient and diverse types of many crops in advanced temperate agriculture, and it is an important objective of agricultural development in the tropics to do the same with all major crops, and so to increase crop production. One consequence of these efforts is that the great genetic diversity now represented by vast numbers of 'unimproved' cultivars in tropical peasant agricultures is in danger of extinction. Fortunately the need to preserve this diversity in collections for present and future use by plant breeders is recognized.

Two kinds of evidence help us to determine, or to speculate about, the origins of agriculture and of crop plants. First, and most important, rare archaeological evidence of the remains of primitive crop plants and of primitive agricultural communities (especially their farming tools and the utensils they used to prepare and to cook their food) provide the only empirical information about the time and place of very early agriculture. Though such evidence has been obtained from many places in the world, two locations are outstanding sources, and have been very carefully studied. One covers an arc in Western Asia extending north from the foothills of the Zagros Mountains in Iran and Iraq to southern Turkey, and then south-west along the eastern Mediterranean basin. In this area, which is known as the 'fertile crescent', agriculture based upon the cultivation of primitive diploid (einkorn) and tetraploid (emmer) wheats, and on barley, began some 9,000 years ago, and eventually gave rise to the great civilizations of the eastern Mediterranean. The other is the Tehuacan Valley in Mexico where ancient remains of plants such as maize, *Phaseolus* beans, cucurbits and chilli peppers indicate that primitive agriculture was practised as long as 7,000 years ago. This kind of evidence seems to suggest that the earliest farming communities, and the earliest civilizations, were based upon the cultivation of cereals, and that they developed in the seasonally dry tropics and sub-tropics. We have relatively little information about the origins of agriculture or of crops in the wet, humid tropics, mainly because the environment there does not favour the preservation of archaeological evidence, especially of fleshy roots and tubers. We assume that agriculture in those regions began with root and tuber crops, but it seems unlikely that such crops were domesticated in the rain forests; rather, it seems more likely that they originated in the transition zone between the wet forests and the savannas. In that environment the short period each year when there is insufficient rain for plant growth would have favoured the evolution of crops with food storage organs which enabled them to survive such adverse conditions.

The second chief source of information about the origin and history of crop plants is genetic, and depends upon studying the distribution of genetic diversity in crops and their wild and weedy relatives, and upon trying to determine which wild species are most closely related to the crops, and if possible which were their ancestors. Such studies suggest that crop plant domestication has been concentrated in several distinct regions of the world, and that these were centres from which many of our crops were dispersed after they were domesticated. These areas were first delineated in the 1920s by the great Russian botanist, Vavilov, who called them 'Centres of Origin' of crop plants. He considered that centres of genetic diversity were probably also centres of origin, but since Vavilov proposed this concept evidence has accumulated which suggests that it must be slightly modified. For example, it seems likely that some crops had a 'diffuse' origin, and were domesticated over a large area within the range of their wild progenitors, and not in any clearly defined or distinct 'centre'. Others may have been domesticated at about the same time in at least two distinct areas. Furthermore, many crops have become genetically diverse in regions which could not have been their centres of origin because they are outside the geographical range of the wild progenitors (unless the wild species have become greatly restricted in distribution since the cultivars were derived from them). Consequently, when we are able to identify the wild ancestor or ancestors of a crop we may be able to distinguish between two kinds of 'centre'. On the one hand there is a 'primary' centre of origin, and often of genetic diversity as well, which occurs within the range of the wild progenitor, and on the other there may also be a 'secondary' centre of genetic diversity, but not of origin, which occurs outside the wild progenitor's range, and to which the crop spread after it was domesticated.

The main centres of origin of crop plants delineated by Vavilov, and some of the important crops which were domesticated in them are:

1. *China*. Soyabean, adzuki bean, orange.

2. *The Indian sub-continent.* Rice, tea, jute, pepper, some *Phaseolus* species.
3. *The Indo-Malaysian peninsular and South-east Asia.* Sugar cane, coconut, banana, some yams.
4. *The Near and Middle East, including Mesopotamia.* Wheat, barley, rye, lucerne, lentil, chick pea.
5. *The Mediterranean region.* Cabbage, lettuce, olive, broad bean.
6. *The Ethiopian region.* Sorghum, pearl millet, cowpea, arabica coffee, finger millet, teff.
7. *South America west of the Andes Mountains, parts of Central America and south Mexico.* Maize, some cucurbits, upland cotton, sisal, chilli peppers, potato, sweet potato, tomato.
8. *South America east of the Andes Mountains.* Groundnut, cassava, cocoa, rubber, pineapple.

Examples of crop dispersal which have not involved human migration or trade appear to be rare. Long before there was any maritime trade between Asia and the New World coconut fruits may have floated across the Pacific ocean from Asia to the western coast of Central America; and perhaps the capsules of sweet potatoes crossed the Pacific in the same way, but in the opposite direction in pre-Columbian times, though an alternative theory suggests that these crops were carried with human migration across the ocean. It is convenient to distinguish several phases in the history of crop plant dispersal by man. At the earliest stage after they were domesticated our ancient crops spread with the expansion and migration of primitive agricultural communities whose movements were restricted to the environments where their crops could grow, and by the physical barriers of mountain ranges, deserts and oceans.

As human communities developed over several millenia, a second and more important phase in crop dispersal was associated with very early trade, with conquest, and with more widespread human migration. By these means crops spread within the Old and New Worlds, but significantly (with the possible exceptions mentioned above) not between them in pre-Columbian times. Important early trade routes were the 'Silk Route' through Central Asia along which spices, silk and other 'Oriental' produce first reached Europe; and the 'Sabaean Lane', a maritime route in the Arabian Sea along which the Sabaean peoples of southern Arabia traded between East Africa and the western coasts of the Indian sub-continent as long as 3,000 years ago. This ancient trade, using vessels called 'dhows' which have changed little over the centuries, continues to the present day. Examples of crop dispersal with human migration are the ancient spread of wheat and barley to the Ethiopian region with migrants from the Caucasus some 5,000 years ago, and much more recently, the dispersal of crops which accompanied the Arab Islamic expansion into north Africa and as far as Spain about 1,200 years ago.

The third, and the most important phase in the history of crop dispersal, began with attempts by European maritime powers (especially the Portuguese and Spanish) to find a sea route to Asia. They wanted to obtain valuable Asian produce which had previously been available only by overland trade along the 'Silk Route' through South-west Asia. The first unsuccessful attempt to do this was made by the Genoese in 1291, but it was not until two centuries later in 1487–8 that the Portuguese were first to sail around the southern tip of Africa from Europe. The discovery of this route to the East enabled them to begin trade with India in 1500, with the Moluccas (the 'Spice Islands') in 1511 and with China in 1516. Meanwhile the Spanish and British searched to the west for a sea route from Europe to Asia. Spain financed an expedition led by the Genoese explorer Columbus to find it, but instead he 'discovered' America in 1492; and a British expedition under the Venetian, Cabot, reached North America in 1497. Then, in 1519, a Spanish expedition commanded by Magellan (who was Portuguese) sailed around Cape Horn and to the Philippines. These important explorations and discoveries were followed by a long period of European settlement in the tropics, and by greatly expanded world trade with which most of our important tropical crops were dispersed widely during the sixteenth and seventeenth centuries. The crop movements of greatest significance during this long period were the earliest ones to and from the New World. During the sixteenth century New World crops such as maize, groundnuts, pineapples, tomatoes and sweet potatoes (which were already in Polynesia in pre-Columbian times) were spread throughout the tropics, cassava had reached West Africa and the 'Irish' potato had reached Europe. Sugar cane, which was taken from the Old World to the Caribbean first by Columbus, had become the major economic crop there by the middle of the seventeenth century. The need for labour to grow it gave rise to the slave trade from West Africa, and with this trade there was an increase in the movement of crops between America and Africa.

During the eighteenth and nineteenth centuries the development of agricultural enterprises in the tropics was stimulated by the demand from Europe for agricultural raw materials for use in industry, and for tropical beverages, foods and spices. Crops spread as a result of these developments, and botanic gardens began to play an important role in their dispersal as early as the beginning of the eighteenth century. The best known examples of their work in this respect were the spread of arabica coffee from Java via botanic gardens in Holland and France, and the very famous role of the Royal Botanic Garden at Kew, England, and the Singapore Botanic Garden when they helped to establish the rubber industry in Malaysia using seed collected in the basin of the Amazon River in Brazil. An example of the influence of European industry on the spread of tropical crops is provided by the dispersal of American 'Upland' cotton into the Old World towards the end of the nineteenth century in response to a demand for high quality lint from the English cotton industry.

The most recent developments in the dispersal of tropical crops have been associated with the expansion of agricultural research in the tropics, and with international cooperation which has included the exchange of seeds, or even of large collections of germplasm, between workers in different parts of the world. Much of this work has been coordinated by international agencies such as the Food and Agriculture Organization of the United Nations, and the more recently established Consultative Group on International Agricultural Research. As our understanding of the nature and distribution of genetic diversity in crops has improved, plant exploration, especially in known centres of genetic diversity, and the collection, evaluation and conservation of seeds and sometimes of vegetative material has become an increasingly important part of international agricultural research. This was given formal recognition in the mid-1970s when the International Board for Plant Genetic Resources was established to collect, conserve and disperse plant genetic resources on a world scale.

Further reading

The tropical environment

Critchfield, H. J. (1960). *General Climatology*, London: Prentice-Hall.
Cooper, J. P. (1970). Environmental physiology, in Frankel, O. H. and Bennett, E., *Genetic Resources in Plants — Their Exploration and Conservation*, Oxford: Blackwell.
Bunting, A. H. (1961). Some problems of agricultural climatology in tropical Africa, *Geography*, 46, 283–94.
Hartley, W. (1970). Climate and crop distribution, in Frankel, O. H. and Bennett, E., *Genetic Resources in Plants — Their Exploration and Conservation*, Oxford: Blackwell.
Harris, D. R. (1969). Agricultural systems, ecosystems and the origins of agriculture, in Ucko, P. J. and Dimbleby, G. W., *The Domestication and Exploitation of Plants and Animals*, London: Duckworth.
Hopkins, B. (1965). *Forest and Savanna*, London: Heinemann.
Huxley, P. A. (1965). Climate and agriculture in Uganda, *Exptl. Agric.*, 1, 81–97.
Kowal, J. and Andrews, D. J. (1973). Pattern of water availability and water requirement for grain sorghum production at Samaru, Nigeria, *Trop. Agric. (Trin.)*, 50, 89–100.
Jameson, J. D. (ed.) (1970). *Agriculture in Uganda*, London: Oxford University Press. (A revised edition of Tothill, J. D., 1940.)
Phillips, J. (1959). *Agriculture and Ecology in Africa*, London: Faber and Faber.
Ruthenberg, H. (1971). *Farming Systems in the Tropics*, Oxford: Clarendon Press.
Webster, C. C. and Wilson, P. N. (1966). *Agriculture in the Tropics*, London: Longman.
Williams, C. N. and Joseph, K. T. (1970). *Climate, Soil and Crop Production in the Humid Tropics*, Kuala Lumpur: Oxford University Press.
Wrigley, G. (1961). *Tropical Agriculture*, London: Batsford.

Crop plant origins

Frankel, O. H. and Bennett, E. (eds) (1970). *Genetic Resources in Plants — Their Exploration and Conservation*, Oxford: Blackwell.

Frankel, O. H. and Hawkes, J. G. (eds) (1975). *Crop Genetic Resources for Today and Tomorrow*, London: Cambridge University Press.
Hawkes, J. G. (1970). The origins of agriculture, *Econ. Bot.*, 24, 131–6.
Hutchinson, Sir Joseph (ed.) (1965). *Essays on Crop Plant Evolution*, London: Cambridge University Press.
Hutchinson, Sir Joseph (1971). Changing concepts in crop plant evolution, *Exptl. Agric.*, 7, 273–80.
Hutchinson, Sir Joseph (ed.) (1974). *Evolutionary Studies in World Crops. Diversity and Change on the Indian Subcontinent*, London: Cambridge University Press.
Purseglove, J. W. (1968). The origin and spread of tropical crops, in *Tropical Crops, Dicotyledons*, London: Longman.
Simmonds, N. W. (ed.) (1975). *Crop Plant Evolution*, London: Longman.
Stebbins, G. L. (1950). *Variation and Evolution in Plants*, New York: Columbia University Press.
Ucko, P. J. and Dimbleby, G. W. (eds) (1969). *The Domestication and Exploitation of Plants and Animals*, London: Duckworth.

Chapter 2

The Cereal Crops

The family *Gramineae*

The cereals are those members of the great grass family, the *Gramineae* of the Monocotyledons, grown for their characteristic fruit, the caryopsis, which has been the most important source of the world's food supply since cereal farming began in the Middle East some 8,000–9,000 years ago with the domestication of wheat and barley. In 1973 world production of all kinds of cereal grains exceeded 1,300 m. tonnes. The *Gramineae* is a very large and specialized family of about 10,000 species. As well as the cultivated cereals, some wild species have edible seeds which are gathered for food in regions subject to famine, and many wild and cultivated fodder grasses are important in agriculture as livestock food. In view of their great importance to man, and their long history in agriculture, it is remarkable that so few cereal species have been brought into cultivation. Oats and rye are the dominant cereals of cold temperate regions, wheat and barley predominate in warmer temperate climates, while in the tropics rice, maize, sorghum and the millets provide the bulk of the diet for large populations. Rice is the most important cereal in the world, though annual production of about 320 m. tonnes is less than that of wheat (about 380 m. tonnes). It is the staple food for about half the world's population, and is especially important in South-east Asia where it constitutes a much greater proportion of human diets than wheat in American or European diets.

Judged by its world distribution and by the number and diversity of its species, the *Gramineae* is the most successful family of flowering plants. Grasses are found all over the world in most kinds of habitat from wet tropical forests (the bamboos) and tropical marshes (where grasses may have originated), to the hot, dry deserts of Africa, Asia and Australia and even the cold deserts within the arctic circle. They form the climax vegetation of the semi-arid prairies of the American continent, the steppes of Asia and the savannas of Africa. Except for some woody bamboos, grasses are herbaceous annual, biennial or perennial plants. Many have growth habits which enable them to survive grazing, drought or even fire. Some form a 'turf', or mat of vegetation, with underground stems (rhizomes) or creeping surface stems (stolons) which produce adventitious roots from their nodes and aerial branches from buds in the axils of their scale leaves. Others, including the cereals, have a tufted growth habit with erect stems and often several erect branches, called tillers, which arise from the axils of leaves at the bottom of the stem; branches may also grow from axillary buds higher up the stems and tillers, but the numbers of both tillers and branches varies between cultivars of a species, and with density of sowing. Long, lanceolate leaves are borne singly at prominent, solid nodes. Each has a sheathing base (but no petiole) which clasps the hollow internode of the stem above its origin more or less closely; a small flap or tongue of tissue called the ligule, often accompanied by a pair of auricles (ears), marks the junction of the leaf sheath and lamina. The stem internode, the leaf sheath and the leaf lamina grow from intercalary meristems at their bases; so the oldest part of the lamina is at its tip. When the grass seed germinates it produces a limited number of seminal roots, but the greater part of the root system in a mature plant is adventitious from stem and tiller nodes below ground. It is much branched and spreads widely in the upper soil, with a few roots penetrating more deeply.

The great success of grasses depends not only upon growth habits which enable them to survive grazing and adverse climates, but also upon their efficient reproductive system. They have evolved a relatively simple flower adapted to wind pollination, and a dry, indehiscent fruit with a rich food reserve which enables annual or ephemeral species to avoid, and so to survive, long periods of cold or of heat and drought, and then to become established quickly as soon as growing conditions once more become favourable. The caryopsis is a source of human food which man has found easy to sow, to harvest and to store; about 70 per cent of its weight is carbohydrate and some 9–14 per cent is protein. The grains of some cereals contain oil, notably those of maize in which some cultivars have 10 per cent or more (most of it in the embryo). The importance of the protein content of cereal grains is emphasized by the fact that it provides some 70 per cent of world protein supply; persons whose staple food is a cereal and who get enough calories from it automatically get enough protein, biological value notwithstanding. Most of the world's cereals are eaten as pancakes, biscuits or gruel, or they are commonly used to make beer or alcoholic spirits. Bread is made only from cereal grains whose endosperm contains the protein gluten, and bread wheat, *Triticum aestivum*, is outstanding in this respect. Gluten in the dough made from wheat flour holds the bubbles of carbon dioxide which are produced by fermenting yeast, and so the dough 'rises' and can be baked to make porous bread.

The grass inflorescence in its simplest form is a spreading, much branched panicle carrying spikelets laterally and at the ends of its ultimate branches. The spikelet, though strictly an inflorescence itself, is the unit of construction of the grass 'inflorescence', not the flower as in other families of flowering plants. Except in the bamboos the flowers are small, inconspicuous and remarkably uniform throughout the family. They have a superior ovary of three united carpels with one locule containing a single ovule, two styles with feathery stigmas and generally three stamens (six in rice and more in some bamboos) on long, slender filaments which are capable of rapid extension in length. As well as hermaphrodite or 'perfect' flowers, most grasses have others which are male or completely sterile. The essential organs of the flower are protected by two bracts, a lower, outer lemma which subtends the flower and which is sometime awned, and a smaller, more delicate palea. These two bracts enclose the anthers and ovary until the flower opens, when they are forced apart by the swelling of two small, fleshy organs, the lodicules, at the base of the flower. The lodicules are thought to have been derived from a change in function of two of the three inner members of an ancestral perianth. The flower, the lemma and the palea are together often called the 'floret'.

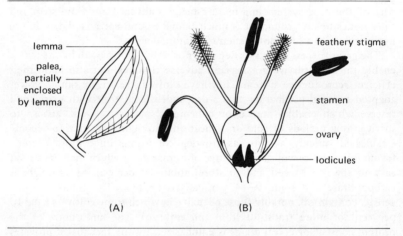

Fig. 2.1 Structure of the floret of the *Gramineae*. (A) Lemma and palea intact. (B) Lemma and palea removed to show the essential organs of the flower.

Whereas the florets have changed little during the evolution of the *Gramineae*, the spikelets have undergone such a variety of modifications that the classification of grasses into sub-families, tribes and sub-tribes has been based largely upon differences in spikelet structure. The simplest spikelet has many florets carried alternately on two sides of its axis, the rachilla, with two bracts, the glumes, subtending the collection of florets. In this primitive spikelet the flowers are protected only by their lemmas and paleas, not by the glumes.

With the evolution of more advanced types the spikelet became modified to provide greater protection for the flowers and fruits, greater economy in the use of water and food and more efficient seed dispersal.

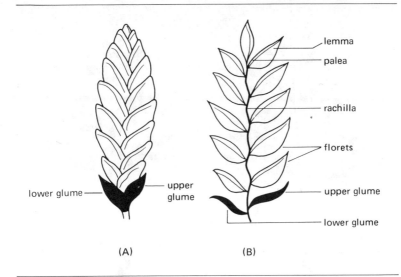

Fig. 2.2 The spikelet of the *Gramineae*. (A) Structure of the primitive spikelet. (B) Spikelet with florets and glumes separated.

The lax, paniculate inflorescence has become concentrated to a more spike-like form (not in the cereals rice, oats and 'shallu' sorghums), the number of florets per spikelet has decreased, the flowers and fruits are more effectively protected by the glumes, while the lemmas and paleas have become less important, and whole spikelets, rather than individual fruits, became the units of dispersal.

During the development of the caryopsis the ovary wall (pericarp) becomes fused with the seed coat (testa) to form one composite covering for the seed. This fruit is characteristic of the grasses and occurs throughout the family though in a few cases (e.g. finger millet and proso millet) the pericarp and testa are less intimately fused. Beneath the testa a single layer of cells, the nucellus, surrounds the endosperm (the main food reserve, rich in starch) which makes up the bulk of the seed. The outer layer of the endosperm is the aleurone in which much of the protein and vitamins of the seed are stored. The rest of the endosperm consists of large cells packed with starch, and is sometimes differentiated into two distinct regions characterized by the presence of large or small starch grains. The embryo occupies a small volume of the seed, lying at the base of the lower surface and in close contact with the endosperm. The single cotyledon has become a large shield-shaped scutellum which remains in the seed when it germinates; food passes through it from the endosperm to the growing embryo. The scutellum encloses or enfolds to varying degrees the apex of the embryonic stem or plumule which is enclosed within a sheath, the

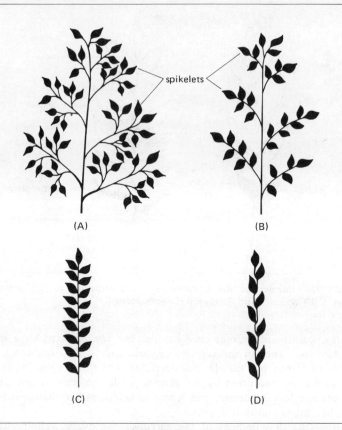

Fig. 2.3 Modifications of the inflorescence of grasses. (A) Much branched open panicle. (B) Contracted panicle with sessile spikelets. (C) Spike of sessile spikelets. (D) Spike of sessile spikelets carried on a notched rachis.

coleoptile, and the embryonic root apex or radicle which is also within a sheath, the coleorhiza. Slightly above the coleorhiza on the lower side of the embryo opposite the scutellum a small flap of tissue called the epiblast is a common feature in some large groups of grasses (notably the sub-family *Festucoideae* – see below), whereas it is rare in others (notably the sub-family *Panicoideae*). There has been much discussion about the interpretation of the epiblast. Recent evidence suggests that it is probably an extension of the coleorhiza, not a rudimentary second cotyledon nor the first true leaf of the embryo.

The embryo is rich in fats, oils, minerals, protein and sugars. Many grass seeds germinate only after a period of rest or dormancy, but most cereal grains will do so almost immediately after they are mature. The coleorhiza emerges from the fruit first, and through it the radicle bursts as the first

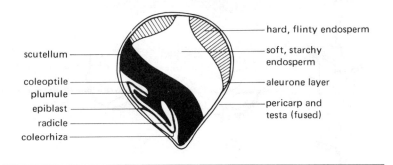

Fig. 2.4 A diagram of a longitudinal section through the caryopsis of *Sorghum bicolor*.

seminal root, followed usually by others depending upon the number of nodes in the embryo and upon the environmental determinants of seedling vigour, though some species have only one. After the coleoptile has emerged through the soil surface the first leaf is extended through its ruptured tip and the seedling becomes established.

The *Gramineae* has been divided into two or more sub-families by various authorities. Within each sub-family one or more tribes of grasses are recognized. In general the classification into sub-families is based upon a knowledge of the evolutionary relationships between major groups (it is a phylogenetic classification). On the other hand, the classification of sub-families into tribes has been largely a matter of taxonomic convenience, and has been less concerned with the evolutionary relationships between groups, though taxonomists work to alter this and to produce a phylogenetic system for the classification of tribes. One current view is that the grasses originated in tropical rain forests, and that the earliest forms were the ancestors of the modern bamboos and their allies. As the grasses were dispersed from the rain forests into seasonally dry climates, and eventually from the tropics into cool temperate regions, distinct types evolved which now form the basis for the classification of grasses into sub-families. For example, the *Oryzoideae* (the rice sub-family) evolved in tropical swamps and marshes, the *Panicoideae* (the sub-family of tropical savanna zone cereals) evolved in seasonally dry tropical and warm temperate regions, and the *Festucoideae* (or *Pooideae* — the sub-family of the temperate cereals) evolved in cool temperate regions.

In an important nineteenth century classification of the *Gramineae* Bentham recognized only two great sub-families, the *Festucoideae* and the *Panicoideae*. Though the modern classification differs from Bentham's, and includes six sub-families, the *Festucoideae* and the *Panicoideae* are retained (in a restricted sense compared with Bentham), and a comparison of their important characteristics illustrates the two general evolutionary trends outlined above; viz., decreased numbers of florets per spikelet,

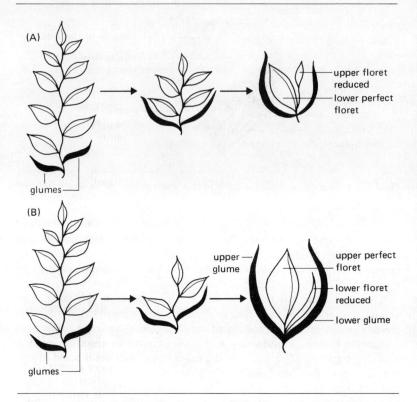

Fig. 2.5 Stages in the reduction of spikelets in the *Festucoideae* and *Panicoideae*. (A) Reduction of the spikelet in the *Festucoideae*. (B) Reduction of the spikelet in the *Panicoideae*.

increased protection of the flowers by the glumes and concentration of the inflorescence. In the *Festucoideae* the upper florets of the spikelet tend to be suppressed so that they are male, sterile or absent, while the lower flowers are perfect (that is they contain both ovary and anthers). The opposite has occurred in the *Panicoideae*; the lower flowers tend to be suppressed and are either male, sterile or absent while the upper flowers are perfect. The spikelets may be compressed or flattened in both sub-families, but in the *Festucoideae* the compression is lateral, while in the *Panicoideae* it is dorsal.

In the *Festucoideae* fruits are dispersed because the axis of the spikelet, the rachilla, fractures so that individual florets with their fruits are shed. In the *Panicoideae* whole spikelets, including the glumes, are shed. Cytological evidence suggests that the two sub-families are significantly separate from one another; the basic chromosome number of the *Festucoideae* is seven while that of the *Panicoideae* is nine or ten.

In the modern classification of the *Gramineae* morphological and anatomical characteristics of the roots, stems and leaves and biochemical criteria (especially those of food reserves in the endosperm) are used as well as those very important features of the reproductive organs described above. Of the six sub-families now recognized two, the *Bambusoideae* (bamboos and their allies) and the *Arundinoideae*, contain no economically important crop grasses. The remaining four sub-families, with their economic tribes and species are:

1. The *Festucoideae*

Grasses of cool temperate regions. Vascular bundles of the stems usually in concentric rings; epiblast present in most species.

(*a*) Tribe *Aveneae*

Panicle lax and typically spreading; very large, prominent glumes completely enclose whole spikelets. As few as two florets per spikelet.

Avena sativa, oat. Oats are an important crop in cool, moist temperate climates, especially in North America, northern Europe and Russia where they are grown principally to feed livestock. The wild oat, *Avena fatua*, is an important weed of wheat, barley and oats in many parts of the world.

(*b*) Tribe *Triticeae*

Temperate or warm temperate annual grasses with sessile spikelets usually more or less sunk in depressions of the rachis of the terminal spike. Each spikelet has few flowers, in some species only one.

Triticum durum, macaroni wheat; *Triticum aestivum*, bread wheat; *Hordeum vulgare*, barley; *Secale cereale*, rye. Though wheat is generally considered to be a crop of temperate climates, locally important quantities are produced in the tropical highlands (for example in Kenya and Ethiopia), and with irrigation in the cool dry season of the semi-arid tropics. As the demand for wheat flour increases in tropical countries efforts are being made to breed suitably adapted cultivars to grow in them. Major advances in this direction has been made by breeders in Mexico who, partly from Japanese ancestors, have produced dwarf, reproductively day-neutral cultivars which respond well to large applications of fertilizer to give increased yields. These cultivars have been notably successful in India, but they are also important in breeding wheat for tropical climates closer to the equator.

2. The *Panicoideae*

A large and economically important sub-family of tropical and sub-tropical grasses. Vascular bundles of the stems scattered throughout the cortex in many species; epiblast absent. All members have two florets per spikelet.

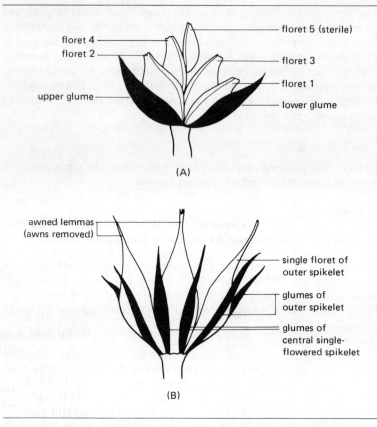

Fig. 2.6 Spikelets of the *Triticeae*. (A) Single spikelet of wheat with five florets. (B) Group of three spikelets of six-rowed barley — each spikelet with only one floret.

(a) Tribe *Paniceae*
In this tribe the glumes are often very small so that the flowers are protected only by the lemmas and paleas. The panicle is often very condensed and spike-like. Though most of the wild species are found in the humid tropics, cultivated species are best known as crops of semi-arid regions.

Panicum miliaceum, common millet, proso.; *Panicum sumatrense* (syn. *P. miliare*), little millet; *Pennisetum americanum* (syn. *P. typhoides*), bulrush or pearl millet; *Echinochloa frumentacea*, Japanese barnyard millet; *Digitaria exilis*, hungry rice; *Panicum maximum*, guinea grass; *Pennisetum clandestinum*, Kikuyu grass; *Pennisetum purpureum*, elephant or Napier grass.

(*b*) Tribe *Andropogoneae*
In this tribe the spikelets are borne in pairs, one sessile and one stalked (pedicelled), except at the tips of inflorescence branches where one sessile spikelet is accompanied by two which are pedicelled. The sessile spikelets contain one perfect, upper floret and one lower floret which is sometimes represented only by a lemma. Pedicelled spikelets normally have two male florets, though one or both of them may be sterile. The spikelets are well protected by the glumes. The outer glume is large and coreaceous (hard) and encloses the whole spikelet. The tribe includes many of the common wild species of savanna grassland, among which the most widespread belong to the genera *Themeda*, *Andropogon* and *Hyparrhenia*.

Sorghum bicolor, sorghum; *Saccharum* spp., sugar cane.

(*c*) Tribe *Maydeae*
A small tribe that in structure and many other respects closely resembles the *Andropogoneae* to which it is related and from which it may have been derived. Indeed, in some modern classifications the tribe *Maydeae* is not recognized and its members are included in the *Andropogoneae*. Within the *Maydeae* there is a tendency for the sexes to be separate on the same plant, and this is most pronounced in maize, which is the only crop. Maize is monoecious, with male flowers in a terminal panicle, the 'tassel', and female flowers in a spike borne on a modified lateral branch, the 'cob' or 'ear'.

Zea mays, maize or corn; *Euchlaena mexicana* (syn. *Zea mexicana*), teosinte.

3. The *Oryzoideae*

A sub-family of only one tribe, the *Oryzeae*, in which there is only one flower in each spikelet, and six stamens in each flower. It is an interesting, isolated sub-family of grasses which are adapted to very wet habitats. The inflorescence is a panicle; the glumes are minute, or may be absent, and the flowers are protected by the lemma and palea which together form the 'husk' of the harvested grain.

Oryza sativa, rice; *Oryza glaberrima*, African rice.

4. The *Eragrostoideae*

A sub-family intermediate in many respects between the *Festucoideae* and the *Panicoidea*. The structure of the spikelets is similar to that in the *Festucoideae*, while in morphological and anatomical characteristics of the leaves, stems and embryos, and in chromosome numbers the *Eragrostoideae* resembles the *Panicoideae*. The sub-family is distinct from both of these others in having a three-nerved, not a five or more nerved lemma.

(a) Tribe *Eragrosteae*
In this tribe the inflorescence is a spreading or contracted panicle of one to many-flowered spikelets, most of them with hermaphrodite flowers. The glumes are small and inconspicuous.

Eragrostis tef, tef, teff or ingera.

(b) Tribe *Chlorideae*
Among the economic genera in this tribe the inflorescence is a digitate collection of spike-like racemes.

Eleusine coracana, finger millet; *Chloris gayana*, Rhodes grass; *Cynodon dactylon*, Bermuda grass, star grass. Both finger millet and tef are good examples of the effects on inflorescence structure of evolution with human selection for grain production, which depends ultimately upon the number and size of fruits produced. It has led to a general tendency among cereals for grain cultivars to have inflorescences with more branches, and therefore more spikelets and grains than their wild relatives. For example, there is a marked contrast between cultivars of finger millet with their bulky, compact inflorescences and relatively large grains on the one hand, and the wild forms of the species with their slender, delicate inflorescences and small grains on the other. Similarly, the degree of branching in the panicle of grain cultivars of teff grown in Ethiopia is much greater than in types grown for fodder in South Africa or for forage in India.

Rice: *Oryza sativa*

Rice is an annual swamp plant that has been cultivated for several thousand years as the principal cereal of South-east Asia, from India to Japan. In Asia it is still grown by peasant and subsistence farmers, who use traditional farming methods that seem to have changed little since rice cultivation began. In other parts of the world, especially in the United States and in Australia, rice production is highly mechanized. Annual world production of rice increased from less than 200 m. tonnes in 1950 to around 250 m. tonnes in 1961–5, around 280 m. tonnes in 1966–70, and to more than 320 m. tonnes in 1973. Mainland China produces most with a crop of more than 110 m. tonnes annually; other countries with huge crops are: India (around 68 m. tonnes each year); Indonesia (20 m.); Bangladesh (18 m.); and Japan and Thailand (each with about 15 m. tonnes in 1973). Annual production in all of Africa is around 7 m. tonnes. The largest crops in the New World are from Brazil (7.4 m. tonnes) and the United States (4.2 m.). Little of world production enters international trade, and except in Thailand and Burma, most rice in Asia is eaten by those who grow it.

Oryza sativa is thought to have been domesticated in India more than 4,000 years ago from the wild species *O. perennis*. The only other culti-

Fig. 2.7 *Oryza sativa*: Rice   Harvesting an experiment conducted in Indonesia by the International Rice Research Institute. (By courtesy of the FAO.)

vated species in the genus is *O. glaberrima*, African rice, which probably originated around the swampy headwaters of the Niger River in West Africa. According to the most recent authority *O. barthii* was the wild progenitor of African rice; previously *O. barthii* and *O. perennis* were considered synonymous. Since rice was domesticated many thousands of cultivars with diverse morphology and physiology have been selected by man in this self-fertilized crop, to suit local tastes and environments. They have been grouped according to the characteristics of their seeds and according to the conditions under which they are grown, but it has been found difficult to produce a satisfactory classification of so many cultivars. Three sub-species are recognized in *O. sativa*, and they correspond with geographical races of the crop.

Oryza sativa subsp. *indica* is 'indica' rice, which consists of a large group of reproductively photosensitive, short-day cultivars of the tropical monsoon region of South-east Asia. They are grown in warm, humid regions where rain during the growing season, or irrigation water, is plentiful, and commonly on poor, infertile soils which are flooded as the crop grows. They tend to be tall, with long straw which lodges easily if growth and grain yield are increased by the use of nitrogenous fertilizers. Compared with the 'japonica' rices discussed below, their grain yields are small. The 'indica' rices flower in response to shortening days, on a more or less fixed calendar date at any place, regardless of when they are sown. Sowing and harvest dates at any place are closely associated with the seasonal distribution of rainfall, and its duration (the 'growing season' — see Chapter 1), and local cultivars are physiologically adapted to the length of the growing season so that they flower close to the end of it. Consequently cultivars well adapted to the climate of their own area are unsuitable for cultivation where there is a different seasonal pattern of rainfall distribution, or north or south of their own location in latitudes with different day-lengths.

The 'indica' rices commonly have grains longer than 9 mm; when they are overcooked they remain separate, and do not become 'mushy'.

Oryza sativa subsp. *japonica* is 'japonica' rice which is both sensitive and insensitive to photoperiod according to the cultivar. With insensitive cultivars, flowering date, and consequently the time of harvest, are determined by sowing date so that a single cultivar may be, at least physiologically, suitable for cultivation in more than one environment. 'Japonica' rice is grown extensively in Japan, Korea, north China and elsewhere outside the tropics (California and Louisiana in the United States, and in Australia and Italy). It tends to have shorter straw than 'indica' types, and responds better to the application of fertilizer to give large yields without lodging.

Typically, the grains of 'japonica' rice are short (about 7 mm long), and do not resist overcooking, but tend to become 'mushy' and stick together.

Oryza sativa subsp. *javanica* is 'javanica' rice which is insensitive to photoperiod, but which has a long vegetative phase. It is well adapted to the equatorial climates of Java and Indonesia, but is little grown elsewhere.

Cultivars from each of these sub-species are commonly referred to one of three classes according to the conditions under which they are grown and to which they are best adapted.

1. Swamp rice or wet paddy ('paddy' is also the name given to rice grain before the persistent lemmas and paleas ('husk') have been removed by milling) accounts for by far the greatest part of world rice production. It is grown in fields surrounded by bunds or levees built to retain flood water and to control its flow. In parts of Asia where the population is very dense the fields are even built on steep hillsides to form terraces. The land is cultivated and levelled, and fertilizers (animal manures in Asia) are applied during a dry part of the year, after which the seed is drilled or broadcast by hand. Where rainfall at sowing time is uncertain, or where there is insufficient irrigation water, seedlings are raised in a nursery, and then transplanted 5—7 weeks later when they are 20—30 cm tall. Before seedlings are transplanted the soil of the rice field is 'puddled' by working it while it is wet. This destroys the soil structure and prevents water from draining away. It also destroys weeds. The fields are often quite deeply flooded by the time the seedlings are transplanted into them. Ideally the

Fig. 2.8 *Oryza sativa*: Rice. Recently transplanted paddy in an experiment at the International Institute of Tropical Agriculture, Ibadan, Nigeria.

crop is grown in 15–30 cm of very slowly flowing water until just before it matures, then the fields are drained, but good water control is essential in this. In parts of Asia two, or even three crops of wet paddy are grown on the same land in one year, but this can be done only when non-photosensitive cultivars are used.

2. Floating rices are cultivars adapted to conditions where flood waters are uncontrolled and become very deep (up to 5 m); the internodes elongate quickly as the flood rises, and the leaves and inflorescences float on its surface.

3. Upland rice or dry paddy is grown like other dry land cereal crops, and is dependent upon rain irrigation though it is sensitive to dry weather.

Another important classification of rice cultivars depends upon the kind of starch stored in the endosperm. The starch of 'glutinous' or soft rices is mostly amylopectin; the broken surface of the grains has a chalky appearance and they are said to have a soft, opaque fracture. When they are overcooked soft rices become gelatinous and sticky, and on subsequent cooling set to a gel. On the other hand starch in the endosperm of hard rices is about one-quarter amylose and three-quarters amylopectin. The grains have a vitreous (glass-like) fracture and do not become sticky when overcooked. Hard rices are grown more extensively, and are more important in world trade than soft rice.

Rice has a hollow, erect stem 30–150 cm tall (up to 5 m in floating rice), depending upon differences between cultivars and the environments

30   The Cereal Crops

Fig. 2.9 *Oryza sativa*: Rice.   Mature paddy in an experiment at Ibadan, Nigeria.

in which they are grown. The internodes are fairly uniform in diameter, becoming longer up the stem. The nodes are solid and swollen, and there is often a coloured, pink to deep purple pulvinus at the base of each internode. The stem is protected by a small celled epidermis which may be silicified; its anatomy is typical of an aquatic plant with large air cavities in the cortex. Consequently the stems are weak, and often unable to support the weight of leaves and the inflorescence. Tillers grow from buds in the axils of lower leaves on the main stem. There are three seminal roots and an extensive adventitious root system whose extent and depth varies with soil conditions, though most of it is in the surface soil to a depth of about 25 cm. As the adventitious roots become older they change in colour from white to brown. Extensive aerenchyma develops in the cortex of old roots and extends also into the stem and leaf sheaths. The central stele of the root is eventually protected by a well developed endodermis and a few layers of parenchymatous cortical cells.

Each stem bears 10–20 leaves with long sheathing bases which are split along their entire length. A characteristic long ligule, and usually auricles, mark the junction of the sheath and the long, narrow lamina (30–50 cm

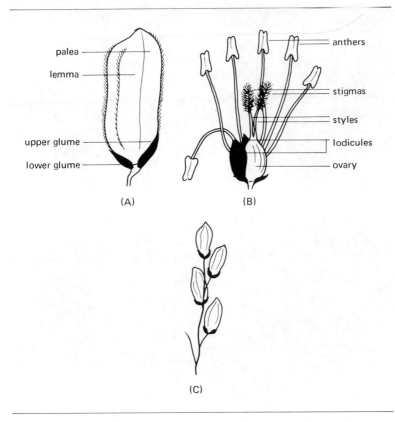

Fig. 2.10 *Oryza sativa*: Rice. (A) A spikelet. (B) Lemma and palea removed to expose the flower with six stamens. (C) A portion of the rice panicle.

long, 1.2–2.5 cm wide) which has a well defined mid-rib, and which may be slightly hairy.

The lax or compact panicle is terminal, and is protected during its early development by the sheath of the 'flag', or uppermost, leaf. The main axis of the panicle is angular with numerous primary branches from its nodes. These in turn carry secondary branches, each of which bears from one to seven spikelets. Each spikelet is carried on the flattened end of a short pedicel which is sometimes extended into two small membranous margins whose interpretation is uncertain (perhaps they represent reduced florets). The spikelets are laterally compressed, each with two small, narrow glumes which are 2–3 mm long. The glumes subtend a single floret which has a large, often coloured, boat shaped, five-nerved, hairy and pointed or awned lemma, a firm, three-nerved palea, two broad lodicules, and six stamens inserted around the ovary in two whorls.

Fig. 2.11 *Oryza sativa*: Rice. A bunch of mature panicles harvested in Indonesia. (By courtesy of the FAO.)

Once flowering begins anthesis proceeds from the apex of the panicle downwards over a period of a week or more, though individual flowers open quickly, mostly just before mid-day. The anthers dehisce just before or at about the same time as the flowers open and self-pollination is the rule, though a small and variable amount of natural crossing does occur. Part or all of the rice inflorescence is often sterile because pollen is defective, because temperatures at anthesis are cool or because the stem has been so much damaged by pests and parasites that the development of the panicle is affected. Partial or complete sterility also occurs in the offspring of crosses between some cultivars, especially if they come from different geographical races.

The grain matures 30–40 days after flowering and is enclosed closely by the heavily indurated lemma and palea, which form the 'husk' or 'hull'. The pericarp is usually white, but it is sometimes red or brown. When rice is milled the husk and the bran (which includes the embryo, pericarp and much of the nucellus and aleurone), with its nutritious protein and vitamins, are removed. Consequently, vitamin deficiencies are common where milled rice is the staple food, especially beriberi, which is a symptom of thiamine (vitamin B_1) deficiency. The water-soluble vitamins of the aleurone and embryo are partially transferred to the endosperm if the grain is soaked in water, then steamed and dried, a process called parboiling. If the grain is subsequently under-milled the nutritive value of rice is greatly improved.

Rice is inbreeding and cultivars consist of homozygous pure lines which breed true. Some early improvement in the crop was achieved by selecting superior cultivars from very large collections, or superior lines from mixed accessions. More recently the greatest advances have been made by selecting from the segregating progeny of crosses between carefully chosen parents. Since the mid-1960s notable success has been achieved by this means at the International Rice Research Institute (IRRI) in the Philippines, and the new cultivars bred there have had a substantial impact on rice production in perhaps 70 per cent of the rice area of Asia. The best known early product of this work was the short strawed cultivar IR8 which is able to utilize large applications of N fertilizer to give grain yields of more than 6 tonnes per hectare. Unfortunately, IR8 is susceptible to a range of important pests and diseases, its grain was not popular as food with consumers, nor did it keep well in store. IR8 was eventually superseded by IR20 with much better grain quality and resistance to some pests and diseases, as well as a potential for large grain yields on soils of relatively low fertility. The most recent products of breeding at IRRI are the short season (as few as 105 days to maturity), day-neutral cultivars IR28, IR29 and IR30 which yield as well as cultivars bred earlier, but are superior to them in their tolerance of pests and diseases and of adverse soil conditions such as salinity or alkalinity and nutrient deficiencies.

Maize, corn: *Zea mays*

Maize is a large, annual, monoecious grass of the small tribe *Maydeae*, closely related to two wild grasses of Central and South America, *Euchlaena mexicana* (= *Zea mexicana*; teosinte) and *Tripsacum dactyloides* (gama grass). It was unknown outside the New World until after Columbus's voyage to America, but archaeological evidence has shown that primitive forms were cultivated there, certainly as long as 5,000 years ago. There is no simple answer to the question of the origin of maize, though it has been the subject of intense study for many years. Though modern maize is not known to occur as a wild plant, its ancestor may have been wild maize of a type similar to modern 'pod corn', which, unlike maize, has prominent glumes protecting the female flowers; or maize may have originated following natural hybridization between a primitive form and both teosinte and gama grass; or as a third, and perhaps most likely alternative, it was domesticated from teosinte. Nor is it certain whether maize was domesticated in only one region, or independently in more than one, but it seems most likely that it originated in southern Mexico or in some nearby part of Central America and spread very early into North and South America. After it was domesticated maize evolved through mutation, gene exchange with teosinte and perhaps also gama grass, and as a result of the spread and geographical isolation of cultivated populations. Once discovered by Europeans in the New World it spread rapidly through southern Europe, and to Asia and Africa, especially with the Portuguese during the sixteenth and seventeenth centuries.

An outstanding feature of the crop is the separation of the sexes into different parts of the same plant, the male flowers in a terminal panicle or 'tassel', the female flowers in a spike on a modified lateral branch, the 'cob' or 'ear'. This arrangement, and the maturation and dehiscence of some anthers before the stigmas of the same plant are mature (protandry), ensures that most female flowers are cross-fertilized. Consequently maize is very heterozygous. As a result the crop is so variable that man has been able to select and breed types suited to a very wide range of environments, and continues to extend its range into cooler and drier regions. As a cereal maize is exceeded in world importance only by rice and wheat. It is grown from latitude 58° N. to 40° S., and at altitudes which range from sea level to 3,300 m (in the Andes Mountains), where rainfall during the growing season exceeds 200 mm and mean temperatures are warmer than 20°C. However, most maize is grown on 400–900 mm of rain in the temperature range 20°–30°C. In Africa and Asia it has replaced ancient local food crops like sorghum and the millets in many areas where the climate is suitable, partly because it is preferred as food, but also because the structure of the female inflorescence confers unique advantages over other cereals. The husks of the cob (see below) effectively protect the grain from damage by birds and from spoilage due to wet weather at harvest, and eliminate entirely the losses which occur in other cereals from shattering. Furthermore the grain of maize can be cooked in a great variety of ways to make very palatable and popular foods. Immature cobs, often of 'sweet corn' (see below), are boiled or roasted to be eaten as a vegetable; mature grain is ground to a coarse or fine flour to make many kinds of pancakes, thick gruel or soup (but not leavened bread). The grain is used to brew beer or to produce distilled alcoholic spirits; starch, oil and a large number of other industrial products are obtained from it. Maize is an excellent fodder plant and is often grown for silage, and some cultivars, the 'sugar maizes', accumulate sucrose in their stems which can be expressed as sugar syrup.

The family Gramineae 35

Fig. 2.12 *Zea mays*: Maize. A standing crop in flower in the Yemen Arab Republic. (By courtesy of the FAO.)

World production is estimated to exceed 310 m. tonnes of grain each year. Almost half of this (143 m. tonnes in 1973) is grown in the United States where 90 per cent of the crop is fed to livestock, and the rest is used chiefly for the manufacture of corn starch, or for the extraction of a semi-drying oil which constitutes around 13 per cent by weight of the embryo.

Maize grows up to 4.5 m tall with few, if any, tillers and no lateral vegetative branches. There are four, sometimes more, seminal roots, but soon after germination adventitious roots develop from nodes of the stem below ground, and extend widely in the top meter of soil, though some roots penetrate deeper. Thick, dark coloured 'prop' roots arise adventitiously from the lowest nodes of the stem above ground and provide some mechanical support when plants become tall. The stem is solid with short, fairly thick internodes at the base of the plant, which become longer and thicker higher up the stem, then taper again to the tassel which terminates the axis. A groove in the stem opposite each leaf blade results from the pressure of buds in the leaf axils on the extending internodes. There are 8—21 leaves inserted alternately at the nodes of the main stem with fewest in short, early maturing cultivars. The leaf sheath is entire at its base, but split above, and clasps the internode above its origin very tightly. A large membranous ligule and small auricles arise at the base of the lamina which

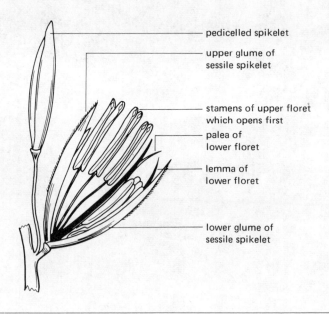

Fig. 2.13 *Zea mays*: Maize. Sessile and pedicelled spikelets of the male inflorescence.

Fig. 2.14 *Zea mays*: Maize.   The male inflorescence or 'tassel'.

is up to 15 cm wide and 150 cm long. The lamina tapers to an acute tip, and is slightly hairy on the upper surface, with a prominent mid-rib; its edges are often noticeably wavy at the base.

The male panicle is fairly compact and much branched. It varies in size, and is either erect or drooping depending upon the length and strength of its main axis and branches. The spikelets are in pairs, one sessile and the other pedicelled, which occur in several (usually four) rows on the main axis of the panicle, but in only two rows on its branches. Both sessile and pedicelled spikelets contain two male florets enclosed by glumes which are longer than the florets. Each floret has an oval, concave lemma enclosing a membranous palea, three stamens and two lodicules. The filaments elongate quickly at anthesis to expose the purple, pink, yellow or green anthers.

The lateral branch which bears the female florets grows from an axillary bud of the main stem. It has compressed, very short internodes, and except for the basal leaf with its well-developed sheath and small lamina, the leaves consist only of broadly sheathing bases terminated by ligules.

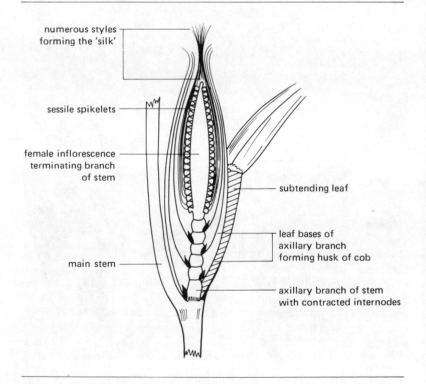

Fig. 2.15 *Zea mays*: Maize. Diagrammatic longitudinal section through the female inflorescence.

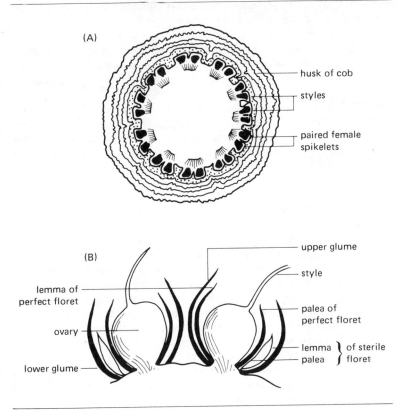

Fig. 2.16 *Zea mays*: Maize.   (A) Diagrammatic transverse section through the female inflorescence.   (B) Diagram of a pair of female spikelets.

These modified leaves overlap each other closely to form the 'husk' of the cob, and they protect the flowers and fruits so effectively that the seeds cannot be dispersed.

The spikelets of the female inflorescence also occur in pairs, but both members of a pair are sessile on the thickened axis. The pairs of spikelets are arranged around the axis in a close spiral. When the number of pairs per revolution of the spiral is a whole number, the spikelets appear to be arranged in longitudinal rows on the cob, but when the number of pairs of spikelets per revolution is uneven they appear to be arranged haphazardly, and not in rows. In each spikelet there are two florets subtended by very small, membranous glumes which give no protection, this function being served by the husk. The 'lower' floret of each pair is sterile and has only a lemma and palea. The 'upper' floret is fertile with a knob-like ovary borne on a short pedicel and surrounded by the short, broad, membranous lemma and palea. The style, or 'silk', grows rapidly from the apex of the ovary, and emerges from the top of the husk, a distance of around 30 cm from the base of the cob in some modern cultivars. The first styles to emerge from the husk are those from the basal florets of the inflorescence which mature first. The style is bifurcated at its tip, but it is stigmatic and receptive of pollen along most of its length, including that part within the husk.

Pollen is produced in prodigious quantities, first from flowers nearest the apex of the tassel, then by those progressively further down. Within each male spikelet the upper flower matures first, and there are two waves of pollen production, which may continue for 2 weeks, though most is shed about the third day after anthesis begins. The styles are receptive as soon as they emerge from the husk, the earliest appearing some 2–3 days after anthesis begins. Pollen is dispersed by wind, or by gravity to female inflorescences beneath the tassel of the same or nearby plants, though maize is more likely to be outcrossed than selfed.

After fertilization the silks wither and the broad, obovate and wedge-shaped fruits appear to develop in rows, or else more or less haphazardly over the cob, depending upon their phyllotaxy. When the fruits develop in rows the number of rows is always even because each fruit was derived from the single fertile floret of a pair. The mature seed rarely has a testa, and the ovary wall is fused directly to the endosperm, which is of two kinds; one which is hard, flinty and opalescent, with a greater proportion of protein (mostly zein) than the other softer, starchy kind which is white and floury. The proportions and distribution of these two kinds of endosperm in the seed vary between cultivars. Those with starchy endosperm at the top of the caryopsis surrounded by flinty endosperm are called 'dent' corns because the soft starch shrinks as it dries, creating a depression at the top of the grain. Cultivars in which the top of the caryopsis is composed entirely of flinty endosperm, which does not shrink, are called 'flint' corns. Several other distinct groups of cultivars are recognized, which also differ mostly in grain characteristics. The grains of sweet maize contain some sugar as well as starch and the mature grain is typically translucent and wrinkled. In waxy maize the starch of the endosperm is mostly amylopectin which becomes glutinous when it is cooked, so waxy maize is popular and common in Asia where people are familiar with, and prefer, glutinous cereals. Flour maize contrasts with 'flint corn' because all of the endosperm is soft and starchy. The interesting phenomenon of xenia is often apparent in the various colours of the mature grains on the cobs of some cultivars. Xenia is the expression in the hybrid tissues of the seed of a character brought by the male gamete. In maize this is expressed in the colour of the hybrid triploid endosperm which is visible through the transparent pericarp of the fruit. (In many cereals the pericarp, a diploid maternal tissue, is pigmented so that if the endosperm is coloured it is not visible in whole grains.) The grain contains up to 15 per cent of protein, which in most cultivars is notably deficient in the essential amino-acids

Fig. 2.17 *Zea mays*: Maize. Young female inflorescences, the 'cobs' or 'ears' with the styles or 'silks' exserted.

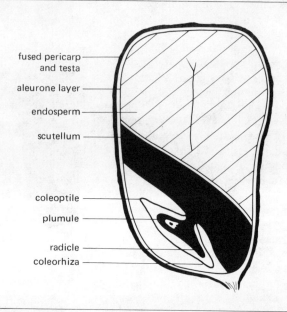

Fig. 2.18 *Zea mays*: Maize.   Diagrammatic longitudinal section of the caryopsis.

lysine and tryptophan. Mutant forms have been discovered in which the protein of the grain has an improved biological value because it is relatively rich in these two amino-acids. Maize with the mutant gene *opaque-2* has relatively more lysine and tryptophan, whereas the mutant *floury-2* has relatively more of another essential amino-acid, methionine, and a smaller increase in lysine.

The range of times from sowing to grain maturity varies widely in modern cultivars from about 90—190 days, and is influenced by both day-length and temperature. Maize flowers earlier in short, warm days than in longer, cooler ones. This is an important reason why maize can be grown in such a wide range of environments, because very early maturing types can avoid adverse growing conditions, whether they be due to cold or drought, and late ones are suitable for regions with very long growing seasons. Maize shows a marked response to applications of fertilizer, especially nitrogen and phosphate fertilizers, and is most productive on deep, fertile well-drained soils, and when modern cultivars are sown densely (up to 50,000 plants/hectare) with large applications of NPK fertilizers. Under the best conditions, and using hybrid maize (see below), more than 5 tonnes/hectare of grain are obtained (up to 10 tonnes with irrigation in Israel), but often less than one-tenth of this amount is produced by peasant farmers in Africa and Asia.

Throughout its history it is likely that maize has been improved by mass selection because farmers have very often kept the best cobs from their crops to provide the next year's seed. During the first half of this century, especially in the United States during the 1930s, the use of 'hybrid maize' gave spectacular yield increases, as great as 15—35 per cent over common open pollinated cultivars of the day. Now most of the maize crop in the United States consists of such hybrids. They are produced by breeding many inbred lines which are rigorously selected to find those which are best adapted to local conditions, and which combine well in crosses to give progeny with large grain yields. Their superiority over open pollinated cultivars is due largely to the greater effectiveness of selection among more or less homozygous inbred lines compared with selection among heterozygous, open pollinated populations. Hybrid seed is expensive to produce when only two inbred lines are used to make a 'single-cross' because grain yields from the inbred female parent are small. Seed for the commercial crop is less expensive when four inbred lines are used to make a 'double-cross' hybrid, because it is obtained from a female parent which is itself a 'single-cross' hybrid. The four inbred lines are first crossed in pairs with two single crosses, (A x B) and (C x D), then the progeny of this cross are hybridized (A x B) x (C x D), to produce seed for the commercial crop. Single and double cross hybrids were originally made by sowing several rows of the female parent for each row of the male in the field, then the tassels were removed from the female rows mechanically to prevent selfing. Now male sterile female lines are used in the commercial production of hybrid seed; because they produce no viable pollen it is unnecessary to incur the expense of detasselling them. The male sterility is caused by a factor inherited through the female cytoplasm. Fertility is restored to the progeny of a cross when the male parent used has genes which suppress the cytoplasmic factor for male sterility. The seed from crops of hybrid maize cannot be sown to produce a second crop because the desirable characters of the hybrid are lost in the second, segregating generation. Consequently hybrid maize is not suitable for introduction into peasant agriculture in countries without well-developed seed production and marketing organizations. In such areas 'synthetic varieties' or 'composite' cultivars consisting of open pollinated mixtures of many carefully selected lines offer better prospects for increased production and the further extension of maize.

As well as increasing grain yields and extending the range of the crop, the work of maize geneticists and breeders has been the source of great advances in the science of plant breeding which are applicable to other crops, and it has given us a better understanding of the genetics of maize than we have of any other flowering plant.

Grain sorghum: *Sorghum bicolor*

Little grain sorghum enters world trade and there are no accurate estimates of the areas sown to the crop. None the less, it is one of the most

Fig. 2.19 *Sorghum bicolor*: Sorghum. Indigenous tall sorghum ('Guineacorn') and recently bred dwarf sorghum in north Nigeria. (By courtesy of the Institute for Agricultural Research, Ahmadu Bello University, Zaria, Nigeria.)

important cereals of the semi-arid tropics, and provides the staple food for large populations in Africa and Asia. It is typically well adapted to the climate of the savanna zones of the tropics because it tolerates heat and drought, being notably superior to maize in this respect. It has a larger root system than maize, a smaller leaf area and xeromorphic characters such as waxy leaves which curl when subject to water stress, and so tend to restrict water loss. Sorghum is well suited to regions with erratic, uncertain rainfall at planting time because, even as a seedling, it soon recovers after limited periods of drought. Furthermore it withstands waterlogging and can be grown on heavy clays or light sandy soils. Sorghum is cultivated most extensively for human food as a rainy season crop in the seasonally dry African and Asian savanna zones, especially in West Africa and India, and typically in areas where the climate is also suitable for cotton and groundnut production. In these areas, because it is often the staple food crop, it receives priority for land and labour over other crops; it is sown early in the season, as soon as the rains begin, and is commonly grown in interplanted mixtures with other crops, especially legumes such as cowpeas.

Annual production for human consumption in Africa and Asia is estimated to be around 18 m. tonnes, whereas in the southern Great Plains region of the United States annual production exceeds 24 m. tonnes, all of it used for livestock feed. As well as its importance as a staple food, sorghum grain is used in peasant and subsistence farming communities to make beer, and its strong stems are valued for fencing and for the construction of temporary buildings. In more advanced tropical agriculture the grain is chiefly used for livestock food, the crop is grown for forage and silage, sugar is extracted from the sweet juicy stems of the 'sorgo' cultivars, brushes are made from the inflorescences of 'broomcorn' and starch is extracted from the grain of yet another group of cultivars.

Cultivated sorghum was probably domesticated from the wild diploid species *S. arundinaceum* in Africa 5,000–7,000 years ago, perhaps by the Cushite people in what is now Ethiopia, though an alternative interpretation suggests that it was domesticated independently in several parts of Africa. During its subsequent evolution in Africa natural hybridization between cultivars and their putative wild progenitor (which is still a weed of some sorghum crops in Africa) has given rise to diverse forms. Though *S. arundinaceum* did not reach Asia with the cultivars, a wild diploid relative, *S. propinquum*, does occur in China, and it too produces hybrids in natural crosses with a distinctive Chinese race of cultivars, the Kaoliangs. Consequently large numbers of closely related forms have evolved, and the taxonomy of the genus appears to be complex.

Sorghum cultivars are classified into several groups depending to a large extent upon the size and shape of their panicles, their grain type and plant size. Large-grained, tall cultivars mostly with lax panicles of the Shallu, Milo and Durra groups, as well as many forms with local names (the Guinea-corns of West Africa for example) are common in most of tropical Africa, but the Kafir group of cultivars with very compact panicles predominates in South Africa. In China sorghums of the Kaoliang group already referred to, with lax panicles and small, brown grains are most common; in India the Shallus are the most important group. In the United States dwarf cultivars with compact heads have been bred for mechanical harvest; they were derived principally from the Milo, Kafir and Hegari groups.

The mature sorghum plant resembles maize, varying in height from 45 cm to 4 m. It is an annual grass with erect, solid stems up to 3 cm in

Fig. 2.20 *Sorghum bicolor*: Sorghum. Mature panicles of the Nigerian 'Guineacorn' cultivar *'Farafara'*.

diameter at the base, and a profusely branching adventitious root system in the top meter of the soil, with a few roots which go much deeper. The single seminal root may function throughout the life of the plant. Supporting 'prop' roots burst through the sheathing leaf bases in whorls from a ring of extra-axillary buds at the base of each internode low on the stem. They are stiff and light-green, with anatomy intermediate between stem and root.

The main stem in some cultivars is solitary; in others many tillers arise from the basal nodes, and sometimes there are branches from axillary buds higher up, but the primary, main stem usually predominates, and its inflorescence is first to mature. The number of tillers produced varies between cultivars and is influenced by density of sowing; it is greatest when plants are widely spaced and on fertile soil. Each tiller has its own adventitious root system and is independent of the main stem for its nutrition. The stem nodes are slightly thickened. The internodes are shortest near the bottom of the stem, but their length also varies between cultivars, and is least in some dwarf types. Other, early maturing types are dwarf because they have few leaves, and the shortest, earliest sorghums are those with short internodes and as few as 7 leaves on the main stem compared with as many as 24 in tall, late cultivars. The sheathing leaf bases closely clasp and protect the stem, they have membranous margins and usually exceed the length of the internode above their insertion. They also secrete wax which forms a white, powdery covering on it. A short membranous ligule and auricles are usually present at the junction of the sheath and the lamina which is 30–135 cm long, bluish-green, glabrous and waxy with a prominent mid-rib. The uppermost leaf is called the 'flag'; its sheath enfolds and protects the developing panicle. The phase of crop development during which the lamina of the flag leaf is visible at the top of the plant, but before the panicle is exserted from the sheath, is called the 'boot' stage in all cereals.

The large panicle may be 20–40 cm or more long. It is carried on a stout peduncle which may be erect, pendent or recurved. Its main axis is deeply furrowed, variable in length and bears numerous primary branches in loose whorls at the hairy nodes, or in a loose spiral. From these arise secondary and tertiary branches. Whether the panicle is dense and compact (as in Milo or Kafir sorghums), or lax (as in West African Guineacorn, or in

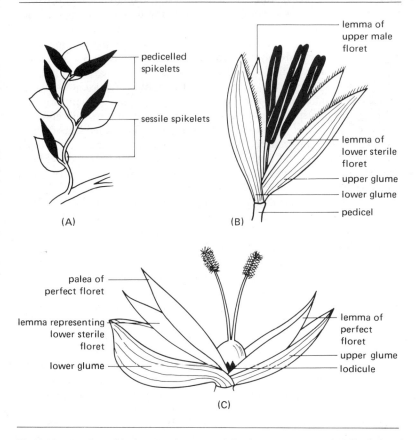

Fig. 2.21 *Sorghum bicolor*: Sorghum. (A) The arrangement of pedicelled and sessile spikelets on the ultimate branches of the panicle. (B) A pedicelled spikelet with one male floret. (C) A sessile spikelet with a single perfect floret from which the stamens have been shed. For the sake of clarity the hairs on the lemmas, palea and glumes have not been included.

Shallu types) depends upon the length of the main axis and its branches. The ultimate branches of the panicle bear the spikelets which are of two kinds, occurring in pairs; one is sessile and hermaphrodite, the other is pedicelled and either male or sterile. This arrangement is characteristic of the tribe *Andropogoneae* and the male inflorescence of maize in the tribe *Maydeae*. At the ends of the ultimate branches spikelets occur in groups of three, one sessile and two pedicelled.

The sessile spikelet of sorghum is broad and relatively large (3–10 mm long), with two glumes of about the same size enclosing the immature floret completely. The glumes persist with the fruit, though they are only one-half to two-thirds of its length. The lower, outer glume of the sessile spikelet is the firmer of the two; it is often stiff and coloured with several prominent nerves. The lower glume overlaps the base of the upper, inner glume which has fewer nerves and tends to be keeled and slightly pointed. The sessile spikelet contains two florets, the upper one perfect, the lower sterile and represented only by a lemma. The fertile floret has a thin, small palea and a narrow hairy lemma which is usually divided at the tip and sometimes has a twisted, kneed awn. These two bracts enclose three stamens, two lodicules and the ovary with its long styles and hairy stigmas.

The pedicelled spikelet is longer and narrower than the sessile spikelet. It is often deciduous. Its two glumes enclose two florets, neither of which has a palea; the lower floret is represented only by its lemma, while the upper has a lemma and three stamens. Sometimes the pedicelled spikelet consists only of the two glumes, or even more rarely, it may contain an upper perfect floret which produces a fruit.

Sorghum is a short-day plant, but the mechanism of its flowering response to photoperiod is not clearly understood. For example, it has been suggested that the date of inflorescence initiation in Nigerian sorghums may depend upon the number of successively shorter days after 21 June, rather than their absolute length, or the advent of a 'critical' short day-length. Whatever the mechanism may be, short-day cultivars in Nigeria are physiologically adapted by their short-day requirement for inflorescence initiation so that they flower close to the average date when the rains end in their own locality. Such adaptation tends to ensure that the grains neither become mouldy by maturing in wet weather, nor fail to mature for lack of water, because for some time after the rains have ended there is water in the soil. After the inflorescence has been exserted from the sheath of the flag leaf the first spikelets to open are those near the apex of the panicle; then flowering progresses downwards over a period of 6–9 days. The sessile spikelet of a pair opens first so that, by the time pollen is shed, the ovary in the sessile spikelet is mature and its stigmas are receptive. Self-fertilization is usual, but up to 5 per cent crossing does occur. Pollen remains viable only a few hours after anthesis whereas the stigmas are receptive before the sessile spikelets open, and may remain so for several days afterwards.

There are differences between cultivars in the shape, size and colour of the mature caryopsis. It is more or less rounded and bluntly pointed, 4–8 mm in diameter, with a black scar marking the point of its attachment to a stalk at one end, and sometimes the shrivelled remains of the two styles at the other. In some cultivars two or three fine lines run from the base to the tip of the grain. Its colour ranges from white or yellow to brown or black, and it is common for white grain to be discoloured by fungi or by sucking insects, especially if it has matured in wet weather. Not all cultivars have grain with a testa, but when present it is brown. Other pigments occur in the pericarp. The grain is ready for harvest 4–8 weeks after flowering. A hard, corneous endosperm surrounds a softer, floury centre, but the amount and distribution of amylose (the starch in corneous endosperm) and amylopectin (floury endosperm) varies between cultivars. The yellow endosperm of the caryopsis of the Nigerian cultivar 'short kaura' contains carotene. The pericarp and testa constitute about 6 per cent of the dry weight of the sorghum grain. It commonly contains around 9–12 per cent of protein, though cultivars with as much as 20 per cent protein have been reported. The flour prepared from sorghum is used to make porridge or batter, cooked either in water or with fat. It contains no gluten and so cannot be used to make bread.

Sorghum seeds will germinate soon after harvest, and retain their viability for considerable periods of time, even when stored at rather high temperatures; indeed, the embryo will germinate after it has been heated to 60°–70°C for several hours. The grain is subject to severe damage from insect pests in storage unless precautions are taken to exclude them, and, like other cereal grains, it soon becomes mouldy and looses its viability if it is stored damp.

Most of the sorghum grown in the United States consists of dwarf F_1 hybrids which are produced using essentially the same breeding techniques as those for the production of hybrid maize. Inbred lines are bred, selected and tested for their ability to combine well in crosses to give high-yielding progeny. Cytoplasmic male sterile female lines are used to produce the commercial seed crop. They are pollinated from male lines carrying genes which restore male fertility to the first generation offspring of the cross. Such high-yielding hybrids are available in parts of Africa and Asia, where breeders have taken care to ensure that the quality of their grain, and of the food made from it, is acceptable to consumers. However, as in the case of maize, it seems more likely that improved yields from peasant farmers' sorghum will be achieved in the near future by using open pollinated mixtures of several carefully selected lines which make up 'synthetic varieties', and from which the farmers can save their own seed.

The millets

Several ancient cereal crops with very small grains and which are, with the exception of pearl (or bulrush) millet, smaller plants than sorghum or maize, are collectively called the millets. They are important tropical and warm temperate cereals because they tolerate drought and intense heat, or avoid these conditions by growing to maturity very quickly, and because some of them will produce a crop on poor, infertile soil where other cereals might fail. The principal millets are *Pennisetum americanum* (pearl or bulrush millet), *Setaria italica* (foxtail millet), *Panicum miliaceum* (common or proso millet), and *Echinochloa frumentacea* (Japanese barnyard millet), all of the tribe *Paniceae*, and *Eleusine coracana* (finger millet) of the *Chlorideae*. All four of these genera from the *Paniceae* include wild species which are widespread in tropical grassland, especially in Central Africa. Some of them provide useful food in times of famine, when their grains are gathered. Indeed, the wild grasses *Echinochloa colona* (jungle rice) and *Echinochloa crus-galli* (barnyard millet) produce grain almost as useful as the cultivated species, though these two are typically swamp plants. Finger millet is the only cultivated species from the wetter tropics, and the only one with a digitate inflorescence. All the others have terminal panicles, which vary among the species from those which are open and lax to others which are very condensed and spike-like.

Millet grain is ground to a flour to make cakes or porridge, it is used to brew beer or to feed poultry and ornamental cage birds. The stems and leaves of cultivars are good forage, and wild members of the *Paniceae* are among the best of tropical forage grasses.

Pearl or bulrush millet: *Pennisetum americanum*

The cultivated *Pennisetums* are all referred to one species, *P. americanum*, which originated in the semi-arid savanna zone of Africa, where many of its wild relatives are now to be found. Pearl millet does not occur as a wild plant, nor has its wild progenitor been identified. There is no general agreement on whether the crop was domesticated in West Africa or in Ethiopia (perhaps with sorghum), or whether it had a diffuse origin at more or less the same time in several parts of the African savanna zone. It does seem likely that the early history of the crop was associated with that of sorghum, and that both reached India together from East Africa along the Sabaean Lane. The present world distribution of pearl millet is much like that of sorghum, but though the two are often grown together in interplanted mixtures in Africa and Asia, the cultivation of pearl millet alone extends into regions with as little as 250 mm of rain during the growing season, which are too dry for sorghum. However, even though pearl millet is less tolerant of drought as a seedling than sorghum, it is commonly grown on sandy, free-draining soils in areas where the early planting rains are unreliable. Young crops sometimes die from lack of water and have to be resown, so it is fortunate that the seeds, and therefore the seed rate for sowing, are so small.

Pearl millet is a freely tillering, tufted annual up to 4 m tall, though recently bred dwarf cultivars may be as short as 1 m. It is protogynous

The family Gramineae 51

Fig. 2.22 *Pennisetum americanum*: Pearl millet.   Condensed panicles with mature grain. In the cultivars illustrated the bristles which subtend spikelet groups are short and not visible. (By courtesy of the FAO.)

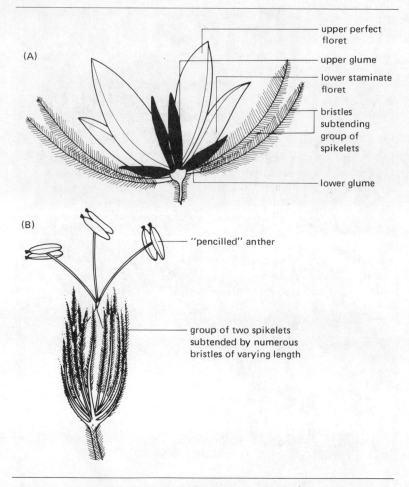

Fig. 2.23 *Pennisetum americanum*: Pearl millet. (A) A group of spikelets with most of the subtending bristles removed. (B) Anthers exserted from one floret.

(stigmas exserted and receptive before anthesis) and predominantly outbreeding so that the crop is heterozygous and variable. Within it there is a range of maturation periods (both photosensitive, short-day cultivars and day-neutral ones occur), of leaf and inflorescence size, and of the length and colour of the bristles which are a prominent feature of the inflorescence. The stem is solid and often much branched as well as producing tillers. Branching occurs from any node, and each branch and all the tillers (and their branches) may produce terminal panicles. The stem is narrow compared with sorghum, and it has prominent nodes marked by rings of long, white silky hairs. There is only one seminal root. Adventitious roots grow from stem and tiller nodes below ground, and 'prop' roots arise from the lowest nodes above ground.

The leaf sheaths clasp the stem closely and have pronounced, often swollen bases which are purple in some cultivars; their inner surfaces are furrowed and waxy. The lamina is up to 1 m long and 5 cm across. It is usually glabrous, and smooth on the lower surface, but rough above.

The terminal, much contracted panicle is usually around 50 cm long, but varies in length from 15—140 cm. It has a stout, finely hairy main axis and many short (2—25 mm), thin branches, each one usually carrying a pair of spikelets, though the number varies from one to four. A characteristic feature of these spikelet groups is the involucre of 25—90 bristles which arise at their base from the swollen tip of the inflorescence branch. The bristles may be longer or shorter than the spikelets. The outer bristles of each involucre are stiff and hairless, but the inner ones are often densely hairy. The group of spikelets and the bristles are shed together when the grain is mature.

Each spikelet contains two florets enclosed by a short, broad outer glume and a longer inner glume. The lower floret is usually male with a pointed, hairy lemma enclosing three stamens; the palea is very small, or it is absent. The upper, perfect floret has a broad, pointed lemma, a very thin oval palea, three stamens and an ovary with two styles fused (connate) at their base. Neither floret has lodicules.

Fig. 2.24 *Pennisetum americanum*: Pearl millet. Panicles from three cultivars in which the bristles subtending spikelet groups are long.

The mature caryopsis is small and wedge-shaped, 3–4 mm long and easily separated from the enclosing lemma and palea. It varies in colour from white to dull grey/blue with a black hilum.

Pearl millet cultivars have not been classified, though we do recognize the physiological distinction between groups of photosensitive, short-day and non-photosensitive, day-neutral cultivars, and distinguish between cultivars bred and grown for forage production in the south-east United States and the grain producers grown elsewhere. Recent breeding methods in the crop include the utilization of cytoplasmic male sterility to produce high yielding (often dwarf) F_1 hybrids, and the development of 'synthetic' mixtures of cultivars for use where agricultural development is not sufficiently advanced for the introduction of F_1 hybrids. Birds, especially *Quelea*, cause large grain losses in African and Asian crops of pearl millet (as they do also in sorghum); such losses may be decreased by growing cultivars with exceptionally long inflorescence bristles which hinder the birds when they try to eat the grain. Such cultivars occur in Senegal and the Gambia in West Africa.

Foxtail millet: *Setaria italica* (Italian, Hungarian, Siberian or German millet)

Foxtail millet was one of the earliest crop plants in Asia and Eastern Europe. It was probably domesticated in Asia more than 5,000 years ago, and has since spread throughout the tropics and warm temperate regions, though it is less important as a cereal now than in the past. It is grown for its grain in India, China, Japan, North Africa and South-east Europe; in the United States it is grown for forage or for grain to feed cage birds (parakeets, canaries, etc.; not poultry).

The plant is a vigorous, freely tillering annual with slender, strongly jointed stems up to 1.5 m tall, though usually much shorter. Lateral branches grow from axillary buds at the upper nodes. The internodes are hollow, the nodes are swollen and solid. All parts of the plant may be purple, but the intensity and distribution of this pigmentation varies widely between cultivars. The seedling has three seminal roots, and in the mature plant there is a diffuse, much branched adventitious root system. The leaf sheaths are loosely overlapping at their bases, but widely open above and longer than the internode they clasp. There is a short, thick and much divided ligule at the junction of the sheath and the lamina, but there are no auricles. The lamina is 30–45 cm long, and up to 2.5 cm wide; it narrows to a pointed tip, and is smooth on the upper surface, but rough below.

The terminal spike-like panicle is erect, curved or nodding. It has a ridged and furrowed, finely hairy (ciliate) main axis 5–30 cm long, and bears many short branches which are also covered with fine white hairs. Each branch of the panicle carries up to twelve almost sessile spikelets, each one subtended by four or five hairy bristles which are about as long as the spikelets. Some bristles occasionally bear a much reduced spikelet (or very rarely a functional spikelet) and are therefore interpreted as modified branches of the panicle. The spikelets are elliptic and have two florets which are partly protected by the glumes; the outer glume is short, pointed and three-nerved, the inner is almost as long as the spikelet, broad, pointed and five-nerved. The lower floret of each spikelet is sterile with a reduced membranous, oval lemma and sometimes a thin palea. The upper floret is perfect and has a stiff, variously coloured, five-nerved lemma and palea which enclose two small lodicules, three stamens with white or yellow anthers, and an ovary with two long styles terminated by brush-like stigmas. Flowering begins at the top of the panicle and continues for 8–16 days, with individual flowers opening late at night or in the early morning, depending upon temperature and humidity; flowering is promoted by cool temperatures and a rise in relative humidity. The crop is self-fertilized, but an average of less than 1 per cent crossing is reported. The mature caryopsis is firmly enclosed by the coloured lemma and palea, but these bracts are not fused to the pericarp of the fruit (as they are for example in rice and barley). The mature grain is shiny and commonly white or yellow, though the colour varies and may be red, brown or black. It is about 2 mm long, on one side flat with a black hilum, on the other convex with the position of the embryo marked by a groove. The spikelets are shed singly without their subtending bristles, not in groups as they are in pearl millet.

Twelve or more rather variable groups of cultivars are recognized, depending upon the size and shape of the panicle, and upon the colour of the grain. Foxtail millet matures 70–90 days after sowing, and though it will succeed on very little rain, it is less tolerant of poor soils than other millets, nor will it tolerate a waterlogged soil.

Common millet; proso: *Panicum miliaceum*

Common millet is one of the most ancient of crops, and is renowned for its quick growth and small water requirement. It will produce mature grain on very little rainfall in as few as 60 days after sowing. It was probably domesticated in Central and Eastern Asia and spread to India, Russia, Europe and the Middle East where it was among the earliest cereals to be cultivated. Common millet has not become well known in Africa, but it has been introduced to the United States where the grain is fed to livestock.

The plants are annuals 30–100 cm tall with a shallow adventitious root system. The slender stems are usually erect, though they sometimes spread at the base before ascending. They are glabrous or slightly hairy, with hollow internodes and few branches or tillers. The leaf sheaths are open (not clasping) and hairy at their junction with the lamina. There is a short ligule, but no auricles. The soft, pointed lamina is up to 30 cm long and 2 cm broad, and slightly hairy on its upper surface. The stem is terminated by a drooping panicle 10—45 cm long which varies in degree of compactness. It is often one-sided, and has a glabrous main axis and ridged branches which bear ovate, pointed spikelets at their swollen tips. The spikelets are about 5 mm long and contain two florets which are partially enclosed by the glumes. The outer glume is short and five-nerved, while the inner glume is longer and has many nerves. The lower floret of each spikelet is sterile with only a lemma and a very small palea, but the upper floret is perfect with a stiff, broad and pointed lemma and palea. The perfect floret has two lodicules, three stamens and an ovary with two long styles and feathery stigmas. The flowers are both self and cross-fertilized (selfing is said to be more common) and produce a nearly globular caryopsis about 3 mm long, enclosed by the hard, persistent lemma and palea. The grain is nutritious with 10—18 per cent protein, and some people find its 'nutty' flavour pleasant.

A second species, *Panicum sumatrense* (syn. *P. miliare*; little millet), is grown extensively only in India but also occurs wild in Burma. The plants are much smaller than common millet and very hardy. They tolerate both drought and waterlogging, and will produce a crop on the poorest, most infertile soils.

The genus *Panicum* is one of the largest in the *Gramineae* with about 500 species which are widely distributed throughout the tropics and the warmer parts of the temperate world. They occupy a variety of ecological habitats, and provide some of the best natural grazing in the tropics. The genus is poorly represented in the drier tropics, apart from a few more or less xerophytic species, such as *Panicum turgidum*.

Japanese barnyard millet: *Echinochloa frumentacea*

The genus *Echinochloa* comprises about 30 species of annual or perennial grasses in the tropics and sub-tropics, some of them locally important as fodder and others as human food in times of scarcity. *Echinochloa colona* (jungle rice) occurs in wet ground and as a weed of rice crops, and was once cultivated in Egypt. It is an excellent fodder plant, especially when in flower and fruit, and its grain is gathered for human food. *Echinochloa pyramidalis* (antelope grass) and *E. stagnina* are among the main constituents of the sudd of the Nile and Niger rivers; they are tall swamp grasses that provide excellent forage at all stages of growth, and grain for humans when food is scarce. *Echinochloa crus-galli* (barnyard millet), from which Japanese barnyard millet may have been derived, is an annual weed in many parts of Africa, Europa, Asia and America. Its grain is also gathered for food.

Japanese barnyard millet is grown for forage in the United States, and for its grain and for forage in many parts of the Far East. It is the quickest growing of all cereals, and produces mature grain in as few as 45 days after sowing. In Asia and Japan it is a crop of minor importance, often sown when the staple rice crop has failed, or where rice does not grow well. The grain is commonly eaten with rice or as a substitute for it, or it is used to brew beer.

The plant grows up to 1 m tall with strongly angled, smooth, glabrous main stems, tillers and branches, and has an extensive, finely branched adventitious root system. The seedling has three seminal roots. The internodes of the stem are hollow, and the solid nodes are slightly flattened. The leaves have glabrous, open sheaths which do not clasp the internodes above their origins, and which are slightly constricted where they join the laminas, but there is no ligule nor any auricles. The lamina is slightly hairy and has a rough, finely toothed margin; it is up to 2.5 cm wide at the base, about 35 cm long and tapers to a fine point at the tip.

The elongated terminal panicle has a rough, angled main axis and fifteen or more branches which are widely spaced and distinct from each other. The branches are rough to the touch and bristly at the nodes. The spikelets, each with a lower sterile floret and an upper perfect floret, are crowded on short, rough pedicels which are bristly at the base. The lower glume is about one-third the length of the spikelet, and is thin, purple and broadly clasping with three to five nerves. The upper glume is about twice as long as the spikelet, broad and acutely pointed, green at the base but purple above, and with soft hairs covering the upper part except along the five rough, spiney nerves. The sterile lower floret has a lemma similar in shape and texture to the upper glume, but with seven nerves, and a small palea. The fertile floret consists of the ovary and stigmas, two lodicules, and three stamens with purple anthers, all enclosed by an ovate, five-nerved lemma and a five-nerved palea. The lemma and palea are both fairly stiff and become shiny as the fruit matures. They enclose the mature grain closely.

Finger millet: *Eleusine coracana*

African millet, ragi, telabun, nagli, marua, korakan, wimbi and bulo are various local Indian and African names for finger millet. It is grown in several parts of Africa and Asia, but it is an important staple food only in parts of India, Uganda and Zambia where it is cultivated in wetter climates and at greater altitudes than other millets. Finger millet is a tetraploid ($2n = 4x = 36$) closely related to two wild species, the diploid ($2n = 18$) *E. indica*, which is found throughout the world's tropics, and the tetraploid ($2n = 4x = 36$) *E. africana*, which is common only in Africa. Several theories for the origin of the crop have been proposed, but it seems most likely that *E. africana* was the wild progenitor; in the progeny of crosses between it and the cultivars a large proportion of plants have normal chromosome pairing at meiosis. This suggests that *E. coracana* and *E. africana* have similar sets of haploid chromosomes (genomes). On the other hand there is no normal chromosome pairing at meiosis in the triploid hybrid between tetraploid cultivars and the diploid *E. indica*. (If *E. indica* was one of two diploid ancestors in a hybrid origin of *E. coracana* we

Fig. 2.25 *Eleusine coracana*: Finger millet. The digitate inflorescence of spikes of two cultivars.

would sometimes expect to see nine paired chromosomes, or bivalents, and nine unpaired chromosomes, or univalents, at meiosis in their triploid hybrid.) So we assume that the crop was domesticated in Africa, perhaps in Uganda, from *E. africana*, and that it spread subsequently to India, probably along the same route (the Sabaean Lane) and perhaps at about the same time as sorghum, pearl millet and cowpeas.

Within the genus *Eleusine* there has been little reduction of the primitive many flowered spikelet; each contains six to twelve florets which are not protected by the small glumes. The spikelets do not fall as a unit, but split up into individual florets with the fracture of the rachilla when the grains are mature.

Finger millet is a tufted annual up to 1 m tall with many tillers and branches. The stems are flattened and closely enclosed by the hairy leaf sheaths which are split along all their length. The ligule is a fringe of short hairs which clasp the stem at the junction of the leaf sheath and lamina. The lamina is up to 75 cm long, but less than 2 cm wide. It tapers to an acute tip and is folded upwards along the prominent mid-rib; its upper surface if often softly hairy.

The terminal inflorescence consists of four or more digitately arranged spikes, often with one carried below the main terminal group. Each spike is 5—15 cm long and bears two overlapping rows of spikelets along its outer side. In some cultivars the spikes are stiffly erect and diverge from the central point of attachment, while in others they tend to curve inward at the top, becoming almost rolled in extreme cases. Compared with their wild relatives the inflorescence of cultivars appears to be massive with more spikelets per spike and larger fruits which when mature seem to be densely crowded along the inflorescence branches. The glumes are small and strongly keeled. Usually all the florets are perfect, but sometimes those near the tops of the spikelets are male or even sterile. They are

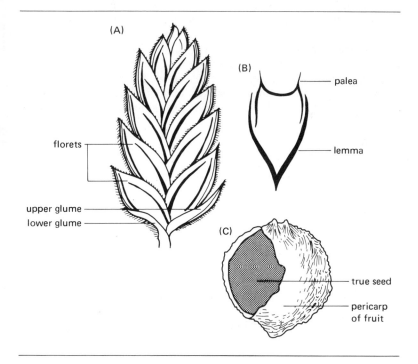

Fig. 2.26 *Eleusine coracana*: Finger millet. (A) A spikelet with many florets. (B) A transverse section through the lemma and palea of a floret. (C) The fruit with its papery pericarp partially removed to expose the seed.

arranged alternately along two sides of the zigzag rachilla, with larger florets at the base of the spikelet than above, and with the lemmas of successive florets slightly overlapping each other so that the spikelet is compact and tapers from bottom to top. The lemma of each floret is boat shaped and pointed at its tip, with a thick, dark green nerve bearing short hairs forming a keel down its centre, and one finer, shorter nerve down each side. The margins of the lemma closely overlap those of the palea, which is smaller with two prominent keels. The palea is pointed and it splits easily; it closely encloses the large ovary, three stamens and two broad lodicules. Short anthers are carried on long filaments. They dehisce just before, or as the flower opens, so that the crop is predominantly self-fertilized.

The caryopsis of finger millet is small and rounded, about 2 mm in diameter, and usually dull red in colour with a black hilum. It is not a typical caryopsis because the wall of the ovary is not fused to the testa of the seed, but forms a loose, light-brown, papery covering surmounted by the shrivelled remains of the two styles when the grain is mature. The pericarp can be removed from the seed easily. The grain is notably free from serious pest damage in store, and it keeps well for several years provided that it is kept dry. It is ground to a flour which is cooked for food, or whole grain is commonly used to brew beer.

Two main groups of cultivars are recognized. Those of the East African highlands resemble *E. africana* and have long spikelets, glumes and lemmas, and fruits which are enclosed by the lemmas and paleas. Those which resemble *E. indica* are grown mostly in India, and have short spikelets, glumes and lemmas, and fruits which are not enclosed by the lemmas and paleas.

Hungry rice: *Digitaria exilis*

There are about 300 species of *Digitaria* in the drier parts of the tropics, many of them, especially *D. decumbens* (pangola grass) of tropical America, providing excellent forage. Several species are cultivated in the savanna zone of West Africa, but *D. exilis* is the most important of them, and it provides the staple food in some restricted areas. It is an ancient crop in West Africa and was probably domesticated there, but it is not cultivated anywhere else, nor is it known to occur wild.

Hungry rice is an annual grass about 45 cm tall with many tillers and branches bearing linear, glabrous leaves around 15 cm long. The terminal inflorescence is made up of several racemes of spikelets which form a digitate panicle. Each raceme may be up to 15 cm long and carries spikelets in a single row along one side of the axis. The upper floret of each pair in a spikelet is hermaphrodite, but the lower one is sterile with a small palea enclosed by a lemma which fulfils the protective function of the very small lower glume. The upper glume is as long as the spikelet (about 2 mm) with three to five nerves.

Hungry rice is grown in areas too barren for other cereals and is given very little care after sowing, save perhaps for bird scaring by children without which it may be badly damaged. The crop matures 90–130 days after sowing, depending upon variation between cultivars. The grain is very small and is used to make porridge or to brew beer.

Teff, Ingera: *Eragrostis tef*

Teff is grown as a cereal only in the highlands of Ethiopia where it was probably domesticated, and where it has been the staple food for several thousand years. In South Africa it is grown for its palatable hay, and in India for green forage. It is a short (40–80 cm) tufted annual with an apical panicle of many long, drooping branches which carry small, five to nine flowered spikelets. The caryopsis is white or brown, and less than 2 mm long. In Ethiopia flour from the grain is used to make very large, thin pancakes called 'ingera'.

Further reading

General

Arber, A. (1934). *The Gramineae*, London: Cambridge University Press.
Clayton, W. D. (1970). *The Flora of Tropical East Africa: Gramineae*, Part 1, London: Crown Agents.
Davies, W. and Skidmore, C. L. (eds) (1966). *Tropical Pastures*, London: Faber.
Gould, F. W. (1968). *Grass Systematics*, New York: McGraw-Hill.
Hector, J. M. (1936). Introduction to the Botany of Tropical Crops, Vol. 1, *Cereals*, Johannesburg: Central News Agency.
Hutchinson, J. (1959). *The Families of Flowering Plants*, vol. 2, *Monocotyledons*, 2nd edn, London: Oxford University Press.
Hutton, E. M. (ed.) (1970). Tropical Pastures, *Agronomy*, 22, Am. Soc. Agron.
Langer, R. H. M. (1972). *How Grasses Grow*, London: Arnold.
Leonard, W. H. and Martin, J. H. (1963). *Cereal Crops*, London: Collier-Macmillan.
Milthorpe, F. L. and Ivins, J. D. (eds) (1966). *The Growth of Grasses and Cereals*, London: Butterworth.
Purseglove, J. W. (1972). *Tropical Crops. Monocotyledons*, London: Longman.
Stanfield, D. P. (1970). *The Flora of Nigeria: Grasses*, Ibadan University Press.
Whyte, R. O. (1969). *Grasses of the Monsoon*, London: Faber.
Whyte, R. O., Muir, T. R. G. and Cooper, J. P. (1959). *Grasses in Agriculture*, Rome: FAO.

Rice

Angladette, A. (1966). *Le Riz*, Paris: Maisonneuve and Larose.
Chandraratna, M. F. (1954). Photoperiod response in rice (*Oryza sativa* L.): effects on inflorescence initiation and emergence, *New Phytol.*, 53, 397–405.
Chandraratna, M. F. (1964). *Genetics and Breeding of Rice*, London: Longman.

Chang, T. T. (1970). Rice, in: Frankel, O. H. and Bennett, E. (eds), *Genetic Resources in Plants — Their Exploration and Conservation*, Oxford: Blackwell.
Grist, D. H. (1975). *Rice*, 5th edn, London: Longman.
Grist, D. H. (1974). Rice production in the past quarter of a century, 1949–1974, *World Crops*, 26, 213–14.
Houston, D. F. (ed.) (1972). *Rice. Chemistry and Technology*, Am. Soc. Cereal Chemistry.
International Rice Research Institute. *Annual Reports*, Los Baños: Philippines.
International Rice Research Institute. (1972). *Rice Breeding*, Los Baños: Philippines.
Ishizuka, Y. (1971). Physiology of the rice plant, *Advances in Agronomy*, 23, 241–315.
Jennings, P. R. (1966). The evolution of plant type in *Oryza sativa, Econ. Bot.*, 20, 396–402.
Owen, P. C. (1971). The effects of temperature on the growth and development of rice, *Field Crop Abstracts*, 24, 1–8.
Shastry, S. V. S. and Sharma, S. D. (1974). Rice, in: Hutchinson, Sir Joseph (ed.). *Evolutionary Trends in World Crops. Diversity and Change on the Indian Subcontinent.* London: Cambridge University Press.
Williams, C. N. and Joseph, K. T. (1970). *Climate, Soil and Crop Production in the Humid Tropics*, Kuala Lumpur: Oxford University Press.

Maize

Brandolini, A. (1970). Maize, in: Frankel, O. H. and Bennett, E. (eds), *Genetic Resources in Plants — Their Exploration and Conservation*, Oxford: Blackwell.
Mangelsdorf, P. C. (1965). The evolution of maize, in: Hutchinson, Sir Joseph (ed.), *Essays on Crop Plant Evolution*, London: Cambridge University Press.
Mangelsdorf, P. C. (1974). *Corn, Its Origin, Evolution and Improvement*, Harvard University Press.
Smith, C. E. (1968). The New World centres of origin of cultivated plants and the archaeological evidence, *Econ. Bot.*, 22, 253–66.
Sprague, G. F. (ed.) (1955). *Corn and Corn Improvement*, New York: Academic Press.

Sorghum

Andrews, D. J. (1972). Intercropping with sorghum in Nigeria, *Expl. Agric.*, 8, 139–50.
Andrews, D. J. (1975). Sorghum grain hybrids in Nigeria, *Expl. Agric.*, 11, 119–27.
Bunting, A. H. and Curtis, D. L. (1966). Local adaptation of sorghum varieties in Northern Nigeria, *Symposium on Methods in Agroclimatology*, UNESCO, Reading University.
Curtis, D. L. (1965). Sorghum in West Africa, *Field Crop Abstracts*, 18, 145–52.
Doggett, H. (1965). The development of cultivated sorghums, in: Hutchinson, Sir Joseph (ed.), *Essays on Crop Plant Evolution*, London: Cambridge University Press.
Doggett, H. (1970). *Sorghum*, London: Longman.
Ross, W. M. and Eastin, J. D. (1972). Grain Sorghum in the USA, *Field Crop Abstracts*, 25, 169–74.
Wall, J. S. and Ross, W. M. (1970). *Sorghum Production and Utilization*, Westport, AVI.
Wet, J. M. J. de and Harlan, J. R. (1971). The origin and domestication of *Sorghum bicolor, Econ. Bot.*, 25, 128–35.

Millets

Burton, G. W. and Powell, J. B. (1968). Pearl millet breeding and cytogenetics, *Advances in Agronomy*, 20, 49–89.

Chennaveeraiah, M. S. and Hiremath, S. C. (1974). Genome analysis of *Eleusine coracana*, *Euphytica*, 23, 489–95.

Ferraris, R. (1973). Pearl millet (*Pennisetum typhoides*), *Review Series*, No. 1/1973. Commonwealth Bureau of Pastures and Field Crops.

Singh, H. B. and Arora, R. K. (1972). Raishan (*Digitaria* sp.) — a minor millet of the Khasi Hills, India, *Econ. Bot.*, 26, 376–80.

Stewart, R. B. and Getachew, A. (1962). Investigations of the nature of injera (Teff — *Eragrostis abyssinica* = *E. teff*), *Econ. Bot.*, 16, 127–30.

Thomas, D. G. (1970). Finger millet (*Eleusine coracana* L. Gaertn.), in: Jameson, J. D., *Agriculture in Uganda*, London: Oxford University Press.

Chapter 3

Sugar Cane: *Saccharum* spp.

Sugar cane is a large, perennial, tropical grass cultivated for its tall, thick stems from which 'cane sugar' amounting to some 60 per cent of the world's annual production of around 80 m. tonnes of sucrose is obtained. The remaining 40 per cent is 'beet sugar', which is extracted from the roots of the temperate crop, sugar beet, *Beta vulgaris* subsp. *vulgaris* of the family *Chenopodiaceae*.

Cane sugar is an important export from several tropical countries, and many developing nations have fairly recently established or expanded their sugar industries to meet internal demand, and eventually hope to produce exportable surpluses. World demand for sugar tends to increase with affluence. Consumption is currently estimated to increase by about 3 per cent each year, though among industrialized, wealthy nations it is already large, as much as 40–60 kg per person per year compared with less than 4 kg per person each year in many developing nations. The national economies of Cuba and Mauritius are entirely dependent upon their sugar industries.

Sugar was not available to Europeans until the beginning of the eighteenth century, and only during the past 100 years has it become a common article in the diets of industrialized nations, though sucrose was first extracted from sugar cane in India more than 1,000 years ago. Few plants store their reserve carbohydrate as sucrose. Before cane or beet sugar were available sweetening materials in common use were honey and plant extracts such as the sap of the sugar maple tree.

Sugar cane is grown between latitudes 35° N. and S. of the equator. It requires mean temperatures warmer than 21°C, and grows best in the temperature range 32°–38°C. Rainfall must exceed 1,500 mm each year in a long growing season or be supplemented by adequate irrigation, and the crop requires a short dry season during which sugar accumulates in the stems. Though sugar cane will grow on heavy or light soils, they must be deep and free draining, whether the crop is grown under natural rainfall or with supplementary irrigation. Commercial sugar production involves very large capital investment in a factory or 'mill' to extract, purify and centrifuge the sucrose from the stems (which are called 'canes'), in facilities to transport their great bulk from the field to the mill, in a large labour force, and often in very large and expensive machines and tillage equipment to cultivate the land. Consequently the commercial production of cane sugar is entirely from large plantations, though the crop grown on them is sometimes supplemented by canes from small farmers nearby. Peasant farmers grow 'soft' canes to chew, and world production of non-centrifugal, dark-coloured sugar amounts to around 12 m. tonnes annually, much of it extracted from the canes in small presses powered by draught animals. The main producers of centrifugal cane sugar are Brazil (around 7.3 m. tonnes annually), Cuba (5.4 m.), India (4.3 m.), China (4.1 m.) and Mexico (2.8 m.). Australia, South Africa and the United States each produce more than 1 m. tonnes every year, and the sugar cane industry is important in many other tropical countries which produce smaller amounts.

As a crop modern sugar cane is remarkable for the large amounts of dry matter it produces. The canes are harvested 1–2 years (most often about 18 months) after planting 12,000–20,000 stem cuttings per hectare, or after regrowth has started following a previous harvest. On good soils with ample water, adequate fertilization (especially during early growth), and the best husbandry more than 200 tonnes of canes per hectare containing around 12 per cent sucrose are obtained. In Hawaii crops of sugar cane grown for 2 years before harvest, and with irrigation, yield up to 22 tonnes of sucrose per hectare. After the first harvest regrowth from the rhizomes can be harvested repeatedly over a period of up to 20 years on the best

Fig. 3.1 *Saccharum officinarum*: Sugar cane. Mature canes in Queensland, Australia (from G. C. Stevenson: *Genetics and Breeding of Sugar Cane*, Longman, 1965).

soils, but it is common practice to replant every 2—3 years, and in some areas the crop is replanted after each harvest.

There are four cultivated species of *Saccharum*, and two wild species which may have been involved in the evolution of the crop. During this century only interspecific hybrids involving *S. officinarum* have been used in commercial sugar estates, though two other cultivated species, *S. barberi* and *S. sinense*, are still grown in restricted areas. The fourth cultivated species, *S. edule*, is confined to New Guinea and is of little importance. All the cultivated species are polyploids with a rather wide range of chromosome numbers from 51—185, though the number is commonly 80 in *S. officinarum*. They rarely produce viable seed and are propagated as clones from stem cuttings.

Saccharum officinarum was probably domesticated from the wild species *S. robustum* in New Guinea, and spread very early into India through Java and Malaysia. Its cultivars are called 'noble' canes because their stems are larger, and contain more sucrose, but less fibre, than the canes of other cultivated species. During the eighteenth century two clones of *S. officinarum*, 'Bourbon' and 'Cheribon' from the South Pacific, were dispersed throughout the main sugar-producing countries, especially to the New World, where they replaced inferior cultivars of *S. barberi* (see below) which had been grown there since the sixteenth century. Towards the end of the nineteenth century 'Bourbon' was very seriously affected by epidemics of virus, fungus and bacterial diseases in the New World, and was consequently replaced, first by 'Cheribon', then by newly bred, disease-resistant hybrids between wild and cultivated species.

Saccharum barberi is indigenous to north India where it may have originated by natural hybridization between the wild species *S. spontaneum* and cultivars of *S. officinarum*. Alternatively, it has been suggested that it arose directly from *S. spontaneum* in India. Its cultivars have shorter, thinner stems than the noble canes, they yield less sugar and have a large fibre content. Though it is no longer important in the sugar industry, except in sugar cane breeding, the cultivation of *S. barberi* in India for chewing or for sugar extraction is ancient. Cultivars of this species were the first to reach Southern Europe, North Africa and eventually, with Columbus, the New World. Indeed, it seems probable that only one clone, subsequently called 'Creole', was involved in this dispersal and used to establish the early sugar industry in the West Indies in the sixteenth century.

Saccharum sinense originated in south China, perhaps as a hybrid between *S. officinarum* and *S. spontaneum*, or by selection from *S. barberi*. It is cultivated in India and Asia, but like *S. barberi* is now important only in sugar cane breeding programmes.

All of the cultivated species and the two wild species are inter-fertile, but the fertility of the offspring of various crosses is very variable. Modern, high-yielding, disease-resistant cultivars have been bred by a process known as 'nobilization', which involves the transfer of disease resistance and growth vigour from a wild species (most often *S. spontaneum*) to the noble canes of *S. officinarum*. This is achieved by hybridizing the noble canes and the wild species, followed by a limited programme of backcrosses to the noble canes in order to preserve their large sucrose yields while incorporating the disease resistance and vigour of the wild species. It is unusual for more than four backcrosses to be made because the progeny of subsequent ones do not express the desired characteristics of the non-recurrent, wild parent. The phenotype of segregating material can be 'fixed' at any stage in the breeding programme because the crop is propagated vegetatively. Perhaps the best known product of nobilization has been the cultivar POJ 2878, which was bred in Java in 1921, but is seldom grown now.

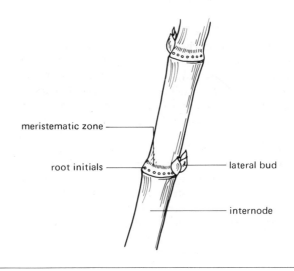

Fig. 3.2 *Saccharum officinarum*: Sugar cane.   Part of a stem (x¼).

Crops are propagated vegetatively using 'setts', which are stem cuttings up to 30 cm long taken from the top of 8—12 month old canes. Care in the selection and preparation of the setts for planting, even to the extent of growing the plants from which they are obtained in a nursery, is an important feature of disease control in sugar cane.

At each node of the cane are an axillary bud and a band of root primordia which develop to produce aerial shoots and adventitious roots when the sett is planted. The roots produced from the sett function briefly until adventitious roots arise from the basal nodes of the new plant and spread widely in the surface soil to a depth of about 30 cm, with a few which reach depths as great as 5 m. Once established the plants have many tillers, each one with its own adventitious root system. Supporting prop roots grow from the lower nodes of the canes above ground. The canes grow as tall as 6 m and are up to 6 cm in diameter, though cane size varies between species and is greatest in *S. officinarum*. They are solid, unbranched and usually erect, but sometimes more or less spreading, with a varying number of distinct nodes. The internodes are 5–25 cm long, the shortest ones at the bottom of the cane, the longest ones about halfway along its length. The canes have a fairly thick waxy covering and are variously coloured green, yellow, pink, red, black or even striped.

The leaves arise from the nodes in two opposite rows along the cane. The thin leaf sheath is closely overlapping at its base but open above, and has margins which become dry and brittle with age. A papery ligule, which is often split into two parts and is sometimes coloured, occurs with two auricles at the junction of the leaf sheath and the lamina. The lamina may be more than 1 m long and up to 10 cm wide with a prominent mid-rib; it is white on the upper surface but green below. It has a fine pointed tip and finely toothed margins which form a cutting edge in some cultivars. The leaves become yellow as the crop matures, and the ease with which they can be removed from the canes at harvest time varies between cultivars. Those from which the leaves fall readily are called 'free trashing', which is a desirable characteristic. The lamina of the uppermost, or flag leaf, is short, but its sheath is large and enfolds and protects the developing panicle.

Saccharum is a member of the tribe *Andropogoneae*, characterized by the occurrence of spikelets in pairs, one sessile and the other pedicelled. The terminal panicle is much branched and up to 50 cm long; it is called the 'arrow'. The main axis of the panicle narrows gradually from the terminal internode of the cane until it merges with the terminal rachis of spikelets. It has a slightly furrowed surface and branches which arise from the nodes in uneven whorls, some above and others below each node. The bases of the panicle branches are swollen and they are thinly covered with short, white hairs. Primary branches are about 15 cm long at the base of the panicle, but shorter above, and they carry secondary branches in two rows; these in turn may carry tertiary branches. The ultimate branches carry the pairs of spikelets, each spikelet of a pair subtended by a ring of long, silky white hairs more than twice as long as the spikelet. These hairs give the sugar cane inflorescence its characteristic and attractive appearance. The sessile and pedicelled spikelets both have two florets. The lower floret is sterile with only a thin, pointed lemma with hairy margins, while the upper one is perfect with a small palea, no lemma (except in *S. spontaneum*), two lodicules, an ovary with large plum-red coloured feathery stigmas, and three large, versatile, purple anthers. The glumes of each spikelet are broad, long and pointed, the lower with a prominent, green central nerve and redish pigment near its base on either side of the nerve. The glumes enclose and protect the florets completely.

Sugar cane is a reproductively photosensitive, short-day plant, and there

Sugar Cane: Saccharum spp. 69

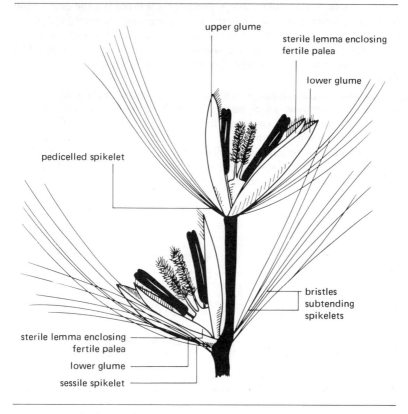

Fig. 3.3 *Saccharum officinarum*: Sugar cane. The paired spikelets.

is a great deal of variation in flowering dates within collections of cultivars grown together at one place; so much so that breeders have difficulty in synchronizing the blooming of plants from different sources for hybridization. The production of an inflorescence by the apical meristem of the cane brings to an end the initiation of new leaves on that stem, but tends to stimulate the development of vegetative branches at lower nodes. Such branches contain less sugar than the main stem. Furthermore, the developing inflorescence consumes sugar stored in the cane, and the amount of fibre in the cane increases after flowering and makes it difficult to express the sap. For the largest sucrose yields it is sometimes necessary to delay flowering, and this is achieved by withholding water from irrigated crops, or by spraying them with chemical defoliants such as 'diquat'. The flowers open during the night and in the early morning, beginning at the top of the panicle and progressing downwards and inwards. The stigmas are exserted about 3 hours before the anthers, so the flowers are protogynous, but the pollen remains viable only a few hours after it is shed, and the anthers fall from the filaments soon after anthesis. The caryopsis is ovate, yellowish-brown and rarely more than 1 mm long, with the shrivelled remains of the styles often persistent at its tip. The seeds soon loose their viability, but they can be stored and remain viable for long periods if they are freeze-dried.

The canes are harvested when they are uniformly mature and have a large sucrose content all along their length. If harvest is delayed after uniform maturity has been achieved the inversion of sucrose into glucose and fructose begins at the base of the cane and gradually proceeds upwards, with a consequent decrease in the quality of the cane which depends largely upon its sucrose content. Samples of canes are taken from the crop for analysis to determine when they are uniformly mature and ready for harvest. In many sugar-producing areas the leaves and trash are burned from the standing crop before the canes are cut by hand or by machine.

Further reading

Barnes, A. C. (1974). *The Sugar Cane*, 2nd edn, London: Leonard Hill.
Fauconnier, R. and Bassereau, D. (1970). *La Canne à Sucre*, Paris: Maisonneuve and Larose.
Humbert, R. P. (1967). *The Growing of Sugar Cane*, 2nd edn, London: Elsevier.
King, N. J., Mungomery, R. W. and Hughes, C. C. (1965). *Manual of Cane Growing*, New York: Elsevier.
Stevenson, G. C. (1965). *Genetics and Breeding of Sugar Cane*, London: Longman.

Chapter 4

The Legumes

The legumes are a very large group of plants second only to the cereals as a source of food for man and his animals. They belong to the family *Leguminosae* sub-family *Papilionoideae* which includes both temperate and tropical trees, shrubs and herbs, and is the largest, most widespread, and by far the most important of the three sub-families of the *Leguminosae*. The other two, the *Caesalpinioideae* and the *Mimosoideae*, consist mainly of tropical trees and shrubs with few economically important species. There are about 18,000 species in the *Leguminosae*, and they are characterized by their fruits, which are pods or legumes, and by their (usually) alternate, compound, pinnate or trifoliate leaves. The three sub-families are distinguished by the structure of their flowers, and by the arrangement of the petals in their flower buds (aestivation). The *Mimosoideae* have regular flowers; in the bud the petals do not overlap, but meet at their edges, and their aestivation is said to be valvate. The *Caesalpinioideae* and the *Papilionoideae* have irregular flowers; in the bud their petals

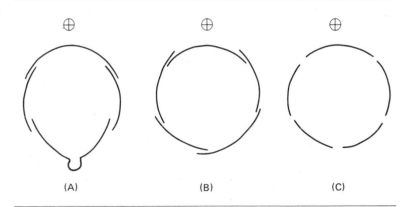

Fig. 4.1   Aestivation of the petals in the three sub-families of the *Leguminosae*.   (A) *Papilionoideae*. Aestivation imbricate, ascending.   (B) *Caesalpinioideae*. Aestivation imbricate, descending.   (C) *Mimosoideae*. Aestivation valvate. (⊕ Axis of the inflorescence.)

overlap and aestivation is said to be imbricate. In the *Caesalpinioideae* aestivation is imbricate-ascending with the upper or posterior petal (the one nearest the axis of the inflorescence) innermost, while in the *Papilionoideae* the opposite is true and aestivation is imbricate-descending with the posterior petal folded outside all the others in the flower bud.

Though the *Papilionoideae* is the chief subject of this chapter, economically important species in the other sub-families deserve brief mention. The *Caesalpinioideae* includes three species of the genus *Cassia* which are shrubs of minor economic importance in the tropics. *Cassia senna* grows wild, and is cultivated in the Sudan and in Egypt, and *C. angustifolia* is cultivated in India; the leaves and pods of both species are the source of the laxative senna. *Cassia auriculata*, avaram, was once important in India, and is still cultivated there for the tannin which constitutes about 20 per cent by weight of its bark. The tannin is used to cure raw animal hides in the manufacture of leather, but most is now obtained from the bark of the black wattle tree which is described briefly below. *Tamarindus indica*, the tamarind, is a tree which grows wild in the semi-arid parts of Africa, and is now widespread throughout the dry tropics. It is cultivated, especially in India, for the edible pulp which surrounds the seeds within the pods. The pulp is used to make sweetmeats, and in curries, preserves and drinks, or as a laxative. The *Caesalpinioideae* also includes several popular ornamental trees (*Bauhinia* spp., *Cassia* spp., and the flamboyant *Delonix regia*), and shrubs or small trees such as *Brachystegia* and *Isoberlinia* which are major constituents of savanna woodland in parts of Africa.

The most important economic species in the sub-family *Mimosoideae* is the Australian tree, black wattle, *Acacia mearnsii*. It was introduced to South Africa at the end of the nineteenth century, and from there to Kenya in 1903 where it was first cultivated as a source of firewood for the steam engines of the Kenya–Uganda Railway, then for its bark which contains around 40 per cent of tannin. It is now grown on a plantation scale in the Kenya highlands around 2,000 m above sea level, and in Natal, South Africa. Black wattle is a good source of tannin for the leather industry, and though it has been introduced to several parts of the world the major producer is still South Africa. The tree grows up to 20 m tall and bears large, bipinnate leaves with about fifteen pairs of pinnae, each pinna with up to 70 pairs of very small leaflets. The tiny yellow flowers are clustered in globose heads which in turn are numerous in axillary or terminal panicles. Each flower has five sepals fused into a calyx tube, five free petals about 2 mm long, and numerous free stamens whose yellow anthers are exserted well beyond the mouth of the corolla. The pods are distinctly constricted between the seeds. The crop is propagated from seed which has a very hard testa and which germinates unevenly unless it is first treated by soaking in hot water. The trees are felled when they are 4–12 years old and yield 4–7 tonnes of fresh bark per hectare. The bark is air dried and the tannin is extracted from it in water which is evaporated off to leave solid wattle extract. Several other species of *Acacia* are common in the arid and semi-arid parts of Africa and Australia, many of them with their stipules modified into protective thorns; but in spite of the thorns Acacias are browsed by animals. Gum arabic is the gum obtained by wounding the bark of *Acacia senegal* which grows wild in Africa. The gum exudes from the wound and solidifies on the bark, where it is collected, mostly for export to Europe and use in adhesives, confectionery and as size in the textile industry. Finally, *Parkia clappertoniana* (syn. *P. filicoidea*) of the *Mimosoideae*, the locust bean tree, is abundant in the savanna zone of West Africa, especially Nigeria. It has been estimated that around 0.2 m tonnes of its protein-rich seeds are eaten each year in north Nigeria. The trees are owned and protected, but they are not cultivated. Their flowers are clustered together in large spherical heads which are pendent on a stalk up to 30 cm long; they are pollinated by bats.

The sub-family *Papilionoideae*

The sub-family *Papilionoideae* includes about 480 genera and 12,000 species which form a natural group with distinct characteristics. Many of them are used as sources of food for man and his animals. The sub-family includes herbs, shrubs and trees, all with alternate (rarely spirally arranged), compound, pinnate or trifoliate leaves with stipules, and often with stipels subtending the pinnae or leaflets. In some species part, or all, of the compound leaf is modified into a climbing tendril. The inflorescence is a spike or raceme, always indefinite, sometimes contracted and sometimes with the flowers widely spaced. Bracts, varying in size and degree of persistence, usually subtend the flowers. With the notable exception of groundnuts among the economically important species, the structure of the flower is remarkably constant and is the characteristic feature of the sub-family. The flowers resemble butterflies and are said to be papilionaceous, from which the sub-family gets its name. They are zygomorphic and usually hermaphrodite, with five more or less fused sepals which partially enclose the base of mature flowers. Of the five petals the largest is the upper, posterior 'standard' which is held erect when the flower opens, and which is folded around the others in the bud. Two smaller lateral petals lie more or less parallel to each other and are called the 'wings', while the lower, anterior pair of petals are fused along their lower, or sometimes both margins, to form the 'keel' which encloses the androecium and gynaecium. There are ten stamens, usually diadelphous with nine filaments fused and the upper (vexillary) stamen free, or monadelphous with all ten filaments fused. The staminal tube encloses the superior ovary which consists of a single carpel with a long style having a broad, often oblique, stigma at its tip. The ovary contains one or few or many ovules, and matures into a dry, dehiscent fruit, the pod or legume,

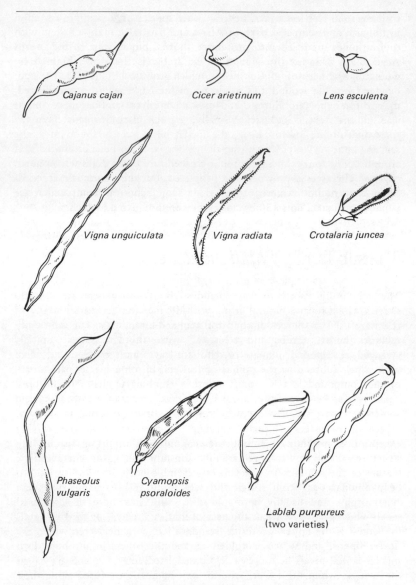

Fig. 4.2 The pods of some tropical leguminous crops (all ×½).

which in wild species and few cultivars dehisces, sometimes explosively, along both ventral and dorsal sutures. The seeds of the *Papilionoideae* vary greatly in size, shape and colour; none of them have an endosperm. Within the sub-family the range of variation in testa colour and colour pattern is similar in several different genera, and is a good example of the 'homologous variation' described by Vavilov.

The flowers are adapted to insect pollination, especially by heavy insects like bees, whose weight forces the keel down to release the anthers and stigmas which spring up against the lower surface of the insects' hairy abdomen (the flower is then said to be 'tripped'). Even so, many important legumes are self-pollinated and inbreeding because pollen is shed and the stigmas are receptive before the flowers open. Such flowers are said to be cleistogamous.

Man very early recognized the value of legumes as food, both for himself and his livestock, and among the large number of species which have been domesticated some are to be found wherever agriculture is practised. Legumes are important because they are rich in protein compared with other crop plants. This is especially true of their seeds, which in many species contain 20—30 per cent by weight of protein, though up to 50 per cent has been reported in some soyabean cultivars. Legumes cultivated especially for their mature seeds for human consumption are called pulses or grain legumes, but their immature seeds, and the young pods and leaves are also eaten as vegetables. The vegetative parts of grain legumes are commonly fed to livestock after their seeds have been harvested; species which are cultivated only to feed livestock are called fodder or forage legumes, or if they are grown in mixtures with pasture grasses, as pasture legumes. Few tropical legumes are grown exclusively for forage. Another group of legumes, the cover crops, is grown in the tropics to smother weeds, restrict soil erosion and to enrich soil nitrogen. They are often grown to cover the ground in plantations of trees such as rubber.

The special value of legumes as food has always been appreciated, but it has recently been emphasised as a result of increasing world population, and our awareness of the need to produce more food, especially more protein. Protein deficiency, or imbalance between protein and carbohydrate intake, is a feature of human diets in the wet tropics where the staple foods are starchy roots and tubers and where very few grain legumes are grown, but it is much less common where the staple foods are cereals. Cereal grains contain around 9—12 per cent of protein, and though increased meat and grain legumes consumption would improve staple cereal diets in the tropics, the most urgent need is to provide more protein in root and tuber diets.

In addition to their importance as food, legumes are important in agriculture as replenishers of soil nitrogen. Nearly all legumes, and most species in the sub-family *Mimosoideae*, but very few in the *Caesalpinioideae*, produce nodules on their roots in which symbiotic bacteria use carbohydrates from their host plant to supply energy for the fixation of atmospheric nitrogen in forms useful to plants. The host plants are able to use nitrogen from the nodules, and some remains in the soil when the plants die, or when they are ploughed in as green manure. The benefits derived in this way from grain legumes in the tropics are uncertain, for it is probable that a large part of the nitrogen fixed in the root nodules is harvested in the seeds. None the less, legumes can play an important role in helping to maintain soil fertility in permanent agricultural systems because soluble nutrients are readily leached from free-draining soils in the tropics, which are characteristically deficient in nitrogen.

Root nodules

The bacteria capable of infecting the roots of legumes and stimulating nodule development belong to several strains (sometimes given the status of species) of the genus *Rhizobium*. Each strain of the bacterium is able to infect the roots of only a particular group of legumes, or in the case of the root nodule bacteria of soyabean, only one species. The bacteria are free living in the soil, but fix nitrogen only in symbiotic association with the host legume. The seedling roots may be infected by *Rhizobium* at any time after root hairs are present on them. The bacteria in the soil are attracted by secretions from the root, especially of tryptophan, which they convert to indole acetic acid (IAA). The presence of this auxin may cause a root hair to grow around the bacteria nearby. In the free living state the bacteria are non-motile or motile (flagellated) coccoid cells, motility being the greatest in soils rich in phosphates. Motile forms pass through the wall of an epidermal cell or root hair, but the mechanism of their entry is not clearly understood. An 'infection thread' or tube containing the motile bacteria is formed by the wall of the host cell they enter; it grows into the cell and eventually to the inner wall where the bacteria pass into a cell of the root cortex. Each new cell entered by the bacteria produces a continuation of the infection thread, and so the process of invasion continues as the thread branches and ramifies across one-third or one-half of the root cortex. The bacteria do not stimulate nodule development unless they enter a tetraploid cortical cell, but whether their presence causes the chromosome number to double is not known. Once in a tetraploid cell, one or more bacteria are released from the infection thread, but remain within a sheath, the 'infection vacuole' formed by the plasma membrane of the infection thread. Now rapid divisions of the bacteria and the tetraploid host cell occur, until after about 2–3 weeks the bacteria become free from the infection vacuole, they lose their mobility and stop dividing; at about this time they begin to fix atmospheric nitrogen, using carbohydrates from the host legume as an energy source for the reduction of N_2 to NH_3 by the enzyme nitrogenase. The nodule increases in size for about 3 weeks after infection, and in it vascular tissues derived from undifferentiated cells of the root cortex develop until they become continuous with the stele of the root. The mature nodule has a characteristic shape, depending upon the species of legume; it consists of an outer layer of small cells containing a vascular trace, and an inner region of much larger cells which become vacuolated

Root nodules 77

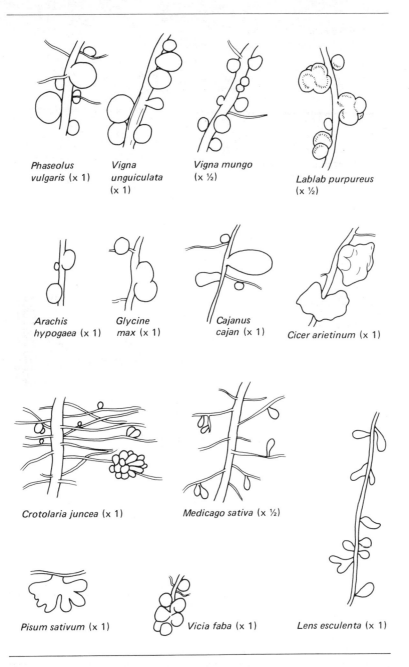

Fig. 4.3 Root nodules of some tropical legumes.

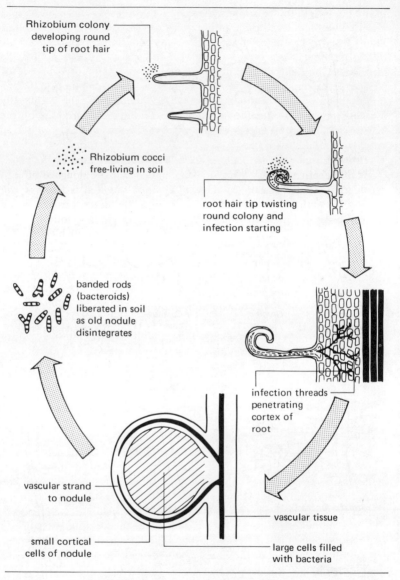

Fig. 4.4 A diagrammatic representation of nodule formation in a legume, and the life cycle of the root nodule bacteria (*Rhizobium* strains).

and are filled with bacteria. After 8—10 weeks the nodule begins to break down, and eventually falls from the root and disintegrates. The bacteria are then released into the soil as free-living coccoid cells whose cytoplasm is not evenly distributed so that, when stained, they appear as 'banded rods'. The symbiotic relationship between the bacteria and the host is a very delicate one; any failure in the supply of carbohydrates to the nodule results in the bacteria becoming parasitic and consuming nitrogen from the host plant instead of fixing it from the atmosphere. Though nodulation may occur normally, there is no nitrogen fixation in the absence of phosphate, calcium, magnesium, molybdenum, cobalt or boron. All functioning nodules contain the red pigment leghaemoglobin, and are usually pink in section. Considerable quantities of carbohydrates are used by the bacteria as a source of energy during nitrogen fixation, and the respiration rate in roots with active nodules is greater than in other roots.

Although some of the nitrogen fixed is undoubtedly used by the plant, the main benefit to the soil accrues when old nodules are sloughed off and disintegrate. Under certain conditions, as when nitrogen is fixed in the nodules faster than it is utilized by the host plant, excess nitrogen in the form of β-alanine or aspartic acid may be excreted into the soil. In mixed cropping systems involving cereals and legumes it is probable that the cereals do derive some nitrogen from this source.

A wide variety of leguminous crops is found in the tropics, but only the important ones are described below, and a great many have been omitted. They are described under two headings, the 'grain legumes', grown mainly for their mature seeds for human consumption, and the 'forage legumes', grown for their vegetative parts for livestock food. Soyabean and groundnuts are well known as oil seeds, but they are both included in this chapter partly because they are legumes and it is botanically convenient to describe them here, and also because they are very important grain legumes.

The grain legumes

Grain legumes are important in human nutrition in the less humid parts of the tropics, where they contribute substantially to total protein intake, particularly in those less arid parts where the main energy sources are starchy roots and tubers which contain little protein. Apart from some species of *Phaseolus* very few grain legumes are adapted to the humid tropical environment (though it may be possible to alter this by breeding) where consequently they are seldom grown. The proteins of grain legumes contain relatively more of the essential amino-acids lysine and tryptophan, and so usefully complement the amino-acids supplied by cereals in which the contents of lysine and tryptophan are relatively small. On the other hand, the proteins of grain legumes contain relatively small proportions of the sulphur-containing amino-acids methionine and cystine/cysteine. Grain legumes are useful sources of thiamine (Vitamin B, cocarboxylase), niacin (Vitamin B6) and of calcium.

The seeds of some species contain substances which inhibit the action of pancreatic proteases (trypsin inhibitors), but there is some doubt whether they act in humans in the same way as they do in experimental animals. They may also contain blood-clotting substances (haemagglutinin), cyanogenetic glucosides and fiatus factors, but the content of these antimetabolites has been substantially decreased in many species by selection and breeding.

In the peasant agricultures of Africa and Asia grain legumes are often interplanted with cereals, and indeed some of them are an integral part of traditional cereal farming systems. Though grain legumes may yield little seed when interplanted with cereals, they do help to smother weeds and protect the soil surface from the impact of heavy rain. The cereal crops may derive some nitrogen from the root nodules and the legume seeds are a valuable supplement to the staple cereal diet of the communities who grow them in this way.

Groundnut, peanut, monkey nut: *Arachis hypogaea*

Groundnuts were domesticated in South America probably in the region of south-west Brazil, northern Argentina, Paraguay and perhaps Bolivia and eastern Peru. This large area is the centre of diversity of the genus and contains perhaps 90 or more wild species though no more than 15 have been formally described. *Arachis hypogaea* is not known to occur wild and is the only cultivated species. The crop was unknown outside the New World in pre-Columbian times, but in the sixteenth century it was taken from Brazil to West Africa by the Portuguese, and from Peru to the Philippines by the Spanish. Since then it has spread throughout the tropics and sub-tropics to latitude 40° N. and S. of the equator, where rainfall during the growing season is greater than 500 mm, and is followed by a distinct dry season during which the subterranean fruits ripen and mature. The crop is grown most on loose, friable soils because it is easier to harvest the 'nuts' from them than from heavy soils. Annual world production of fruits, which are the seeds or kernels 'in shell', is around 17 m. tonnes, of which 60–80 per cent by weight is seed. India produces around 6 m. tonnes (little is produced in Pakistan or Bangladesh), China 2.7 m., the United States 1.6 m., Nigeria 1.2 m. and Brazil 0.8 m.; and Senegal, Burma, Indonesia, South Africa, Sudan, Niger and Argentina each produces more than 0.25 m. tonnes annually.

Arachis hypogaea is a variable annual herb whose chief, and remarkable, characteristic is the production of fruits underground. The many cultivars fall naturally into two distinct botanical groups depending upon differences between them in their branching habit. The groups are regarded as distinct sub-species.

(a) Alternately branching 'Virginia' groundnuts produce no reproductive branches from the buds in the axils of leaves on the main stem, which either give rise to vegetative branches (including the two which arise in the axils of the cotyledons) or remain dormant. On the primary branches from the main stem, and then on secondary and higher-order branches, the first

Fig. 4.5 *Arachis hypogaea*: Groundnut. A young crop growing on ridges made with a tractor-drawn implement in north Nigeria.

two nodes produce vegetative branches, the next two produce inflorescences, the next two branches, and so on in alternating pairs. In those Virginia cultivars which are known as 'runners', vegetative branches are diageotropic, and spread along the ground; in 'spreading-bunch' Virginia types the branches are ascending (more or less negatively geotropic) and produce densely bunched plants; in both groups the branches are commonly longer than the main stem. They are long-season cultivars requiring 120–180 days (rarely much longer) from sowing to maturity depending on temperature and to some extent spacing. Their foliage is dark-green and more or less resistant to the very common leaf spot disease caused by fungi of the genus *Cercospora*. In most cultivars the pods usually contain two seeds with russet-brown testas, which remain dormant for varying periods of time after they are mature, and so do not germinate if the soil becomes wet before they are harvested. On average the seeds of Virginia groundnuts contain less oil than 'Spanish-Valencia' cultivars (38–47 per cent), and about 25 per cent protein.

(*b*) Sequentially branched, 'Spanish-Valencia' cultivars produce vegetative branches from the axils of the cotyledons, and the first two or three leaves on the main stem. These branches rarely exceed the length of the main stem. Some of the higher nodes on the main stem, and most of the nodes on the branches produce inflorescences. Secondary vegetative branches are uncommon. All the Spanish-Valencia cultivars have an erect-bunch growth habit, but because there are so few secondary branches growth is less dense than in Virginia groundnuts. They are short-season

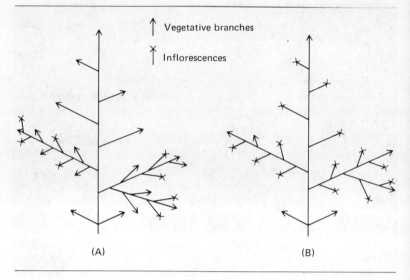

Fig 4.6 *Arachis hypogaea*: Groundnut. Diagrammatic representations of branching habit. (A) Alternately branching 'Virginia' groundnuts. (B) Sequentially branching 'Spanish-Valencia' groundnuts.

cultivars which mature 90–110 days after sowing; their foliage is pale-green and susceptible to the *Cercospora* leaf spot diseases. The pods contain two to six seeds with a range of testa colours including white, but never dark russet-brown; the seeds have no dormancy and will germinate in the pod before harvest if the soil is wet. The Spanish group of cultivars normally have two seeds per pod, the Valencia group normally more than two. The seeds contain much oil (47–50 per cent by weight), and about 25 per cent protein.

Groundnuts with a branching habit which is apparently intermediate between these two major groups were described for the first time in 1974. Such forms are extremely rare, but they have desirable characters which may be useful in breeding; they have flowers and branches on the main stem like sequentially branching groundnuts, and many secondary branches like alternately branching forms.

Groundnuts grow to a height of up to 50 cm, the Virginia bunch types generally taller than the Spanish-Valencia forms. They have robust primary roots which branch freely in the soil, and adventitious roots may arise from the hypocotyl, the zone of transition between stem and root. There are no roots hairs, and nutrients and water are mainly taken up in the region behind the tip of the finer rootlets. Provided that the correct strain of *Rhizobium* is present in the soil, or the seeds are inoculated with it before they are sown, root nodules are abundant on the main and lateral roots. They are globular, never lobed. The stems have short internodes, cylindrical in section and covered with short hairs; they become hollow with age. The four-pinnate leaves have petioles up to 7 cm long and are borne in a spiral sequence, though at lower nodes they seem to be alternate. They are subtended by a pair of long, pointed stipules which are fused together and to the petiole at their base. There is a marked pulvinus at the point where the petiole becomes free from the stipules; other pulvini occur at the point of attachment of each leaflet. The four leaflets are borne in two opposite pairs; each leaflet is obovate, softly hairy and 3–7 cm long, with a mucronate tip. Nastic (sleep) movements occur at the pulvini in which the petiole folds downwards and the leaflets fold upwards until they touch, the lower pair overlapping the upper pair.

Groundnuts are reproductively day-neutral, non-photosensitive plants. They begin to flower 4–6 weeks after sowing, with a peak of flower production 10–12 weeks after sowing. The flowers are carried in axillary inflorescences of condensed spikes containing several flowers which open successively at intervals of one or several days. At first sight the open flower appears to be typically papilionaceous, with bright yellow standard, wing and keel petals; but it is in fact most unusual in that the lower parts of the calyx, corolla and filaments are fused to form a hollow tube, the hypanthium, 4–6 cm long, which has often been mistaken for a peduncle. The superior ovary is borne within and at the base of this long tube, within which the style rises free. At the top of the hypantheum the calyx, corolla and filaments become free from each other, and appear to be inserted there. Four fused sepals form a large calyx-lobe behind the standard petal; the fifth is beneath the keel. Two stamens are sterile. The eight functional stamens are fused below, but free in their upper third, and carry anthers of three kinds; four anthers, of which one is uniloculate and three are biloculate, are oblong and they alternate with four which are smaller, uniloculate and globose. The cleistogamous flowers are self-pollinated as they open. After fertilization the hypanthium and other flower parts wither and are shed. Rapid cell divisions in a meristem beneath the ovary produce a long, negatively geotropic stalk, called the 'peg', which carries the ovary at its tip down, and into the soil to a depth of 2–6 cm. Most of the fruits on a groundnut plant are produced from flowers at the lower nodes on the branches, and in sequentially branched forms, on the lower nodes of the main stem. This is largely because these are the earliest flowers; many late flowers fail to produce fruit because, for several reasons, there may be insufficient assimilates available to support their development to maturity. A large number of fruits may even be from flowers produced beneath the soil surface as a result of cultivations, which throw earth towards the bases of the plants. The pegs from flowers above ground will grow as far as 15 cm to reach the soil, but those from flowers higher than this do not do so, and the ovaries at their tips die. Once in the soil the peg turns into a horizontal position where the fruit develops with its long axis parallel to the soil surface. The peg beneath the soil, and the

Fig. 4.7 *Arachis hypogaea*: Groundnut.   A plant lifted to show immature fruits, the 'nuts'. (By courtesy of the FAO.)

ovary, become hairy, and although the function of these hairs is uncertain it has been shown that the fruit absorbs water and some of its nutrients, including calcium, directly from the soil. The mature fruit is an oblong, indehiscent pod, 3—4 cm long (exceptionally as long as 8 cm). It is more or less constricted between the seeds, and has a fibrous, rugose, often beaked pericarp, or 'shell'. The seeds are ovoid, 1—2 cm long, without endosperm, and sometimes only loosely enclosed by the papery testa which has an epidermis of irregularly thickened cells. On germination the cotyledons are carried to the surface of the soil, but no further; so germination is neither epigeal nor hypogeal. The press cake remaining after the oil has been expressed from the kernels is a very valuable stock feed, with around 50 per cent by weight of protein. Mature groundnut seeds are liable to infection by the fungus *Aspergillus flavus* if they become damp in the soil or in store. A very toxic by-product of the metabolism of the fungus often produces a series of very toxic substances, the aflatoxins, which remain in the press cake, and have caused the death of animals, including poultry, fed with it. The production of toxic cake can be avoided by ensuring that, as far as possible, groundnuts are harvested, stored and transported in dry conditions. Groundnuts for human consumption are hand picked and selected to eliminate discoloured seeds which may have been infected with *A. flavus*.

Groundnuts are tetraploid ($2n = 4x = 20$). The regular pairing of chromosomes at meiosis (bivalents, no multivalents or univalents) suggests that they are allotetraploids, but the diploid ancestors are unknown. One of the main objectives of current plant breeding is to improve the resist-

Fig. 4.8 *Arachis hypogaea*: Groundnut. A mature crop which has been lifted and gathered into stacks to dry before picking the 'nuts' from the haulms. North Nigeria.

ance of cultivars to serious fungus (especially *Cercospora* spp.) and virus (rosette disease) pathogens, and it is likely that some recently discovered wild species from South America will play an important role in achieving these ends. At least one wild species important in this respect is diploid ($2n = 10$), and the progeny of its crosses with cultivars are infertile triploids ($2n = 3x = 15$). It may be possible to produce fertile, disease-resistant hexaploids ($2n = 6x = 30$) by doubling the chromosome number of such hybrids using colchicine.

Soyabean; Soybean: *Glycine max*

The genus *Glycine* is divided into three sub-genera with *G. max*, the soyabean, *G. ussuriensis* and *G. gracilis* in the sub-genus *Soja*, which has its centre of diversity in east China, Korea, Taiwan and Japan; of the other two sub-genera, one is endemic in Australia, the other in Africa and India. The three species in sub-genus *Soja* are interfertile, and some authorities suggest that they constitute a single species. *Glycine ussuriensis* is a wild species, and is the putative ancestor of soyabeans, which do not occur wild anywhere. Whether the second wild species, *G. gracilis*, was intermediate in the evolution of soyabeans from *G. ussuriensis*, or whether it arose after the cultivated species is not known. The most recent estimate suggests that the crop was domesticated in north-east China as recently as the eleventh century B.C., very much later than previous estimates, which suggested domestication earlier than 2800 B.C.

Annual world production of soyabeans is estimated to be around 63 m.

tonnes, mostly from areas north of latitude 30° N. which have temperate climates. The seeds have not become popular as human food outside Eastern Asia, but they are a major source of protein in diets there with annual production around 13 m. tonnes, 11.5 m. from China (mostly the north-east of the country). The United States is the leading producer with a crop of about 43 m. tonnes annually, all used in industry for the extraction of oil and protein; slightly more than half of soyabean production in the United States is from the 'corn belt' centred around the State of Illinois in the east-central part of the country 40° or more north of the equator. Before the Second World War soyabeans were important in the United States chiefly as a forage crop, but the great expansion of the crop there in recent decades has been as an oil-seed. The seeds contain 13—25 per cent of oil, and soyabean meal contains up to 50 per cent of protein which can be extracted to make artificial meat for human consumption. Soyabean flour is used as a component in human foods. Corn-soy-milk (CSM) is a compound of maize flour, soyabean flour and non-fat milk solids which has been found to be valuable as a food for infants in developing countries. The crop may also be grown for fodder, hay or silage.

Unfortunately, soyabeans are unpalatable to communities not accustomed to eating them, and who have traditionally eaten other grain legumes; and they are short-day plants not well suited to cultivation in low tropical latitudes. These are major reasons why the crop has not spread more widely in spite of its great value as a source of vegetable protein and oil. The seeds have the largest protein content of all cultivated legumes, some cultivars as much as 50 per cent, and the protein has a high nutritive value, especially after it has been heated to inactivate antimetabolites. Their nutritive value, and the expectation that breeders will produce cultivars well adapted to the humid tropics, stimulate continued attempts to introduce soyabeans as a grain legume into new regions where people suffer from dietary protein deficiency; and it is for these reasons that this very important, though predominantly temperate crop, deserves a place in this book.

Glycine max is a hairy annual with an extensive root system, mostly in the top 15 cm of the soil. The tap root, which is little different in diameter from its many branches, may grow as deep as 2 m. Adventitious roots grow from the hypocotyl. The root nodules are small and globose, but sometimes lobed. The strain (or species) of *Rhizobium* (*R. japonicum*) symbiotic in soyabean roots is specific to the crop, and seeds must be inoculated when soyabeans are introduced for the first time to a new area. Modern cultivars are erect, bushy plants 20—180 cm tall, usually with few primary branches, and no secondaries; but prostrate, freely branching forms do occur. Plant habit is variable, and within any cultivar may depend in part on variation in the duration of growth before flowering, because soyabeans are reproductively photosensitive, short-day plants (and were used by Garner and Allard in their early classical experiments on photoperiodism). If soyabeans are sown close to the date when days are short enough to induce flowering, vegetative growth is of short duration and the plants are small; on the other hand branches are more numerous, and longer, if the crop is sown a long time before it flowers. Cultivars have been classified into maturity groups, but the classification is useful only if the latitude where the crop is to be grown, and the day-length changes at that latitude, are known. Furthermore, both vegetative and reproductive growth are influenced by temperature. The minimum temperature for growth is 10°C, and though vegetative growth is promoted by warm temperatures, it is retarded by temperatures warmer than about 38°C, especially in young plants. In experiments with constant day-length cool night temperatures (19°C) have a marked effect in delaying flowering in some cultivars compared with warmer night temperatures (24°C).

The trifoliate leaves are alternate with long petioles, and small stipules and stipels; the leaflets are ovate to lanceolate and have a more or less mucronate tip. The flowers are in short axillary, or sometimes terminal, racemes on the main stem and branches, each raceme with 2–35 small, white or pale-purple flowers. Each flower is subtended by two bracteoles, and has a hairy calyx of five pointed sepals united for about half their length; two upper calyx lobes are behind the standard petal and three are below the keel, the middle one of these exceeding the length of the keel. All the calyx-lobes are sharply acuminate and may be tinged purple. The flowers are small (around 6–7 mm long), with a notched standard petal which is auricled at the base and hinged to the two narrow wing petals. The standard is white or pale purple, with dark purple veins at its base. The keel is short and broad, its two petals fused only along their lower margins, and open above, exposing the staminal tube which encloses the short, hairy ovary with its curved style and apical stigma. The ten stamens are usually diadelphous, with the vexillary stamen free, but they are sometimes monadelphous. The flowers are normally self-pollinated (sometimes cleistogamous), but around 1 per cent of cross-pollination by insects does occur. Many of the flowers produced by soyabeans abort, or are shed at various stages of development after they have opened, because there is insufficient food to support the full development of them all. The percentage abortion varies from 20–80 per cent, and is greatest in those cultivars which have the largest number of flowers in each inflorescence. The pods are carried on short stalks, and occur in groups of three to fifteen. Their colour and size varies, but commonly they contain two to four seeds and are 3–7 cm long, hairy, light brown when mature, and slightly constricted between the seeds. The seeds vary greatly in size, shape and colour, but are most often round and yellow or yellowish-green, brown or black. Germination is epigeal.

Phaseolus species

Several species of the Central and South American and Asian genus *Phaseolus* are among the most important grain legumes in the tropics and in temperate agriculture. The best known and most widespread of them is *P. vulgaris*, the common bean, whose fruits and seeds are those most of us refer to when we speak of 'beans'. *Phaseolus* species are annual or perennial, twining or bushy herbs with large trifoliate leaves, stipules and stipels, and typically papilionaceous flowers in axillary or terminal racemes. A distinctive character of the flower is the spirally coiled keel, but since several Asian species have been transferred from *Phaseolus* to the cowpea genus *Vigna*, even though their keels are coiled, it is no longer diagnostic of the genus. *Phaseolus* belongs to the tribe *Phaseoleae* of the *Papilionoideae*, and is morphologically similar to *Vigna* (cowpea) and *Lablab* (hyacinth bean) in the same tribe. One group of species with white or purple flowers and large seeds is of New World origin (*P. vulgaris*, *P. lunatus*, *P. acutifolius* and *P. coccineus*), while another with yellow flowers and small seeds, whose true taxonomic affinities are less certain, is Asian (*P. aconitifolius*, *P. angularis* and *P. calcaratus*). The very important Indian grain legumes, green gram or mung (*Vigna radiata*), and black gram or urd (*Vigna mungo*) are among those recently transferred from *Phaseolus* to *Vigna* (*Vigna radiata* was *Phaseolus aureus*; *Vigna mungo* was *Phaseolus mungo*).

The genus *Phaseolus* differs from *Vigna* in that its stipules do not have appendages below their point of attachment to the stem; its pollen grains are smooth, not with an open reticulation of raised walls; and all its species have coiled keels whereas most *Vigna* species do not.

Common bean; haricot, French, kidney or snap bean: *Phaseolus vulgaris*

Phaseolus vulgaris was domesticated in Central and South America more than 6,000 years ago. Wild forms with small black seeds are now to be found in tropical America. It was probably first cultivated with maize, and it seems likely that the two crops evolved together in a cereal/legume farming system in much the same way as cowpea, sorghum and pearl millet in West Africa, and later in India. Common beans were spread to Europe, Africa and Asia by the Spanish and Portuguese, and are now grown throughout the cooler tropics, but not in hot semi-arid or wet humid regions. They are susceptible to drought, and set little seed, or shed many flowers in areas with very high temperatures or year-round heavy rainfall and high humidity. Like soyabeans, *P. vulgaris* grows well in drier areas which are suitable for the cultivation of maize. In the tropics it is typically a short duration crop, in Asia often grown in rotation with rice. Although it is a very variable, annual herb with hundreds of modern cultivars, only two main kinds of common bean are recognized. 'Bush' cultivars are day-neutral, early maturing, dwarf plants 20—60 cm tall with lateral and terminal inflorescences, and consequently determinate growth. 'Climbing' or 'Pole' cultivars are indeterminate, and may grow 2—3 m tall if they have supports to climb by twining; among them are day-neutral and short-day types.

The tap root is large with many branches in the upper soil. The stem is slender, twisted, angled and ribbed, more or less square in section, and

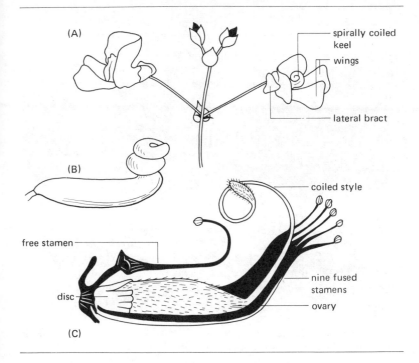

Fig. 4.9 *Phaseolus vulgaris*: Common bean. (A) The inflorescence. (B) A diagram illustrating the coiled keel. (C) A diagram of the essential organs of the flower.

often streaked with purple. The alternate trifoliate leaves are large, rarely with leaflets as long as 15 cm and 10 cm broad; the terminal leaflet is subtended by a pair of tiny stipels, the lateral asymmetrical leaflets by one each. The flowers are white, pink or purple with a reflexed standard about 1 cm broad, and diadelphous stamens; they are self-pollinated. The pods are slender, 10–20 cm long, straight or curved, and terminated by a prominent beak; they contain four to six seeds, rarely more. The seeds are non-endospermic, and vary greatly in size and colour, from small black 'wild types', to the large white, brown, red, black or mottled seeds of cultivars which are 7–16 mm long. Germination is epigeal. Common beans are grown for their mature, dry seeds, for their immature green or yellow pods ('snap beans'), and occasionally in Africa and Asia for their leaves used as vegetables.

Lima, sieva or butter bean: *Phaseolus lunatus*

Phaseolus lunatus is another ancient New World crop which was domesticated in Central and South America, perhaps independently in Guatemala and Peru. It has spread throughout the tropics where it is one of the few

Fig. 4.10 *Phaseolus lunatus*: Lima beans.   'Pole' or climbing cultivars in an experiment at the International Institute of Tropical Agriculture at Ibadan, Nigeria.

species important as a grain legume in wet and humid rain forest zones. It occurs wild as an escape from cultivation in many countries. Two more or less distinct types are recognized. Naturally perennial (but cultivated as annuals), late maturing types usually with large, white, flat seeds, the 'lima beans', are thought to have originated in Peru. Earlier maturing, naturally annual types with small flat or rounded, variously coloured seeds, the 'sieva beans' probably originated in Mexico. Among modern cultivars of both types there are determinate 'bush' cultivars and indeterminate 'pole' or climbing cultivars. The large seeded lima beans have spread more widely, and are more important in the humid tropics, than sieva beans, which are more tolerant of heat and drought. The flowers of *P. lunatus* are produced in large numbers on the racemes; they have a greenish standard petal which contains chlorophyll, and white wing and keel petals. The curved, flat pods are 5–12 cm long, markedly flattened, papery to the touch and slightly hairy. The seeds contain a cyanogenetic glucoside which breaks down to produce hydrocyanic acid. It is most abundant in coloured seeds which are safe to eat only after boiling in an open vessel which allows the HCN to escape. Germination is epigeal.

Tepary bean: *Phaseolus acutifolius*

Tepary beans were domesticated in the region of north Mexico and in the arid south-western United States, and this has remained the only area where they are cultivated on a large scale. Their chief characteristics are their very early maturity and their tolerance of heat and drought, so they are potentially valuable grain legumes for the semi-arid tropics. The plants are small, glabrous or slightly hairy, and either bushy or with trailing and twining stems. The leaflets are broad, long and strongly pointed with veins prominent on their under surface. The flat pods are 5–9 cm long, straight or curved with a prominent beak, hairy when young, and with a rimmed margin. They contain around five dull, round or oval, sometimes flattened, white, yellow, brown, black or mottled seeds. Germination is epigeal.

Mat or moth bean: *Phaseolus aconitifolius* (now *Vigna aconitifolia*)

Mat beans probably originated in the semi-arid parts of India, and they are now cultivated there and in Asia as a grain legume, often in interplanted mixtures with sorghum and pearl millet, and as a forage crop in India and the south-western United States. They grow best in hot climates on free-draining soils where annual rainfall is about 600–800 mm. Mat beans are short-day plants, much branched, widely spreading and hairy, with a more or less erect main stem up to 30 cm tall. Their leaflets are characteristically deeply lobed, the terminal leaflet largest and five-lobed, the lateral leaflets four-lobed. The leaves are subtended by large stipules, up to 12 mm long. Very small, yellow flowers are clustered on axillary peduncles 5–10 cm long; they are self-fertilized and produce thin, narrow, hairy, beaked pods 2–6 cm long. The pods contain four to nine small, rectangular seeds which are grey, black or mottled. Germination is epigeal.

Adzuki bean: *Phaseolus angularis* (now *Vigna angularis*)

Phaseolus angularis is an Asian member of the genus native to Japan. It has never been cultivated widely outside Japan and China, where it is second in importance as a grain legume only to soyabean. Adzuki beans are erect, bushy, annual plants with clusters of bright yellow flowers in short racemes. The fruits are around 10 cm long and contain up to twelve small oblong seeds which are red, brown, straw-coloured or black. Germination is hypogeal. The crop has been introduced to Zaire, South America and the United States.

Vigna species

Vigna is a pantropical genus of about 170 species with the largest number endemic in Africa, but several in India, Australia and the New World. Only three species are important as grain legumes, cowpea (*V. unguiculata*), green gram (*V. radiata*) and black gram (*V. mungo*), and cowpeas are an important forage crop. The morphological similarity between *Vigna* and *Phaseolus* has already been mentioned. As well as less obvious differences between them, *Vigna* is distinguished by distinct projections from the large stipules below their point of attachment to the stem.

Cowpea, catjang, 'yardlong' bean, blackeye pea: *Vigna unguiculata*
There are three cultivated and two wild sub-species of *V. unguiculata* which are all interfertile. The wild sub-species occur only in Africa, and one of them, subsp. *dekindtiana*, is the putative wild ancestor of the cultivars which were domesticated in the Ethiopian region, or in West Africa, or perhaps widely throughout the African savanna zone, more than 4,000 years ago. The earliest cultivars in Africa were probably spreading, reproductively photosensitive, short-day types of subsp. *unguiculata* which may have evolved in cultivation with sorghum and pearl millet in the cereal farming system of the West African savanna zone, where they are now the principal grain legume. In Nigeria alone annual cowpea seed production from sub-species *unguiculata* is around 0.75 m. tonnes, almost all for local consumption; this is some 75 per cent of the estimated annual world production of 1.0 m. tonnes of cowpea seed. In the traditional cereal farming system of West Africa, spreading, short-day cowpeas are interplanted with sorghum or pearl millet about 6 weeks after the cereals were sown, and after they have been weeded and soil has been thrown up around their roots. The spreading growth of the cowpeas soon smothers weeds, and makes further weeding unnecessary; they protect the soil from the impact of heavy rainfall, and it is likely that the cereals derive some nitrogen from the cowpea root nodules, especially towards the end of the growing season. At this time old nodules may be sloughed off and decompose to release nitrogen into the soil, and nitrogen may be excreted from active nodules if the cowpeas are so heavily shaded by the cereal canopy that they cannot use all that is available from their nodules for their own retarded growth. Towards the end of the growing season the cereal inflorescences begin to develop, though there is least nitrogen in the soil at this time because it has been leached or used by the crops; so any derived from the cowpeas will make an important contribution towards cereal grain yield. Like sorghum in Nigeria, spreading, short-day cowpeas interplanted with cereals flower close to the end of the rains in their own locality so that their seeds mature in dry weather while there is stored water in the soil. This local adaptation to the length of the growing season appears to be based upon their short-day requirement for the development of inflorescences which may have been initiated earlier, in longer days.

A distinct group of early maturing, more or less upright, day-neutral forms of subsp. *unguiculata* probably evolved in the transition zone between the West African rain forest and savanna, where rainfall is bimodally distributed, and there is a marked dry period during the 'rainy season'. They are grown there now, commonly as 'sole' crops or interplanted with yams, and are sown at the beginning of the rains for harvest in relatively dry weather 3—4 months later (July/August) when food of all kinds is scarce. These day-neutral forms of subsp. *unguiculata* are the chief forage cultivars in the advanced agriculture of other parts of the tropics.

Sub-species *unguiculata* probably reached India with sorghum and pearl millet from East Africa around 1500 B.C. In India the two other cultivated

The grain legumes 93

Fig. 4.11 *Vigna unguiculata*: Cowpea. An upright cowpea cultivar with pods held erect on long peduncles. (By courtesy of the FAO.)

sub-species were selected from it, the upright, determinate cultivars of subsp. *cylindrica* (cat-jang) with small seeds originally for forage, and spreading cultivars of subsp. *sesquipedalis* ('yardlong', 'asparagus' bean) for its pods which are 30–100 cm long and used as a vegetable. These two Indian sub-species are rare in Africa as relatively recent introductions. Cowpeas reached Southern Europe from Asia before 300 B.C., and the New World from West Africa and Europe in the seventeenth century.

Cultivated cowpeas are very diverse, usually glabrous, annual herbs with a strong, deep tap root and many branches from it in the surface soil. The root nodules are smooth and spherical, about 5 mm in diameter; they are numerous on the tap root and its main branches, but sparse on the smaller roots. The terminal leaflet of the trifoliate leaves is commonly around 12 cm long (rarely as long as 16 cm) and larger than the asymmetrical lateral leaflets. The stipules are large and spurred at the base, the stipels are inconspicuous. The large, showy flowers are commonly white or purple, rarely yellow, with a standard petal 2–3 cm across. They occur in alternate pairs on a long axillary peduncle, from 2 cm to rarely more than 30 cm long; there are large extrafloral nectaries between each pair of flowers. Many flowers may be produced in each inflorescence, but only two to four produce fruit. The rest are shed. The five sepals are fused for

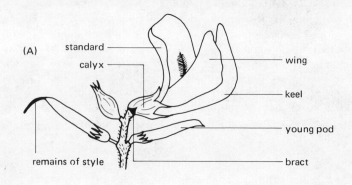

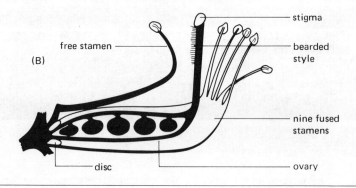

Fig. 4.12 *Vigna unguiculata*: Cowpea. (A) A flower and young pods (x1½). (B) A diagram to illustrate the essential organs of the flower.

more than half their length, but free above as five, pointed calyx-lobes. The stamens are diadelphous, and the ovary is straight with a style bent at right angles, hairy on the inner side, and extending as a beak beyond the lateral stigma. The flowers are cleistogamous, and although 1–2 per cent crossing occurs, cowpea cultivars are usually inbreeding, more or less homogeneous pure lines. The pods of cultivars are indehiscent, and vary greatly in colour and size. Indeed, pod and seed size are the chief diagnostic characters of the three cultivated sub-species as follows:

Subsp. *unguiculata*: Pods 10–30 cm long, pendent. Seeds 5–12 mm long, very rarely shorter than 6 mm

Subsp. *cylindrica*: Pods 7.5–13 cm long, usually erect. Seeds 5–6 mm long

Subsp. *sesquipedalis*: Pods longer than 30 cm, flabby. Seeds usually 8–12 mm long

The two wild sub-species have scabrous, dehiscent pods and small, dark speckled seeds. The pods of the cultivars vary in colour from pale straw to brown, red or dark purple, and may be straight, curved or rarely coiled like a spring. The seeds are white, brown, red, black, or variously mottled and spotted; some short-day cultivars of subsp. *unguiculata*, but not day-neutral forms, have seeds with rough testas. Germination is epigeal.

Cowpeas are tolerant of heat and drought and are most important in the semi-arid tropics where they are grown mainly for their mature seeds, but also for their immature pods and leaves for human food. Their vegetative parts are good fodder. On average the seeds contain 23 per cent of protein, but there is much variation in protein content so that breeders may succeed in producing cultivars with seeds containing much more. A notable feature of crops of cowpeas is their susceptibility to damage by insect pests. The use of insecticides in field experiments in Nigeria has given ten-fold increases of yield, and it is another objective of breeders to produce cultivars with hairy or scabrous pods which are more resistant to insect pests.

Black gram, urd: *Vigna mungo*

Vigna mungo originated in India where it is the most widely grown and highly esteemed grain legume. It occurs throughout Asia, and in Africa and the West Indies, but nowhere is it so important as in India where it is often grown in mixtures with other crops, or in rotation with rice. The young pods are used as a vegetable, and the mature seeds are cooked whole, split to make dhal, or crushed to make flour which is cooked in many ways. Black gram is also grown for forage, or as a cover crop.

It is a spreading, densely hairy annual with growth habits which vary from erect to more or less prostrate and spreading. The tap root produces a much branched root system with smooth, rounded nodules which vary greatly in size on the smaller roots. The stems bear alternate, trifoliate leaves with large (up to 10 cm long), pointed leaflets subtended by stipels; the leaves are subtended by stipules with basal appendages. The inflorescences are axillary and are sometimes branched. They carry clusters of five to six yellow flowers on short, hairy peduncles which elongate as the pods mature. The bracts subtending each flower are longer than the five linear, pointed calyx lobes which are about as long as the flower. The keel is spirally coiled with a terminal appendage. There are ten diadelphous stamens, and a short superior ovary with a spirally twisted style which has a lateral stigma near its tip. The pods are narrow, cylindrical and up to 6 cm long with a terminal, hooked beak, and conspicuously covered with short hairs. The mature pods are held erect. Each one contains six to ten small seeds which are square or oblong, and black with a pronounced white, concave hilum. Germination is epigeal.

Green gram, mung: *Vigna radiata*

Green gram is widely grown in India and Burma as a highly esteemed grain legume, and as a cover crop, green manure or for forage. The plant is a native of India, but is also grown in China, Iran, Japan and on a small scale in parts of Africa, the West Indies and the United States. It is not known to occur wild. Green gram is an annual much like *V. mungo*, but more erect, less hairy and taller, reaching a height of more than 1 m. The stout stems carry alternate dark-green trifoliate leaves on long petioles subtended by broadly ovate stipules. The small yellow or purplish-yellow flowers resemble those of *V. mungo*, but are in larger clusters of ten to twenty at the ends of long, hairy peduncles; each flower is subtended by a pair of bracts. Numerous slightly hairy, straight pods around 10 cm long without a beak are produced on each flowering axis; they are often pendent and each contains ten to fifteen green or golden-yellow seeds which are smaller than those of black gram, and have a flat, white hilum. The seeds are used in much the same way as black gram, but in addition they are germinated to produce 'bean sprouts' which are eaten as a vegetable.

Hyacinth bean: *Lablab purpureus*

Lablab purpureus is an Asian crop, and was probably domesticated in India where it occurs wild and as a cultivated grain legume. It is also grown in Africa, especially the Sudan and Egypt, but it is not an important grain legume outside Asia. It is discussed here, not because it is more important than the grain legumes which follow, but because it is closely related to *Vigna* and *Phaseolus*, and is included with these genera in the tribe *Phaseoleae* of the *Papilionoideae*. *Lablab purpureus* has previously been called *L. niger*, *L. vulgaris* and *Dolichos lablab*. It is a variable perennial, usually cultivated as an annual, with very long, twining, rounded hairy stems, and alternate trifoliate leaves on grooved petioles with swollen bases. The leaves are subtended by a pair of small pointed stipules, and have three large, slightly hairy leaflets. The terminal leaflet is subtended by a pair of stipels, and is symmetrical, ovate, pointed and widening from the base, whereas the lateral leaflets are strongly asymmetrical, and each is subtended by only one stipel. The flowers are in long axillary racemes on which they occur in groups of four to five from thickened outgrowths along the peduncle. They are white or purple, about 1.5 cm long, and subtended by two large bracts which are longer than the calyx-tube. The five sepals expand from the tube into only four lobes because the upper pair remain fused. The ten stamens are diadelphous. The keel and style are incurved at a right angle; the style is hairy on the inner side with a terminal stigma (not oblique as in *Vigna*). The flowers are commonly cross-pollinated by insects, and produce a flat, broad pod, 1.5 cm or more across and 5 cm or more long with a pronounced beak, and the persistent remains of the style. The pods contain three to six oblong, flat seeds which are white, reddish-brown, black or mottled with a prominent white, raised hilum, the aril, around one-third of their edge. Germination is epigeal. The first pair of true leaves are simple, opposite, heart shaped and prominently

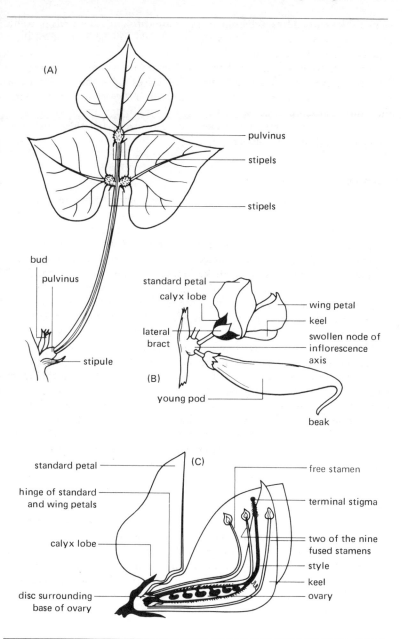

Fig. 4.13 *Lablab purpureus*: Hyacinth bean. (A) A trifoliate leaf illustrating the arrangement of stipules and stipels ($\times \frac{1}{3}$). (B) A flower and a young pod ($\times 1\frac{1}{2}$). (C) A diagram to illustrate the essential organs of the flower.

veined with hairy edges; subsequent leaves are trifoliate. The seeds of some races have a cyanogenetic glucoside, and should not be eaten until they have been boiled or roasted.

Hyacinth bean is well suited to arid climates where rainfall during the growing season is around 600 mm. The leaves of some cultivars are retained, and remain green, for several weeks into the dry season following the rains. The crop is consequently valued for forage and cover in hot, semi-arid regions. Three sub-species are recognized, and there are both long-day and short-day cultivars.

Chick pea, gram: *Cicer arietinum*

Chick pea is an ancient crop thought to have originated in Western Asia, and now of major importance as a grain legume in north India, Burma, the Middle East and the Mediterranean basin, including Southern Europe, and Ethiopia. The crop has spread in recent times in Africa and to South America and Australia. The mature seeds are eaten whole, or they are ground into flour, or split to make 'dhal', and the immature pods and young leaves are used as vegetables. Annual world production of mature seed is estimated to be about 6 m. tonnes, 4.5 m. of them from India, 0.5 m. from Pakistan, but only 0.3 m. from Africa (mostly from Ethiopia and Morocco).

Chick peas are small herbaceous annuals, rarely taller than 60 cm, with erect, much branched stems. All the plant is covered with glandular hairs. There is a strong tap root, and well developed lateral roots in the upper layers of the soil, with many large, much branched and lobed nodules. The large, alternate, pinnate leaves are about 5 cm long with nine to fifteen pairs of leaflets, and a single terminal leaflet. Each leaflet is ovate or oblong, 1—2 cm long and strongly pointed with a serrate margin. The leaflets of each pair are not always opposite. White, pink, pale blue or purple flowers are carried singly on long axillary peduncles. Five sepals are united to form a persistent calyx-tube with pointed lobes. The broad standard is clawed, the ten stamens are diadelphous, and the style is glabrous and curved inwards. The flowers are normally cleistogamous. The fruit is an inflated pod, about 2 cm long and 1 cm broad, containing one or two spherical, wrinkled seeds with an oblique, pointed beak. They are commonly brown, but sometimes white, red or black. Germination is hypogeal. Chick peas are notably tolerant of soil salinity and drought, but do not grow well in warm, humid climates. They are most widespread in seasonally dry regions as a cool season crop on free draining medium to heavy soils, often as a follow-on crop after rice.

Lentil: *Lens esculenta*

The lentil is one of the oldest grain legumes, and has been cultivated in Italy, Greece, Egypt and India since ancient times. It is presumed to be a native of Western Asia and Southern Europe; when grown in the tropics it is cultivated, like chick pea, as a winter or cool season crop in dry areas, or at high altitude. It will tolerate drought, but not waterlogged soils. Lentils are grown most extensively in the Middle East and India, and are one of India's most important and nutritious grain legumes. Annual world production is around 1 m. tonnes, 0.7 m. from Asia and 0.2 m. from Africa. In India the mature seeds are split to make 'dhal', which is used in soup. The young pods are also eaten as a vegetable, and the whole plant is excellent fodder. *Lens esculenta* is a small, light-green annual herb with much branched, square stems. It grows quickly with a low, bushy habit rarely taller than 40 cm, and matures 3—4 months after sowing. It has a slender tap root, and a mass of fibrous lateral roots with very small, round or elongated nodules. The alternate, pinnate leaves have up to seven pairs of leaflets, and a terminal bristle or tendril; they are subtended by upward pointing, linear stipules, but there are no stipels. Each sessile leaflet is oval or lanceolate, and about 12 mm long; the pairs of leaflets are not always opposite. There is a small pulvinus at the base of each leaflet which causes the leaflets to fold up together when plants are subject to water stress. The small white flowers are tinged with purple, and occur in groups of two to three at the ends of long, very slender peduncles. The calyx-tube divides near its base into five narrow, pointed lobes which curve over and beyond the corolla. The standard is white with blue markings, while the wings and keel are white. The stamens are diadelphous, and the ovary, containing one to two ovules, is surmounted by a short, curved style which is hairy along its inner surface. The pod is flattened and broad, rarely more than 12 mm long, tipped by a minute curved beak, and with the persistent calyx around its base. The seeds are lense shaped, light red, brown or grey, and often speckled with black. The hilum is minute. Germination is hypogeal.

Pigeon pea: *Cajanus cajan*

Pigeon pea is an ancient African grain legume which has been cultivated in the Nile Valley for more than 4,000 years. It is now one of the most important grain legumes in India and Pakistan where the greatest diversity of cultivars occurs, and though the crop is also grown in the rest of Asia, in Africa (mostly in Uganda, Malawi and Tanzania) and the West Indies, India produces 1.75 m. tonnes of the total estimated annual world production of 1.9 m. tonnes of dry seed. Pigeon pea is grown for its mature seed, which in India is a very popular source of 'dhal', and for its immature seeds and pods, for forage, or as a cover crop, a wind break or a nurse crop for young cocoa, kola and oil-palms. It is notably tolerant of heat and drought and free from pests and diseases, and grows well on poor soils with little attention; yet it has not become widely popular outside India, perhaps because even early maturing cultivars do not produce a first crop for as long as 8 months after sowing, and because the mature seeds are hard and difficult to prepare as food. The crop is very variable, with both tall, long season, short-day types, and dwarf, day-neutral types. It seems likely that it will become more widely cultivated wherever consumers learn to accept its seeds.

Cajanus cajan is the only cultivated species in the genus. It occurs wild in Africa. It is a woody perennial living 3–4 years, and bears well in the first year, profusely in the second and third years, but progressively less thereafter. It has a very deep tap root and a mass of fibrous lateral roots with almost spherical nodules which are sometimes lobed. The main stem is thick and erect, about 1 m tall in dwarf, day-neutral cultivars, but up to 4 m in tall, short-day ones; it bears numerous, usually ascending branches which give the plant a bushy growth habit. Young stems are angled and hairy with trifoliate, often deciduous leaves inserted spirally. The petioles are short, slender, grooved and subtended by small stipules. The leaflets are lanceolate and entire, slightly hairy and dark-green above but pale grey below.

The terminal leaflet is up to 15 cm long, and is larger with a longer stalk than the almost sessile laterals. There are pulvini at the bases of the petiole and the leaflets. The axillary (or sometimes terminal) panicles are usually shorter than the leaves which subtend them; they bear yellow or reddish flowers singly on short, stiff branches. The calyx is hairy and four-lobed,

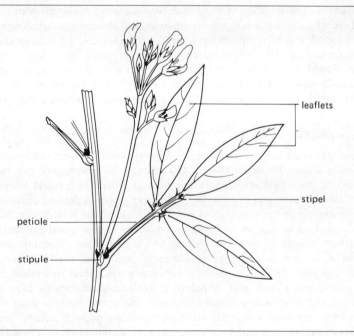

Fig. 4.14 *Cajanus cajan*: Pigeon pea. A trifoliate leaf and its axillary inflorescence ($\times \frac{1}{3}$).

the upper lobe made up of two fused sepals and larger than the lower three. The large standard petal is usually bright golden-yellow, but may be streaked with red, or have red veins; the wings and keel are yellow and about the same length. The ten stamens are diadelphous, but the free parts of the filaments are of unequal length. The ovary and the style, which tapers to a knob-like stigma, are covered with short, shining brown hairs. Although the anthers dehisce before the flower opens there is considerable cross-pollination by bees and other insects, and up to 20 per cent cross-fertilization occurs. The pods are straight and flattened, slightly hairy, usually 4—5 cm long, rarely as long as 10 cm, and have a pronounced, pointed beak. They are constricted between the three to four seeds (rarely as many as eight) they contain. There is much variation, especially among Indian cultivars, in the shape and colour of the pods, and the size of the seeds which are usually spherical or oval, and white, brown, red or speckled with a white hilum. Germination is hypogeal.

Bambarra groundnut: *Voandzeia subterranea*

Bambarra groundnut is wild in isolated locations in the savanna zone of West Africa, and may have been domesticated around the headwaters of the Niger River. It is an ancient crop in that area and in Central Africa, and has spread more recently to Madagascar, Asia and to South America. The fruits of bambarra groundnut mature underground, but they are not buried at the end of a peg like the fruits of *Arachis*. After groundnuts were

Fig. 4.15 *Voandzeia subterranea*: Bambarra groundnut.   A yield trial in north Nigeria.

introduced to West Africa by the Portuguese in the sixteenth century it seems likely that they replaced bambarra groundnuts in many areas because the husbandry of the two crops is similar and they fill the same role in the traditional farming system of the savanna zone. Bambarra groundnuts are not an oil-seed and have never become an export crop.

Voandzeia subterranea is a bunched herbaceous annual with contracted, much branched stems which root at the nodes forming a short crown from which a cluster of leaves arise on long petioles. There are uncommon cultivars with widely spreading branches. A thick tap root grows from the crown, free of lateral roots for part of its length, then producing a profusion of them. Small rounded and sometimes lobed nodules occur on the tap root and the laterals. Trifoliate leaves are carried on stiff, erect, grooved petioles up to 15 cm long. The lanceolate to narrowly oblong leaflets are on very short stalks with pulvini at their bases; the terminal leaflet is 5—10 cm long, the pair of laterals about two-thirds as long, all of them glabrous. Whitish-yellow flowers are usually in pairs on short axillary peduncles. They have a glabrous calyx-tube with short, blunt lobes. The flowers are cleistogamous, and may not open at all, the broad standard remaining folded around the wings and keel. There are ten diadelphous stamens, and the ovary is surmounted by a short bent style, hairy on its inner surface with a short, lateral stigma. Beneath each flower, at the top of the peduncle, a smooth glandular swelling grows down and into the soil after fertilization, dragging the ovary behind it. The pod develops in the soil, and grows to about 2.5 cm in diameter, with a very hard, wrinkled pericarp when mature. The fruit contains one or rarely two large, round seeds up to 1.5 cm across, which are commonly white, reddish-brown or black, but sometimes mottled or spotted. The white hilum may be surrounded by a black ring, or 'eye'. The immature and mature seeds are eaten whole, or mature seeds are ground to flour before cooking. They contain 4—6 per cent of oil, 16—21 per cent protein and 50—60 per cent carbohydrate.

It is interesting to mention here a third grain legume which produces subterranean fruits, *Kerstingiella geocarpa*. It has the same natural distribution as bambarra groundnut, but is much less important. Its fertilized ovaries are forced into the soil at the ends of pegs like those of *Arachis*.

Sword bean, jack bean: *Canavalia* spp.

Two species, *C. gladiata*, the sword bean from India, and *C. ensiformis*, the jack bean from Central America, cultivated principally as fodder or cover crops, are of minor importance as vegetables and grain legumes throughout the tropics. Sword bean is widely cultivated in India and Asia for its immature pods which are eaten as a vegetable. The two species are morphologically similar, but while the jack bean is a much branched, bushy annual up to 2 m tall, sword bean is a taller perennial, twining plant with larger leaflets, flowers and seeds. In both species the stems are

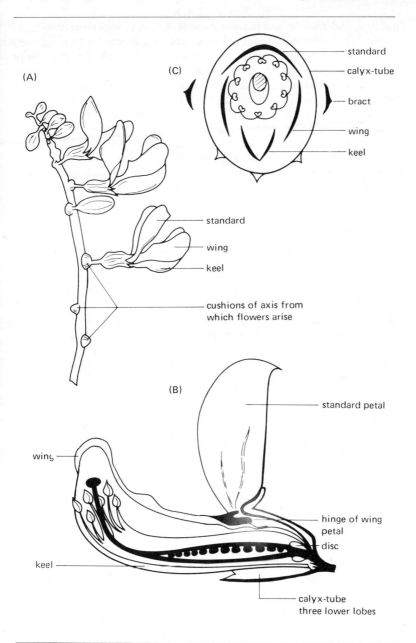

Fig. 4.16 *Canavalia ensiformis*: Jack bean. (A) The inflorescence (×$\frac{2}{3}$). (B) A diagram to illustrate the essential organs of the flower. (C) A floral diagram.

glabrous and tend to become woody with age; in some cultivars they are purple. The alternate, trifoliate leaves have long grooved, often coloured petioles with small stipules and a pulvinus at their base. The petioles of sword bean are longer than the leaflets, those of jack bean shorter. The leathery, ovate leaflets of jack bean are about 12 cm long, those of sword bean slightly longer; in both species there is a large pulvinus at the junction of the leaflets and the petiole; on the under surface of the leaflets veins are prominent.

The inflorescences are axillary racemes, usually produced from extra-axillary buds at the same nodes as vegetative branches. The flowers arise, often in pairs, from knob-like projections on the long, stout peduncles. The calyx has two large upper lobes, reflexed and closely adpressed to the standard petal, and three small lower lobes. The flowers of *C. ensiformis* are reddish-purple, those of *C. gladiata* white or pink. The standard petal is large, with recurved edges and a white throat. The filaments of the ten stamens are all united for most of their length, but the vexillary stamen tends to become free from the staminal tube. The ovary contains twelve to twenty seeds, and when mature is a long, flattened, often curved woody pod with a pointed beak and pronounced ridges along the upper suture. *Canavalia ensiformis* has pods about 25 cm long, and white, flat-sided seeds about 2 cm long and 1 cm broad; *C. gladiata* has more curved pods about 20 cm long, and with more prominent ridges, and larger, red, pink or brown seeds. The mature seeds are among the largest produced by legumes; they may be poisonous and must be boiled in several changes of water before they are eaten.

Cluster bean, Guar: *Cyamopsis tetragonolobus*

The cluster bean is probably indigenous to India and has a long history of cultivation throughout Asia. It has been introduced to Africa and the United States. The seeds and young pods are a source of human food, and the whole plant is used as forage. The gum called guar obtained from the mucilaginous seeds is used in food products and in the paper and textile industries. Cluster bean is a bushy perennial 1–3 m tall, with a long tap root and lateral roots on which are many large, lobed nodules. There are numerous stiff, erect branches with alternate, trifoliate leaves on long, grooved petioles; the leaflets are ovate, and slightly pointed with serrate margins. At the base of the petiole there is a marked pulvinus and two narrow, linear stipules. Small flowers are in dense clusters on axillary racemes; each flower is sessile with a five-pointed, often coloured and hairy calyx-tube, a white standard and keel, but pinkish-purple wings. The ten stamens are monadelphous around a long ovary with its short, bent style and terminal stigma. The stiff, erect pods are in clusters, each about 8 cm long with a double ridge along the upper side, a single ridge below and two lateral ridges along the flat sides. The pods are slightly constricted between the seeds and have a terminal beak. They contain eight to twelve small, oval seeds which are white, grey or black. Ciuster beans are hardy and drought resistant. They are commonly grown in mixed cultivation with other crops.

Winged bean, Goa bean: *Psophocarpus tetragonolobus*

The genus *Psophocarpus* includes five species of large, herbaceous, tropical plants with twining stems and tuberous roots which resemble those of the sweet potato, but which are tougher. *Psophocarpus tetragonolobus* is the most important species. It originated in tropical Asia and is cultivated, especially in Indonesia and Burma, for its edible tubers, its young pods, which are a delicious vegetable, and for its seeds, which are eaten after they have been parched. Winged bean is remarkable as one of the few grain legumes well adapted to the humid, lowland tropics, and for the nutritive value of its seeds and tubers. The seeds contain up to 37 per cent of protein with a large content of essential amino-acids, and the tubers contain up to 24 per cent of crude protein (not all of its digestible). Furthermore, winged bean nodulates very freely provided that the necessary strain of *Rhizobium* is in the soil, and it is consequently valued as a cover and green manure crop. The crop has recently received much attention from research workers because of the potential for increased production in the tropical rain forest zones, especially in Africa.

Winged bean is a perennial, and produces new growth each year from its persistent, tuberous roots. It is cultivated as an annual. The stems are 2–3 m long, twining about any support or trailing along the ground. They bear alternate, trifoliate leaves on long petioles, subtended by divided stipules. The leaflets are subtended by small stipels, and they are broad, ovate and acuminate, from 8–15 cm long. The blue or white flowers are in terminal groups on axillary racemes. They have a two-lobed calyx, a large standard petal with auricles at the base, narrow wings and an incurved keel. The stamens are diadelphous, and the long ovary with its thick style and densely hairy stigma contains many ovules. The pod is 15–25 cm or more long, roughly square in section, with four prominent, jagged wavy wings all along its length. The seeds are embedded in the solid tissues of the pod. They are round, smooth and brown with a very small hilum.

A second species, *P. palustris*, is found growing wild or semi-cultivated in West Africa, parts of Central and East Africa and has been introduced elsewhere in the tropics. Its tubers and young pods are eaten.

Forage and pasture legumes, cover and shade crops

The cultivation of legumes and grasses only to feed livestock, and of legumes as cover crops or to provide shade for young tree crops, are

Fig. 4.17 Leguminous forage and cover crops. *Left* Foreground — *Stylosanthes gracilis*, 'stylo'. *Right* Background — *Pueraria phaseoloides*, 'kudzu'.

features of advanced mixed or rangeland farming and plantation agriculture. Although livestock are kept in relatively undeveloped agricultural systems, land and labour in them are rarely devoted to growing crops which do not provide human food directly, or which cannot be sold. Instead most crop residues, but especially those of the grain legumes, are fed to livestock. Legumes are important as livestock food because they contain more nitrogen than grasses, commonly 15—25 per cent of their dry matter is 'crude protein' (which is the nitrogen content of the plant material, protein or not, which livestock can use to synthesize animal protein). A large number of tropical legumes are grown for forage, and many of them are also used as cover crops and green manures to enrich the soil with nitrogen from their root nodules, prevent soil erosion and smother weeds. Most are quick growing, spreading and twining, drought-resistant plants, many of them perennials with very deep tap roots. They vary greatly in their soil and water requirements, in their tolerance of shade, heat, drought, grazing and of competition from grasses with which they may be sown in pastures. In forage species the age of crops when they are used as well as differences between cultivars determine their nutritive value, digestibility and palatability (which is a measure of how much animals like to eat them). The best forage species tend to be glabrous because animals find hairy plants unpalatable.

Legumes grown to provide shade or as wind breaks for tender young tree crops are small trees or shrubs, and are preferred to other shrubby plants because the bacteria in their nodules fix nitrogen. Woody species may also be used as cover crops in plantations, and their foliage is sometimes harvested as fodder. In the past leguminous trees have been used to provide permanent shade in coffee and tea plantations, but this is no longer common practice.

The following lists contain only a few of the better known forage, cover and shade legumes used in the tropics.

Forage species:
  Several *Phaseolus* species
  Glycine max (soyabean) and *G. javanica*
  Vigna unguiculata (cowpea) and *V. vexillata*
  Lablab purpureus (hyacinth bean)
  Stizolobium deeringianum and *S. atterrimum* (Velvet beans)
  Centrosema pubescens and *C. plumieri*
  Pueraria phaseoloides (kudzu)
  Medicago sativa (lucerne)
  Trifolium alexandrinum (berseem or Egyptian clover)

Lucerne is especially suitable for cultivation with irrigation in the cool, dry tropics or at high altitude. Berseem is not grown in the hot tropics.

Pasture species:
  Trifolium semipilosum (Kenya white clover). Grown only at high altitude in the tropics.
  Clitoria ternata (butterfly pea)
  Glycine javanica
  Stylosanthes gracilis ('stylo')
  Pueraria phaseoloides (kudzu)
  Centrosema pubescens
  Indigofera spp. (Indigo)
  Desmodium intortum and *D. uncinatum* (tick clovers)
  Alysicarpus vaginalis (Alyce clover)

Cover species:
  Pueraria phaseoloides
  Centrosema pubescens
  Calopogonium mucunoides ('Calopo')
  Glycine javanica
  Stylosanthes gracilis
  Indigofera spp.
  Crotolaria juncea (sunn hemp)

Shade species:
  Tephrosia candida
  Glyricidia sepium
  Cajanus cajan (pigeon pea)
  Albizzia spp.
  Erythrina spp.

Other legumes

Yam bean: *Pachyrrhizus* spp.

Pachyrrhizus is a small genus native to tropical South and Central America with two important species cultivated for their root tubers which are eaten both raw or cooked. Neither species is grown for its seeds, so they are not grain legumes. *Pachyrrhizus erosus* reached Asia in post-Columbian times and is now cultivated throughout the humid tropics; it is most important in Asia, from India to south-east China. As well as its root tubers, the immature pods of *P. erosus* are eaten as a vegetable. *Pachyrrhizus tuberosus* is less widespread. It is cultivated mainly in South America and the West Indies; its pods are not edible, but its tubers are larger than those of *P. erosus*. Both species are large, perennial, twining herbs with slightly hairy stems and leaves. They are cultivated like yams (*Dioscorea* spp.), and produce tubers for harvest about 6 months after their seeds have been sown. Their stems bear alternate trifoliate leaves subtended by linear stipules. Large ovate or more or less lobed leaflets are subtended by stipels. White or purple flowers are clustered in small groups along the axis of large axillary racemes. They are papilionaceous with diadelphous stamens, and a hairy style which is curved upward and which has a subterminal stigma. The flat pods are around 12 cm long and 1–2 cm broad; they contain up to twelve square or oblong seeds which are up to 1 cm in diameter in *P. erosus*, slightly larger in *P. tuberosus*.

Derris, tuba: *Derris elliptica*

Derris elliptica is a woody, perennial legume native to the rain forests of Bangladesh and New Guinea which is cultivated throughout South-east Asia for its roots. They have been used for centuries in Asia as a fish poison, and in 1902 the active ingredient, rotenone, was extracted and found to have valuable insecticidal properties. 'Derris' insecticides are not toxic to mammals and so are useful for the control of pests in vegetable crops and of insect parasites like ticks on livestock. As a fish poison the pounded roots are thrown into rivers where the fish eventually float helpless to the surface; though they are poisoned they are wholesome for humans to eat. Three species of a similar, and perhaps related, genus, *Lonchocarpus*, are cultivated in South America for their roots which also contain rotenone, and are used as fish poisons or the manufacture of insecticide.

Derris elliptica (and *Lonchocarpus* spp.) is propagated by stem cuttings which grow to produce a small bush with a short, woody trunk and long vine-like branches which bear alternate, imparipinnate leaves (four to six pairs of opposite leaflets and a single terminal leaflet). The inflorescence of pink, papilionaceous flowers is an axillary raceme. The crop grows 2–3 years before the roots, each weighing around 1 kg when fresh, are harvested, dried and exported (chiefly to the United States). Powdered roots may be mixed with inert ingredients to make insecticidal dusts or wettable powders, or the rotenone is extracted from them in solvents.

Further reading

General

Aykroyd, W. R. and Doughty, J. (1964). Legumes in Human Nutrition. *FAO Nutrition Studies*, Rome: FAO.
Hallsworth, E. C. (ed.) (1958). *Nutrition of Legumes*, London: Butterworth.
Kachroo, P. (ed.) (1970). *Pulse Crops of India*, New Delhi: Indian Council of Agric. Resc.
Rachie, K. O. and Roberts, L. M. (1974). Grain legumes of the lowland tropics, *Advances in Agronomy*, 26, 1–132.
Smartt, J. (1976). *Tropical Pulses*, London: Longman.
Stanton, W. R. *et al.* (1966). *Grain Legumes in Africa*, Rome: FAO.
Verdcourt, B. (1970). Studies of *Leguminosae–Papilionoideae* for the Flora of East Tropical Africa IV, *Kew Bull.*, 24, 507–69.
Whyte, R. O., Nilsson-Leissner, G. and Trumble, H. C. (1953). *Legumes in Agriculture*, Rome: FAO.

Groundnuts

Anon. (1973). Peanuts, Culture and Uses. *A Symposium*, American Peanut Research and Education Association.
Arant, F. S. *et al.* (1951). *The Peanut – The Unpredictable Legume*, Washington: National Fertilizer Association.
Bunting, A. H. (1955). A classification of cultivated groundnuts, *Emp. J. exp. Agric.*, 23, 158–70.
Bunting, A. H. (1958). A further note on the classification of cultivated groundnuts, *Emp. J. exp. Agric.*, 26, 254–8.
Gibbons, R. W., Bunting, A. H. and Smartt, J. (1972). The classification of varieties of groundnuts (*Arachis hypogaea* L.), *Euphytica*, 21, 78–85.
Gillier, P. and Silvestre, P. (1969). *L'arachide*, Paris: Maisonneuve and Larose.
Lepik, E. E. (1971). Assumed gene centres of peanuts and soyabeans, *Econ Bot.*, 25, 188–94.
Krapovickas, A. (1969). The origin, variability and spread of the groundnut (*Arachis hypogaea*), in: Ucko, J. and Dimbleby, G. W. (eds), *The Domestication and Exploitation of Plants and Animals*, London: Duckworth.

Soyabeans

Caldwell, B. E. (ed.) (1973). Soybeans. Improvement, Production, Uses, *Agronomy*, 16, Am. Soc. Agron.
Hymowitz, T. (1970). Domestication of the soybean, *Econ. Bot.*, 24, 408–21.
Norman, A. G. (ed.) (1963). *The Soybean. Genetics, Breeding, Physiology, Nutrition and Management*, New York: Academic Press.
Leng, E. R. (1968). Soybean. Potential for extension into areas of protein shortage, *Econ. Bot.*, 22, 37–41.

Other grain legumes

Sellschop, J. P. F. (1962). Cowpeas, *Vigna unguiculata, Field Crop Abstracts*, 15, 259–66.
Summerfield, R. J., Huxley, P. A. and Steele, W. (1974). Cowpea (*Vigna unguiculata* L. Walp.), *Field Crop Abstracts*, 27, 301–12.
Gentry, H. S. (1969). Origin of the common bean. *Phaseolus vulgaris, Econ. Bot.*, 23, 55–69.
Kaplan, L. (1965). Archaeology and domestication in American *Phaseolus* (beans), *Econ. Bot.* 19, 358–68.
Doku, E. V. and Karikari, S. K. (1971). Bambarra groundnut, *Econ. Bot.*, 25, 255–62.
Karikari, S. K. (1971). Economic importance of bambarra groundnut, *World Crops*, 23, 195–6.
Hepper, F. N. (1970). Bambarra groundnut (*Voandzeia subterranea*), *Field Crop Abstracts*, 23, 1–6.
Gooding, H. J. (1962). The agronomic aspects of pigeon peas, *Field Crop Abstracts*, 15, 1–5.
Masefield, G. B. (1973). *Psophocarpus tetragonolobus* – A crop with a future?, *Field Crop Abstracts*,
Zohary, D. (1972). The wild progenitors and place of origin of cultivated lentil, *Lens culinaris, Econ. Bot.*, 26, 326–32.
Schaaffhausen, R. (1963). *Dolichos lablab* or Hyacinth bean. Its uses for feed, food and soil improvement, *Econ Bot.*, 17, 146–53.
Williams, W. (1954). The Natal wattle industry, *World Crops*, 6(10), 417–19.
De, D. N. (1974). Pigeon Pea, in: Hutchinson, Sir Joseph, *Evolutionary studies in World Crops. Diversity and Change on the Indian Subcontinent*, London: Cambridge University Press.

Chapter 5

Root and Tuber Crops

A large proportion of the inhabitants of the wet rain forest zones of the tropics, where cereals are difficult to grow (because there is no dry weather in which their grains can ripen), depend upon various roots and tubers for their staple food. Although these crops rarely appear on world markets, and are grown mostly for home consumption in gardens and smallholdings, they are of immense importance. Estimates of annual (1973) world production of the important species are: sweet potatoes 133 m. tonnes, cassava 106 m., yams 18 m., taro 4 m., all others 1.5 m., compared with annual production of about 128 m. tonnes of the chief temperate root crop, Irish potato (*Solanum tuberosum*). The tropical roots and tubers can be grown on a variety of soil types, and over a range of climates and altitudes; they may receive little attention after planting and still produce good yields; some of them can be left in the soil without deteriorating until they are needed for eating, and then are a valuable food reserve for use in times of seasonal food shortage. Their chief faults as food crops are their generally poor storage qualities once harvested, their tendency to deplete soil nutrients, and their small protein content, which is commonly less than 3 per cent of their fresh weight. Malnutrition, and especially symptoms of dietary protein/carbohydrate imbalance like kwashiorkor, is most widespread in regions where root crops are the staple food. Apart from indigenous tropical species, several temperate root crops have been introduced to the tropics, and though they contribute little to food production they can be grown successfully where part of the year is cool, or where good soils occur at fairly high altitudes. Irish potatoes are extensively cultivated in China and India, and on a small scale in most tropical countries, often to meet demand from expatriate European populations. Indeed, carrots, beetroots, radishes, turnips and other temperate roots and vegetables are grown wherever Europeans can contrive their cultivation.

Sweet potato: *Ipomoea batatas*

The sweet potato originated in tropical America and has been cultivated there for thousands of years. It is a hexaploid ($2n = 6x = 90$), and though there is no certainty about the identity of its wild progenitors, *Ipomoea tiliacea*, which is wild throughout the world's tropics, has been suggested as one parent in a hybrid origin of the cultivars, while *I. trifida* from Mexico is suggested as a direct hexaploid ancestor. The crop spread very early from the New World to the Pacific islands, and more recently from them to Asia; it was introduced to Europe by Columbus and to Africa by the Portuguese, and is now grown throughout the tropics. Sweet potatoes yield well under a wide range of climates and soil conditions, but do best on light, sandy soils where annual rainfall is around 1,000 mm. They are not tolerant of drought during the growing season, and in dry areas are grown with irrigation.

Ipomoea batatas is the only economically important species among the 400 or so tropical members of the genus, and the only important member of the family *Convolvulaceae*. This is a large family of about 1,000 species of mostly herbaceous, creeping and climbing plants, with latex cells in their stems, bicolateral vascular bundles, and regular flowers with a typically large, bell or funnel-shaped corolla of five fused petals. The sweet potato, in its long history of cultivation, has become very diverse with a wide variety of cultivars, but is not known to occur wild. Most cultivars have vine-like or trailing stems up to 5 m long with long internodes, while some are more erect, with shorter stems and short internodes. Sweet potatoes are self-incompatible, out-breeding, short-day plants. They do not flower in days longer than about 13½ hours, and consequently rarely do so in temperate latitudes more than 30° N. and S. of the equator, where days during the growing season are long. In tropical latitudes they flower readily, and produce viable seed, but the crop is always propagated vegetatively from stem cuttings, or less often from shoots which grow from germinated buds on the tubers. Vegetative propagation of an outbreeding crop which sometimes produces self-sown seedlings favours the accumulation of great diversity. The stem cuttings root easily and produce several long, trailing stems which root at the nodes and eventually produce a dense ground cover. The stems are fairly thin and dark-green, but they are marked with blotches of purple or brown, and are softly hairy when young, becoming glabrous with age. The simple dark-green leaves are arranged spirally in a phyllotaxy of two-fifths, and are very variable in size and shape, even on a single plant. They may be ovate and entire or deeply dissected and lobed, more or less hairy, and often with purple pigmentation, especially along the veins. The petioles are swollen at the base, 5—30 cm long, and either longer or shorter than the lamina; those arising from the lower surface of the creeping stem bend up as they grow so that all the leaves are carried at about the same height. The petioles are usually grooved, and have two lateral nectaries just below their junctions with the lamina; there are no stipules. Large leaf scars are left on the older parts of the stem where leaves have been shed.

The large, reddish-purple flowers are solitary or in cymose inflorescences on peduncles which are usually longer than the petioles of the

Fig. 5.1 *Ipomoea batatas*: Sweet potato. Variation in the leaf form of sweet potatoes.

leaves which subtend them. The calyx-tube is made up of five sepals free from each other except at the base. The corolla-tube is a funnel of five united petals up to 5 cm long and 4 cm across; there are five free epipetalous stamens inserted on the base of the corolla, with white anthers and filaments of varying length, the two longest as long as the style. The spherical pollen grains are covered with numerous minute papillae. The superior, two- or four-loculed ovary has two ovules in each locule, and is surrounded at the base by an orange nectary. The flowers open in the early morning and wither a few hours later. They are cross-pollinated by bees, but pollen tube growth and cross-fertilization occur only after pollinations between cross-compatible cultivars; almost all cultivars are self-incompatible. The fruit is a globular, dehiscent capsule in which false septa may develop. It contains up to four small, black, flat-sided seeds with one smooth and one angular surface, and with a deep micropylar hollow just above the hilum on the flattened surface. The testa is thick, very hard and almost impermeable to water, so that the seeds germinate at irregular

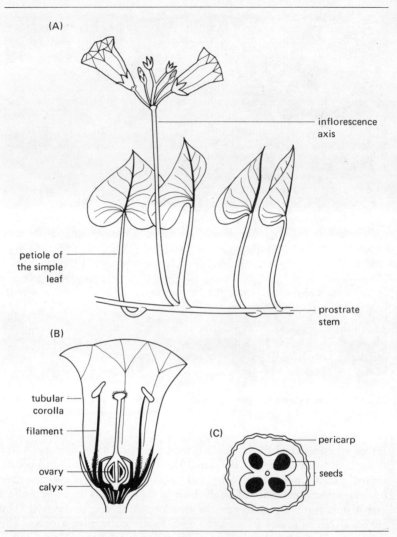

Fig, 5.2 *Ipomoea batatas*: Sweet potato. (A) A portion of the stem illustrating the arrangement of leaves and the axillary inflorescence. (B) A longitudinal section of a flower. (C) A section of the ovary.

intervals, depending upon the time taken for them to absorb water. Regular, more rapid germination can be induced by scratching the testa to permit the entry of water. Germination is epigeal.

The root tubers are formed in the top 25 cm of the soil by secondary thickening in adventitious roots, either those produced by the original stem cutting, or those from the nodes of the creeping stems. Thus a single plant produces several tubers, rarely 50 or more, and since sweet potatoes are perennials, successive crops can be taken by earthing up the stems after each harvest. The tubers vary greatly in size, shape and colour, from long and spindle shaped to almost spherical, and from white through yellow, pink, purple and red to brown. Individual tubers vary in weight from 0.1 kg to 0.5 kg, but exceptionally they may weigh several kilograms. In their early development the vascular cambium of an adventitious root proliferates a mass of secondary vascular tissues and storage parenchyma. Later secondary meristematic tissues, in the form of cylinders or sheets of unrelated tissue laid down in the parenchymatous centre of the root, give rise to small vascular elements and a great deal of storage parenchyma. With the expansion of the tuber the root cortex and epidermis are ruptured, and the tuber is eventually protected by a periderm which arises from a phellogen in the pericycle. The mature tuber consists of a mass of storage parenchyma in which lie the scattered primary and secondary vascular tissues and numerous latex vessels. These vessels contain oil, and the cells from which they are formed retain their transverse walls. The approximate composition of a mature tuber is: water 50–80 per cent, carbohydrate 10–40 per cent, protein 1–2.5 per cent, fat 2–6 per cent; the sugar content varies from 0.5–6 per cent. This wide variation in

Fig. 5.3 *Ipomoea batatas*: Sweet potato. Indonesia. (By courtesy of the FAO.) Carrying sweet potatoes to market in

chemical composition depends upon differences between cultivars, maturity of the tubers at harvest, duration of the storage period and climatic variation during the growing period. The tubers may have a smooth surface, or they may be variously ridged; they bear rows of roots which arise from the grooves of ridged types, and adventitious buds.

Three main groups of cultivars are recognized, depending upon differences between them in the texture, colour and palatability of their tubers. One group has tubers with hard, dry flesh which is often yellow; another has tubers with coarse, fibrous, unpalatable flesh; while the third has soft tubers with white or orange, watery flesh which is sweet when cooked. As well as their importance as a source of carbohydrate in diets, yellow fleshed tubers are rich in vitamin A, and all sweet potato tubers contain small amounts of thiamine, riboflavin, nicotinic acid and ascorbic acid. The crop is important throughout the tropics, but by far the greatest part of annual world production of 133 m. tonnes is from Asia (about 122 m.), and especially mainland China (about 111 m.), whereas Africa produces only 6 m. tonnes each year. As well as mainland China, countries where annual production exceeds 1 m. tonnes are Brazil (2.3 m.), Japan (2.0 m.), Indonesia (2.1 m.), the Korean Republic (about 2.0 m.), India (1.8 m.) and Burundi (1.2 m.). In the United States the crop is grown to be eaten fresh or canned, and is also used for the manufacture of starch, flour (as a supplement to wheat flour in bread), alcohol and glucose. The young leaves are used as a vegetable in parts of Africa and Asia, and the vines are fed to livestock. Sweet potato tubers are liable to be infected by several fungus species which cause them to rot and to deteriorate rapidly in store.

Cassava, manioc: *Manihot esculenta*

Cassava is second in importance only to sweet potato as a tropical root crop, with world production estimated to be around 106 m. tonnes in 1973. The plant was domesticated in South or Central America and has been cultivated there for several thousand years, but it is not known to occur wild. Like so many other New World crops, it was taken to West Africa by the Portuguese in the sixteenth century, but its spread throughout all the world's tropics has been a relatively recent event, and it is only during this century that cassava has achieved its present great importance as a food crop. Annual production in Africa is about 42 m. tonnes, in South America 37 m. and in Asia about 25 m. Brazil, which may have been the centre of origin, remains the world's chief producer with more than 30 m. tonnes each year, followed by Indonesia, Zaire and Nigeria each producing about 10 m. tonnes. Thailand's production is around 6 m. tonnes, but it is the chief exporter of cassava products (more than 0.5 m. tonnes annually). Under the best conditions cassava gives very large yields of tubers, up to 50 tonnes per hectare, but it is often grown with very little attention on poor or exhausted soils, and average yields are around 8 tonnes per hectare. It is easy to propagate from stem cuttings, and the

Fig. 5.4 *Manihot esculenta*: Cassava. Recently harvested cassava tubers (foreground) and cassava plants in Nigeria.

tubers keep so well in the ground that cassava is a most important famine relief crop.

The genus *Manihot* is a member of the economically important family *Euphorbiaceae* (rubber, castor, tung), with several thousand species, many with latex vessels in their stems, and curious flowers and inflorescences. The flowers are small and unisexual, in many species much reduced; the superior ovary is typically trilocular producing a schizocarp which splits at maturity into three one-seeded sections (though the fruit of cassava is a capsule). *Manihot esculenta* is the most important species in the genus which also includes *M. glaziovii*, ceara rubber. The genus has two centres of diversity in the New World, one in Brazil, the other in south Mexico, and includes herbs, shrubs and trees. *Manihot esculenta* is a shrubby, woody, short-lived perennial growing to a height of 3 m or more, with erect, glabrous stems and varying degrees of branching. Dwarf cultivars rarely exceed 1 m in height. In some cultivars branches are produced only from the base of the stem to give an erect bunch growth habit; in others the branching pattern and branch growth produce widely spreading plants. The stem and branches vary in colour with cultivar, from silvery-grey through various shades of reddish-brown to dark-brown, often streaked with purple. The leaves tend to be clustered towards the tops of the stems, as those below are shed, leaving prominent leaf scars. The large, deeply lobed leaves are arranged spirally, and have long petioles subtended by small, deciduous stipules. Each leaf commonly has three to seven (sometimes more) obovate—lanceolate, acute lobes up to 20 cm long, which are broadest about one-third of the distance from their tip, and taper gradually to their base. The leaves vary in shape, colour and size, in degree of dissection and width of the lobes; the petioles vary in length, colour and hairiness, and the stipules in colour and size. The leaves are usually dark-green, but red, yellow, and various shades of purple pigmentation occur in the foliage.

The unisexual flowers are in racemes in the axils of leaves near the ends of branches, with both sexes in one inflorescence, the female flowers near its base. The male flowers are about 1 cm long on short pedicels, and shorter than the females. Five sepals are united for about half their length, and are variously coloured yellow or red, the veins being differentially marked. There are no petals. Male flowers have ten free stamens with small anthers in two whorls of five. Female flowers have a three-loculed ovary borne on a glandular disc, and they open 7—8 days before male flowers in the same inflorescence. Cross-pollination is by insects. The fertilized ovary grows into a globular, explosively dehiscent capsule about 1.5 cm in diameter with six longitudinal, angled wings. Each locule contains a single seed.

Cassava is always propagated vegetatively from stem cuttings which root easily, and produce an extensive, fibrous adventitious root system. The root tubers develop as a result of secondary thickening in adventitious roots, and in well-grown plants a cluster of five to ten tubers is produced near to the base of the stem. In some cultivars the tubers are ready for harvest about 1 year after planting, but in most they take 18 months or 2 years to mature; they are harvested individually for home consumption, because though they keep well in the ground, they spoil quickly after harvest. The tubers are commonly unbranched, about 50 cm long and 10 cm in diameter, but they are sometimes branched and as long as 1 m. Numerous small lateral roots arise from them, and they are surrounded in the soil by the unthickened adventitious roots of the plant. A mature tuber consists of three distinct anatomical regions. The outer periderm, which may be thick and rough or thinner and smooth, varies in colour from white to pink, red or brown. It encloses a thin cortex, usually white, but streaked with brown in some cultivars. The large white or cream-coloured pith is the main storage region; it consists of a mass of parenchyma in which the chief storage material is starch, with a few xylem elements and lactiferous tubes. When the tubers are prepared for eating the periderm and cortex are peeled off. The tubers tend to become lignified as they grow old, sometimes so much that they become unpalatable.

All cassava plants and tubers are to a certain degree poisonous because they contain various amounts of the cyanogenetic glucoside linamarin, which breaks down to give prussic acid. Linamarin is most concentrated in the periderm and cortex, and prussic acid is produced from it in largest amounts when cassava is grown on infertile soils and during dry weather. The cultivars of cassava are classified in two groups, the 'sweet' types in which the prussic acid tends to be confined to the outer rind, and the bitter types in which it is more generally distributed throughout the tissues of the tuber. The distinction between these two groups is not clear cut, because the prussic acid content depends to some extent on the conditions under which they are grown. In all cases care must be taken in eating the tubers, and wherever the plant is grown local methods of removing the poison have been devised. The 'sweet' cultivars are scraped free of rind and boiled, while the bitter types, which are more nutritious, are cut into pieces and boiled, or grated and squeezed to expel the sap and prussic acid, or ground to a powder and pressed. Cassava meal may be fermented briefly in the preparation of foods, such as the West African 'garri'. The tubers cannot be stored after harvest unless they are first cut into slices and dried. Fresh peeled tubers contain about 65 per cent water, 35 per cent carbohydrate and usually less than 2 per cent protein. Apart from their use as a food crop in the tropics, there are several industrial uses, and cassava starch is a commodity in world trade. Tapioca is prepared from the fine starch which is obtained from the tubers when they are washed and pressed. The starch grains are heated on a flat plate until they swell and collect into pellets which are marketed as tapioca. During heating some of the starch is converted into sugar. The starch is also used in adhesives, and as a source of sugars, alcohol and acetone.

Yams: *Dioscorea* spp.

Yams are grown throughout the wetter tropics for their stem tubers, especially in West Africa, where they provide the staple food for many people in the rain forest zone and its northern fringes; they are also important, though grown on a much smaller scale, in South-east Asia and the Caribbean. Annual world production of tubers is difficult to estimate, but is thought to be about 19 m. tonnes, of which some 14 m. are produced in Nigeria, 1.5 m. in the Ivory Coast and about 1 m. in Ghana and in Togo. The eleven cultivated species of *Dioscorea* differ in their climatic requirements, but the greater part of yam production in West Africa is in areas with a short, but distinct dry season, and annual rainfall around 1,500 mm; but the crop is grown where annual rainfall is only 1,000 mm, or where it is as much as 3,000 mm, and one species, *D. opposita* (Chinese yam), can be grown in temperate climates. Yams are grown on deep, well-drained soils, often as the first crop following a period of several years fallow. The land is worked with hand hoes into large mounds or ridges in which small 'seed' tubers or cut pieces of larger tubers are planted to propagate the crop vegetatively. The tubers of the most important cultivated species mature 8—10 months after planting, and though they vary in storage quality and the length of time they remain dormant, yams are generally difficult to store. Yam farmers often take care to make dry, airy buildings to store their crop to decrease the chances of the tubers germinating or going rotten. The tubers of most wild species and of many cultivated ones are bitter, or even poisonous when fresh because they contain alkaloids or oxalic acid and oxalates; the cultivars are not bitter after they have been peeled, then boiled or roasted. They are more nutritious than cassava because they have more protein, and are an important source of vitamin C in diets. Their composition when fresh is approximately 60—70 per cent water, 30—40 per cent carbohydrate (starch) and 4—8 per cent protein. The tubers of some wild species contain compounds valued in the pharmaceutical industry for the preparation of the drug cortisone and the hormones in oral contraceptive pills.

Dioscorea is the principal genus in the monocotyledonous family *Dioscoreaceae*, which is taxonomically close to the *Liliaceae*. The family includes six genera of broad-leaved climbing plants with rhizomes or tuberous food storage organs, inconspicuous, often unisexual flowers, an inferior ovary of three carpels with three styles, and a fruit which is a capsule, or a samara or a berry. *Dioscorea* is a large genus of 600 or so tropical or sub-tropical, mostly monoecious species, in which the rhizome is modified to produce an annual tuber. In wild species buds at the hard, rhizomatous tip of the tuber germinate to produce new aerial shoots at the beginning of the growing season; the new shoots consume the food reserves in the tuber, or the planted pieces of tuber in cultivars, and it eventually shrivels and dies, to be replaced by new tubers during the growing season. Each tuber usually produces only one main stem which may branch. The stems vary in length from 3—12 m, depending upon differences between species; they are weak, and climb supports by twining, either to the right (with the steam apex turning round the support in an anti-clockwise direction) or to the left (clockwise). The direction of twining is characteristic of each species. The stems may be spineless or armed with spines, they may be glabrous or hairy, and carry simple (rarely compound) cordate or deeply lobed, opposite or alternate leaves on long petioles. Some species produce bulbils in the axils of their leaves which are organs of vegetative reproduction; in *D. bulbifera*, the potato or aerial yam (the only cultivated species which is also wild in both Asia and Africa), the bulbils are very large, and are the edible produce for which the crop is grown, though it also has small, bitter tubers. The inflorescences of male plants are axillary racemes, panicles or cymes, those of female plants are axillary spikes, or the female flowers are single or in pairs in the leaf axils. The individual flowers are rarely more than 4 mm across, with a perianth of three sepals and three petals alike in size and colour, white, green or brown. The male flowers have six stamens, three of which may be sterile; the three-locular ovary develops into a dehiscent capsule containing six flat, winged seeds which are dispersed by wind. Even though the flowers are tiny, insects are attracted by their sweet scent, and effect cross-pollination.

The cultivated yams were domesticated independently in four centres, but those which are now of greatest importance are indigenous to West Africa (*D. rotundata* and *D. cayenensis*, and others less important), and South-east Asia (*D. alata* and *D. esculenta*, and others); they were widely cultivated in and near their centres of origin long before the Spanish and the Portuguese spread them around the world in the sixteenth century. *Dioscorea esculenta* belongs to the Section *Combilium* of the genus in which the stems twine to the left (clockwise); the other three more important species, *alata*, *rotundata* and *cayenensis*, belong to Section *Enantiophyllum* in which the stems twine to the right (anti-clockwise). *Dioscorea opposita* (Chinese yam) is the only species from the south China centre of origin, and *D. trifida* (cush-cush yam) the only one to have been domesticated in the New World.

Dioscorea alata, the greater yam, gives the largest yields of tubers of all cultivated species, and with *D. rotundata* accounts for most of world yam production. Its tubers are very variable in shape and size, and may be very large, exceptionally as long as 3 m and as heavy as 60 kg, though they normally weigh about 5–10 kg. Some cultivars produce long, cylindrical tubers which grow deep in the ground, while others produce shorter, almost globular tubers near the surface; or the tubers may be flat and branched, lobed or fan shaped. They mature in 8–10 months and keep better in store than those of many other species because they remain dormant for several months. The flesh of greater yam tubers is commonly white or cream coloured, but sometimes pink, or even purple. The stems are weak and spineless; they are square in section, with corners extended as wings (giving another common name to the plant, 'winged yam'), and bear large, cordate, opposite leaves, occasionally with small bulbils in their axils. The male flowers are in large axillary panicles, the female flowers in axillary spikes.

Dioscorea rotundata, the white guinea yam, is the most important cultivated species in West Africa, and has been cultivated in the West Indies since the sixteenth century, but not in Asia. It is more tolerant of

Fig. 5.5 *Dioscorea rotundata*: White guinea yam. A white guinea yam twining up a stake in Nigeria.

drought than *D. alata*, and so extends further into the dry savanna zone of West Africa than most other species. Its cylindrical stems are up to 12 m long, very spiny in some cultivars, spineless in others, with simple, opposite, cordate leaves up to 12 cm long. The tubers may be very large, but commonly weigh about 5 kg. They become dormant when mature 8—10 months after planting, and consequently store well compared with the tubers of species with no dormancy. Their shape is very variable, but they are often long, smooth and cylindrical; they have a white mealy flesh. The flowers are borne singly or in pairs in the leaf axils.

Dioscorea cayenensis, the yellow guinea yam, grows wild in Africa. It is much like *D. rotundata*, and has been suggested as the ancestor from which the white guinea yam evolved. *Dioscorea cayenensis* is important in the wet forest zones of West Africa, but not in Asia; it is not so tolerant of drought as *D. rotundata*, and is not grown where there is a long dry season. Its tubers do not become dormant when they are mature; consequently the yellow guinea yam does not store well. The large, yellow-fleshed tubers can be harvested most of the year; provided that their rhizomatous tops are left in the ground, and the plants are carefully earthed up after harvest, a second crop of tubers will be produced. The stems are spiny with opposite or alternate, cordate pale-green leaves, and flowers which are solitary or in pairs in the leaf axils as in *D. rotundata*.

Dioscorea esculenta, the lesser yam, is cultivated mainly in Asia and the Pacific islands, but also to a small extent throughout the tropics. The lesser yam produces clusters of five to twenty small round or ovoid tubers which contain no toxins, nor are they bitter; and because they are also less fibrous they are more palatable than the tubers of most other species. The stems are cylindrical and very spiny with small, simple, alternate leaves. The species flowers rarely, but when it does its flowers open widely and are large compared with other species. The tubers sprout soon after they are mature, and bruise easily so that they do not store well.

Dasheen, Eddoe, Cocoyam, Taro: *Colocasia esculenta*

The genus *Colocasia* is a member of the large monocotyledonous family, the *Araceae*, which includes 1,500 or more species, most of them tropical herbaceous plants with rhizomes, tubers or corms containing latex, and very large simple leaves on long petioles. Some species are woody, others are aquatic or epiphytic. The few cultivated species are confined to the wet, humid tropics, or permanently wet soils in the drier tropics. The chief characteristic of the family is the peculiar inflorescence, the spadix, which is a simple spike with a much thickened, fleshy axis bearing numerous small flowers, the whole subtended by a large bract, the spathe, which enfolds and more or less encloses the inflorescence. The flowers are unisexual or hermaphrodite, usually without a perianth; the fruits are small berries which are packed closely together on the spadix.

Colocasia is the most important economic member of the family with seven species native to Eastern Asia and Polynesia. There is some doubt about the true relationships of the various cultivated forms, and the account given here is that followed by Purseglove in his *Tropical*

Fig. 5.6 *Colocasia esculenta*: Cocoyam.   A cocoyam growing at Ibadan, Nigeria.

Crops. He refers all cultivars to *C. esculenta*, but recognizes two botanical varieties, var. *esculenta*, the dasheen, taro or cocoyam domesticated in South-east Asia, and var. *antiquorum*, the eddoe, which may have been selected from var. *esculenta* in China or Japan. The dasheen is the staple food in the tropical rain forests of some Pacific islands, and is widely grown in West Africa, Japan and the West Indies. The eddoe, though of

Asian origin and still cultivated there, is now perhaps equally important in the West Indies. Annual world production of both varieties is estimated to be almost 4 m. tonnes, more than 3 m. of them from Africa (Nigeria 1.8 m., Ghana 1.1 m.), and less than 1 m. from Asia (Japan 0.5 m.). Like other root crops dasheens and eddoes grow best on deep, well-drained soils rich in organic matter. They can be grown under a range of climatic conditions, but require supplementary irrigation where rainfall during the growing season is less than about 1,750 mm, and are most productive on freshly cleared rain forest soils where the annual rainfall exceeds 2,500 mm. Eddoes are more tolerant of drought and cold than dasheens.

The stem is a large globular or elongated corm marked with rings representing nodes, and bearing axillary buds and a few adventitious roots. The corms of var. *esculenta* are up to 30 cm long and 15 cm in diameter with few much smaller lateral corms around the single large one produced by each plant. The main corm of var. *antiquorum* is smaller, and surrounded by numerous lateral corms or cormels. The crop is propagated vegetatively with the tops of main corms or whole side corms. From the apex of the corm arises a whorl of very large heart-shaped leaves on petioles 1–2 m long which have a sheathing base, and which are attached to a point near the middle of the lamina (peltate), not at its basal edge. By the peltate form of their leaves the eddoes and dasheens can be distinguished from the other important tuber-bearing member of the family, tannia (*Xanthosoma sagittifolium*). The dark-green laminas are thick and glabrous, more or less rounded or ovate, and indented at the base (hastate), with three main veins and numerous prominent lateral veins. They are up to 0.5 m long, and hang from the petiole with their tips almost touching the ground. When flowering occurs, which is rare in most clones and unknown in others, the inflorescence is carried on a short peduncle, not exceeding the length of the petioles. It is subtended by a widely spreading yellow spathe 15–30 cm long. The inflorescence carries both male and female flowers, female at the base separated from the male flowers by a region producing only staminodes. At the top of the inflorescence there is a sterile appendage which is short in dasheens, but longer than the male section of the inflorescence in eddoes. In the male flowers the stamens are fused (connate) into a mass, and on the very rare occasions when seed is produced the one celled ovary contains many ovules.

The corms mature 6–9 months after planting, and though they may be harvested and eaten before they are mature, they do not store well unless harvested fully mature after the leaves of the crop have yellowed. The fresh corms contain 60–80 per cent water, 13–30 per cent carbohydrate, mostly small and very easily digested starch grains, 1.5–3 per cent protein and vitamin C, thiamine, riboflavin and niacin. After the flesh of the peeled tubers has been cooked it looses its bitter taste, and has a pleasant, nutty flavour. The young laminas and petioles are sometimes used as a vegetable. A fermented, edible product called 'Poi' is manufactured from the flesh of the corms in Hawaii, and may be canned.

Tannia: *Xanthosoma sagittifolium*

Tannia is a crop of the New World tropics in the family *Araceae*, much like eddoes and dasheens in appearance, with a short stem producing large leaves on long petioles from its apex and a large corm at its base; but the leaves of tannia are never peltate. The petioles grow to heights of 2—3 m with very large, thick, pale-green laminas 50—90 cm long, and arrow-shaped (sagittate) with two large basal lobes (hastate). In mature plants the main corm is surrounded by numerous lateral cormels, as in eddoe. The inflorescence is never carried above the leaves. The spathe is rolled around the spadix at the base, but expands above into a flattened or boat-shaped bract 15—25 cm long, and longer than the spadix. The flowers are closely packed on the spadix, with the male flowers at the top and the female flowers below, the sexes separated by a sterile region bearing only staminodes. There is no sterile terminal appendage on the inflorescence as in *Colocasia*. The male flower has six stamens united into an angular column. In female flowers the ovary is divided into several cells with numerous small ovules in axile placentation, and it has a disc-like, yellow stigma.

The young leaves are used as a vegetable, and the corms are more nutritious than dasheens or eddoes though their starch grains are larger and less easily digested. The flesh of the corms varies in colour from white to pink or yellow, and there are also differences between clones in the amount of reddish-purple anthocyanin pigmentation in the leaves, especially along the veins. Tannia has been cultivated since ancient times in South America and the West Indies, but it was not spread by the Portuguese and Spanish with other New World crops in the sixteenth century. It has become widespread in the tropics only as recently as the mid-nineteenth century. Tannia is grown under similar conditions to those suitable for eddoes and dasheens, but extends into drier areas without irrigation (1,000—1,500 mm of rain), and will not tolerate waterlogged soils.

Arrowroot: *Maranta arundinacea*

Arrowroot is an erect, herbaceous perennial native to the tropics of northern South America, and now chiefly important in the West Indies, especially St Vincent. It is a member of the monocotyledonous family *Marantaceae*, and *Maranta* includes about 25 species all with fleshy rhizomes, but only *M. arundinacea* is economically important. It produces white, thin, cylindrical rhizomes around 30 cm long; they are fibrous and not useful for human consumption when fresh, but are the source of very fine-grained starch which constitutes about 17 per cent of the fresh weight of the rhizome. The starch is easy to digest and is imported into the United Kingdom and the United States to make infant foods. It is also used to make carbonless paper for use in printing machines attached to computers. To extract the starch mature rhizomes are washed, skinned and grated, then the pulp is washed over sieves until all the fibres are removed. The starch is allowed to settle out from the washing liquid, or it is separated by centrifuging; it is then dried, pulverized and packed for export. Over recent years annual production of arrowroot starch in St Vincent has been 1,500–3,000 tonnes.

The plants grow to 1.5 m tall with large lanceolate leaves in two rows (distichous) along the erect stems; the laminas are around 12 cm long on long petioles with clasping bases. The inflorescence is a terminal panicle with slender branches bearing very few flowers in pairs, each flower subtended by a deciduous bract. The flowers are zygomorphic with three free calyx-lobes about half as long as a corolla tube of three fused, white petals. The upper petal is expanded into a broad hood. The flowers are hermaphrodite with one fertile stamen which has a one-celled anther, several staminodes, the outer pair of which are large and petalloid, and an inferior one-celled ovary which develops into a dehiscent fruit containing a single seed. In cultivation the inflorescences are removed as they appear in order that all possible food material is diverted to the growing rhizomes. The crop is propagated vegetatively with rhizome pieces.

Another member of the *Marantaceae, Calathea allouia*, topee tambu, is cultivated to a small extent in northern South America and the West Indies for its small ovoid potato-like tubers. The tubers develop at the ends of fibrous roots in the upper layers of the soil and mature about 10 months after planting. The plants are perennials, about 1 m tall, and produce cone-like inflorescences of white flowers resembling those of arrowroot, but with only one outer staminode and an ovary of three cells with three ovules. The plant is propagated vegetatively from suckers because the tubers have no buds.

Canna edulis, the edible canna or Queensland arrowroot, is an ancient New World crop in the monocotyledonous family *Cannaceae*. It has been cultivated for more than 4,000 years in the Peruvian region of South America for its large, branched rhizomes which are used for food, and has spread through Africa, Asia and the Pacific islands as a minor food crop, and in some areas as a food for livestock, especially pigs. In the West Indies and Australia it is grown for the commercial extraction of coarse-grained starch from the rhizomes which is used in infant and invalid foods. The crop is propagated vegetatively with the small ends of the rhizomes which grow to produce a cluster of fifteen to twenty erect, fleshy stems bearing large, simple, oblong leaves on long petioles with clasping bases. The stems grow to heights of 2–3 m, and both the stems and leaves are often purple. When the plants are mature after about 8 months they have a clump of branched rhizomes up to 60 cm long, ringed by the scars of scale leaves, and bearing thick, fibrous roots. The terminal inflorescence carries pairs of large, red zygomorphic epigynous flowers subtended by bracts. Each flower has three small green sepals, and three larger, red sepal-like petals which are fused at the base and to the staminal column. The androecium is the most striking part of the flower, and has five large petalloid staminodes and one fertile stamen in which the single anther is attached to the side of a broad, coloured petal-like filament. The ovary has three locules and numerous ovules in axile placentation. The fruit is a capsule with many hard, round seeds.

Tacca leontopetaloides, the East Indian arrowroot, is a member of the small family of monocotyledons, the *Taccaceae*. It occurs wild in Africa and Asia, and was once an important food crop in the Pacific islands where it is still cultivated, and has been used on a very small scale as a commercial source of starch. The tubers of wild plants are sometimes gathered and eaten during times of food scarcity in parts of West Africa. The aroid-like plants produce small, globular starchy rhizomes similar in size and shape to Irish potatoes. They are very bitter when fresh, but less so after they have been boiled.

Sago: *Metroxylon* **spp.**

The commercial pellets of starch known as sago are obtained from several tropical plants, the most important of which is the palm *Metroxylon sagus*. A second species, *M. rumphii*, differs little from it; both are indigenous in New Guinea and nearby islands where they grow in freshwater swamps. They are now cultivated in Indonesia and Malaysia. The sago palms have stout, erect trunks rising to heights of 10—12 m; suckers grow up from the main roots so that wild plants grow in clumps, but in cultivation the suckers are removed and used to propagate the crop. The trunk is topped by a magnificent crown of large, paripinnate leaves. Each leaf arises from a broad clasping leaf base which may reach a considerable length, but it does not encircle the stem as in some other members of the family *Palmae*. The leaf has a broad, stiff mid-rib carrying two rows of numerous straight, pointed leaflets, the longest of which may be 1.5 m or more in length.

The sago palm is a monocarpic perennial, and flowers after about 15 years' vegetative growth, then dies. The inflorescence is a terminal spadix with several main branches, each subtended and partially enclosed by a spathe. The whole infloresence may be very large, and secondary branches which bear pairs of flowers hang down well below the crown of leaves. Each pair of flowers is subtended by a broad membranous bract, and the flower pairs are borne spirally around the axis, the bracts of successive pairs tending to be fused together. In each pair of flowers one is male and the other apparently hermaphrodite, but the stamens of the latter do not function and it behaves as a female. Each flower is subtended by a number of bristles or small scales. The male flowers have a bell-shaped calyx-tube which is deeply three-lobed, three inner perianth segments (petals) longer than the calyx and divided almost to the base and six stamens. Each stamen has a versatile anther on a short filament, the filaments being partially united at the base and fused there to the base of the inner perianth segments. Female flowers are similar in structure to the male, but in addition have an ovoid unilocular ovary surmounted by a thick, pointed style.

The male flowers open before the females, which do not mature until the males have been shed. The globose fruit has a very thin endocarp enclosing a single seed, a spongy mesocarp and an exocarp of round or conical reddish-brown, shiny scales.

By the time the palm flowers its trunk is packed with starch, and at this time it is cut down and split open to scrape out the white, floury tissues. They are washed repeatedly to separate the starch from the fibre, then the starch is washed, pressed through sieves and dried in the sun or on iron trays in ovens. The resulting starch pellets are marketed as sago. A single palm yields around 300 kg of starch; the largest yields come from palms of *M. sagus*, which have smooth trunks. The trunk of *M. rumphii* is spiny. Several other palms of South-east Asia yield starch from their trunks, and it is also obtained from the cabbage palm, *Roystonea oleracea*, in tropical America.

Further reading

General

Anon. (1971). International symposium on tropical root and tuber crops, *World Crops*, 23, 25–38.

Kay, D. E. (1973). Crop and Product Digest No. 2, *Root Crops*, Tropical Products Institute, London: HMSO.

Purseglove, J. W. (1968 and 1972). *Tropical Crops*, 2 vols, London: Longman.

Yams

Alexander, J. and Coursey, D. G. (1969). The origin of yam cultivation, in: Ucko, J. and Dimbleby, G. W. (eds), *The Domestication and Exploitation of Plants and Animals*, London: Duckworth.

Ayensu, S. E. and Coursey, D. G. (1972). Guinea yams. The botany, ethnobotany, use and possible future of yams in West Africa, *Econ. Bot.*, 26, 301–18.

Coursey, D. G. (1967). *Yams*, London: Longman.

Waitt, A. W. (1963). Yams, *Dioscorea* species, *Field Crop Abstracts*, 6, 145–57.

Sweet potatoes

Cooley, J. S. (1951). The sweet potato. Its origin and primitive storage practices, *Econ. Bot.*, 5, 378–86.

Crosby, D. G. (1964). The organic constituents of food. iii. Sweet potato, *J. Food Sci.*, 29, 287–93.

Haynes, P. H. and Wholey, D. W. (1971). Variability in commercial sweet potatoes (*Ipomoea batatas* (L) Lam.) in Trinidad, *Exptl. Agric.*, 7, 27–32.

MacDonald, A. S. (1963). Sweet potatoes, with particular reference to the tropics, *Field Crop Abstracts*, 16, 219–25.

Yen, D. E. (1970). Sweet Potato, in: Frankel, O. H. and Bennett, E. (eds), *Genetic Resources in Plants — Their Exploration and Conservation*, Oxford: Blackwell.

Cassava

Jennings, D. L. (1970). Cassava in Africa, *Field Crop Abstracts*, 23, 271–8.
Jones, W. O. (1959). *Manioc in Africa*, Stanford University Press.
Rogers, D. J. (1965). Some botanical and ethnological considerations of *Manihot esculenta*, *Econ. Bot.*, 19, 369–77.
Rogers, D. J. and Fleming S. H. (1973). A monograph of *Manihot esculenta* — with an explanation of the taximetric methods used, *Econ. Bot.*, 27, 1–113.

Aroids

Barrau, J. (1957). Les aracées à tubercules alimentaires des Iles du Pacific Sud, *J. Agric. trop. Bot. appl.*, 4, 34–52.
Coursey, D. G. (1968). The edible aroids, *World Crops*, 20(4), 25–30.
Plucknett, D. L., de la Pena, R. S. and Obero, F. (1970). Taro (*Colocasia esculenta*), *Field Crop Abstracts*, 23, 413–26.

Canna

Gade, D. W. (1966). Achira, the edible canna, its cultivation and use in the Peruvian Andes, *Econ. Bot.*, 20, 407–15.

Sago

Barrau, J. (1959). The sago palm and other food plants of the marsh dwellers in the South Pacific Islands, *Econ. Bot.*, 13, 151–62.

Potato

Simmonds, N. W. (1971). The potential of potatoes in the tropics, *Trop. Agric. (Trin.)*, 48, 291–9.

Chapter 6

Vegetable Crops

Vegetables are a distinct group of crops largely because of the way in which they are grown and their produce is used, not because they have botanical features in common. Indeed, the opposite is true, and from one species or another root or stem tubers, corms, bulbs, stems, petioles, leaves, inflorescences and mature or immature fruits are harvested for use as vegetables. They are eaten raw or cooked, commonly as a dish accompanying a staple food or meat, or in stews or salads; they add variety and flavour to diets, and though their nutritive value varies and is usually greatest in those which are eaten raw, vegetables are important sources of iron and calcium, and of vitamins. Except for the leaves and young fruits of legumes, vegetables contain little protein (commonly 1—2 per cent of their dry weight). They provide 'roughage' in the form of cellulose which aids the digestion of other foods, such as meat. Unlike the major staple foods, which are sometimes called 'field crops', vegetables are usually cultivated intensively in 'gardens' (though many are interplanted with staple food crops in Africa and Asia); consequently, like 'fruits', the vegetables are studied as a part of horticulture. A wide range of indigenous and introduced vegetables is grown in the tropics, commonly in home or compound gardens where they are given frequent and careful attention. Where there is a demand for vegetables from large urban populations they are grown in specialized commercial enterprises, in much the same way as vegetables in the 'market' or 'truck' gardens of countries with temperate climates. The limitations of transport in many tropical countries have confined such commercial production to the vicinity of town markets, where cash profits enable growers to use improved methods of husbandry to overcome restrictions imposed by local soils and climates. For example, it is common practice in the urban market gardens of the tropics to provide nurseries and shade for seedlings and young plants, and to use relatively large amounts of organic manures and inorganic fertilizers as well as supplementary irrigation water and mulches to conserve it, all in order to increase production and profit.

The tendency for temperate biennial vegetables (usually harvested at the end of their first season of vegetative growth) to pass through their vegetative and reproductive phases of growth within the space of a few months in the tropics has posed problems not often met with in temperate climates. Furthermore, exotic vegetables in the tropics are liable to be seriously damaged by insect pests and nematodes, and by fungal and bacterial pathogens, especially in very humid regions. None the less, the majority of temperate vegetable crops can be grown in the tropics, most successfully at cool, high altitudes, or with irrigation during the relatively cool, dry months in semi-arid regions. Indeed, at high elevations in the tropics, where mild, almost temperate climates permit, vegetables (and fruits and flowers) can be produced profitably for export by air to temperate countries during their winter season. An increasing number of new cultivars of many vegetables is becoming available which are suited to the lowland, humid tropics, either because they are resistant to important diseases, or because they are physiologically adapted to grow slowly without 'bolting' (early flowering) in low tropical latitudes. Several important vegetables which have been introduced to the tropics are omitted from this account because they are adequately discussed in other literature, but though the same is true of tomatoes and onions, they are included briefly because they are so popular and widespread in the tropics. The immature cobs of maize, which is discussed in the chapter on cereals, are very popular as a cooked vegetable wherever the crop is grown.

Turning now to indigenous tropical species, experience reveals an enormous range of wild, weedy and cultivated plants which are used as vegetables, especially a large number whose leaves are gathered to be cooked as 'pot herbs'. Grain legumes and root crops are important in tropical diets not only for their seeds and tubers, but often for their leaves and young shoots which are eaten as leaf vegetables (for example, cowpeas and sweet potatoes), and immature legume pods are often eaten as a vegetable. Among the many indigenous tropical vegetable crops this chapter deals only with the cucurbits, which are of special importance in the drier parts of the tropics, the egg plant or aubergine, okra and the important pot herbs Jew's mallow and purslane which are fairly widely distributed.

It must be remembered that peasant agriculture in the wet tropics can well be described as gardening. A relatively small area of land is cultivated, one or two staple food crops such as roots or bananas are grown, and a large number of ancillary vegetables are often interplanted with the main crops. In the semi-arid tropics, though vegetables may be interplanted with the staple cereals, they are more often grown at different times of the year in low, swampy areas where irrigation water is available, or during the main growing season in mixed plantings close to the homestead. Of all the crop plants in the tropics indigenous vegetables have for long been most neglected in agricultural research, and much remains to be done to collect and study their diversity, and to improve their yield and quality. This is especially true for the lowland, wet tropics where vegetables are a major and nutritionally very important part of diets.

The *Cucurbitaceae*

The family *Cucurbitaceae* is an extremely interesting, and in many ways unusual, family of dicotyledons, including about 90 genera and more than 700 species which are widely distributed over the warm parts of the world. All the species are killed by frost, but some tolerate very high temperatures and are widely grown in the tropics for their fruits which are eaten as vegetables or for dessert. The seeds of several cultivated species are eaten roasted, and young leaves and shoots may be used as pot herbs. Most members of the family are tendril climbing herbaceous annuals, often growing very rapidly to a large size; the fruits of some species are among the largest known in the plant kingdom. The cucurbits are characterised by a very extensive root system which ramifies in the surface soil; they have hollow angled stems which may be glabrous, hairy or prickly, and their vascular bundles are bicollateral. The stems may reach great lengths, either trailing along the ground, or climbing supports with their tendrils. The large, simple but often deeply lobed, alternate or spirally arranged leaves are on long petioles; in each leaf axil there is a flower bud, a vegetative bud and a tendril. Branching in the *Cucurbitaceae* is difficult to interpret, but it is probable that in the axil of each leaf the axillary bud either remains dormant or develops into a short leafy shoot, while an extra axillary bud develops into a flower. The tendril is probably a modified floral bract. The plants are monoecious or dioecious with unisexual flowers occurring singly in the leaf axils. The hairy calyx-tube is divided at the top into five lobes, and the corolla has five petals which are commonly fused. In male flowers the androecium is variable, but tends to be massive with three connate anthers. In female flowers the inferior ovary consists of three united carpels with numerous ovules carried on three thick, fleshy, apparently parietal placentas, though the placentas are at first axile, later growing out to the wall of the ovary during development, carrying the ovules with them. The ovary has a thick fleshy style topped by three large, forked stigmas. Monoecious plants are protandrous with male flowers produced before the females, and in much greater numbers. The flowers remain open for about 1 day and are pollinated by insects. After fertilization the ovary develops into the curious fruit of the *Cucurbitaceae* which is a kind of berry known as a pepo. In this fruit, which is normally indehiscent, the tissues of the receptacle surrounding the inferior ovary develop into a very hard skin or rind which encloses and protects the fleshy mesocarp.

The three most important genera whose species are cultivated for their fruits in the tropics are *Cucurbita, Cucumis* and *Citrullus* (syn. *Colocynthis*). Their fruits are eaten mainly as vegetables, though some species produce fruits which are eaten as a dessert. Within the same family *Lagenaria siceraria*, the bottle gourd, has fruits with a rind hard enough to make them useful throughout the tropics for a variety of drinking vessels, cooking pots and utensils. Another cucurbit, *Luffa cylindrica*, the loofah, is cultivated for the mesh of vascular tissue which can be extracted from the fruits, and which is used as a bath sponge, in industrial filters and in sound insulation.

The species cultivated as vegetables (or for 'fruits') are of special importance in dry regions because they thrive on sandy soils with irrigation, and some of them can withstand great heat and low humidity. On the other hand, in the humid tropics many of the curcurbits are seriously affected by pathogens, especially downy mildew (*Pseudoperonospora cubensis*), powdery mildew (*Erysiphe cichoracearum*) and fusarium wilt (*Fusarium oxysporum*), and by root-knot nematodes and several insect pests.

Squashes, pumpkins, marrows, gourds: *Cucurbita* spp.

The genus *Cucurbita* is native to America, and includes about 25 species of monoecious, prostrate, trailing, vine-like plants with rough, prickly, angled stems which tend to root at the nodes. Large, bright-yellow unisexual flowers occur singly in the leaf axils, the males often produced before the females. The tendrils of *Cucurbita* spp. are branched, and the large, alternate leaves may be more or less entire or deeply lobed. Five distinct cultivated species, which cannot be hybridized to produce fertile progeny, are cultivated. One of them, the perennial *C. ficifolia*, is restricted to Mexico, Central and South America, and though it has been cultivated there since ancient times it has not spread to other parts of the tropics, and is not economically important. The other four species are widespread throughout the tropics and are described individually later; they are *C. pepo, C. maxima, C. moschata* and *C. mixta*. They all produce fruits which are variously called pumpkins, summer or winter squashes or marrows, but the indiscriminate use of these terms is often confusing. They have been defined by Whitaker and Bohn (see Further reading) as follows.

Pumpkins: The ripe, edible fruits of any of the four species which have strongly flavoured, coarse flesh; rarely eaten as a vegetable, but used in the United States in 'pumpkin pie', or as fodder for livestock.

Summer squash: Edible fruits of any of the four species, but most commonly those of *C. pepo*, eaten when immature as a vegetable. Equivalent to the term 'vegetable marrow' used in Great Britain.

Winter squash: Edible ripe fruits of any of the four species which have mild flavoured, fine-grained flesh; eaten as a vegetable, or in pies, or fed to livestock.

Because the fruits are not harvested until they are mature and have a hard protective rind, pumpkins and winter squashes store well for consumption some time after harvest.

Cucurbita maxima originated in South America, but did not spread into Central America and other parts of the world until after the Spanish conquest in the sixteenth century. It is more tolerant of cool temperatures than the other cultivated species, and is commonly grown as far north as

The Cucurbitaceae 135

Fig. 6.1 *Cucurbita maxima*: Pumpkin or winter squash.   A young pumpkin plant. (By courtesy of the FAO.)

New England (around latitude 45° N.) in the United States; but it also grows well in the hot, dry tropics where it yields heavily, though it may receive little attention. Well known winter squashes obtained from *C. maxima*, with the names by which they are known in the United States, are 'Hubbard', 'Buttercup', 'Marblehead' and 'Golden Delicious'. The fruits of these cultivars are generally considered to have a better flavour, and are more popular than the winter squashes from other species. *Cucurbita maxima* is an annual with long trailing, soft stems which are more or less round in cross-section. Its large leaves are flaccid, dark-green, and kidney-shaped with large basal nodes and entire margins. Its tendrils are fairly stout, and are divided about halfway along their length into many branches. Male flowers are carried on long, upright cylindrical peduncles, and have a calyx of five sepals fused at the base, but expanded into five long, pointed lobes above, petals united into a showy yellow corolla tube with five spreading lobes, and a short massive androecium with three anthers fused into a convoluted group. The female flowers are larger than the males, fewer in number and carried on shorter peduncles. Their calyx-tube is constricted above the ovary where it flattens out into a disc with five fine, spreading points. Female flowers have a corolla similar to that in the males, though at the base of the petals there is a nectary, and female flowers contain rudimentary stamens. A massive style with three large, yellow two-lobed stigmas arises from the top of the inferior ovary. The fruit is variable in size, colour and markings; it may be round, flattened or oval, and has a hard yellow, orange or green, smooth or warty rind enclosing a thick, orange-coloured mesocarp. Numerous seeds are embedded in the torn tissues of the placenta which lies at the centre of the fruit. The white or brown seeds are plump but flattened, and have a slanting scar at the top. They are edible, and contain about 35 per cent of a reddish-brown, semi-drying oil.

Cucurbita pepo probably orginated in north-eastern Mexico where archaeological evidence has been found which suggests that its fruits were eaten there as long as 9,000 years ago. It is now widespread throughout the tropics, though poorly adapted to wet, humid regions, and grows best in cool, dry environments. *Cucurbita pepo* is grown chiefly for its immature fruits which are summer squashes (though some cultivars of this species produce pumpkins and others winter squashes). The plants are trailing annuals reaching very large size, and with an extensive root system. They have angular, prickly stems, and large, roughly triangular leaves, usually more or less deeply divided into five palmately arranged, pointed lobes. The tendrils of *C. pepo* are more or less branched from the base. Male flowers are carried on long, grooved peduncles, and have a calyx-tube which is slightly constricted before it divides into five pointed lobes. The corolla lobes are more pointed than those of *C. maxima*, while the general form of the androecium is similar in the two species. Female flowers are similar to those of *C. maxima*, but are borne on shorter, angular peduncles. The fruits vary greatly in size, colour and shape among different cultivars, some being long, round and smooth, while others are flattened or more or less spherical. They are usually harvested before they are fully mature, and the flesh is eaten as a cooked vegetable. The seeds of *C. pepo* are slightly smaller than those of *C. maxima* and are easier to separate from the tissues in which they are embedded; the scar at their tip is horizontal or rounded, not oblique as in the case of *C. maxima*. The fruits are borne on stiff, five-angled peduncles which are not broadened where they join the fruit.

Cucurbita moschata occurred some 7,000 years ago in Mexico and Central America and spread in ancient times into North and South America. Of the four commonly cultivated species of *Cucurbita, C. moschata* grows best in the wet tropics, and will thrive not only in humid climates, but also in very hot ones. Consequently it is more widespread in the tropics than the other species. The best known named cultivars from the United States are winter squashes such as 'Butternut', and others erroneously (according to the definition given earlier) called pumpkins such as 'Kentucky Field', 'Large Cheese' and 'Dickenson', which also have mildly flavoured, fine grained flesh. *Cucurbita moschata* has soft stems without prickly hairs, and palmately divided leaves with rounded lobes

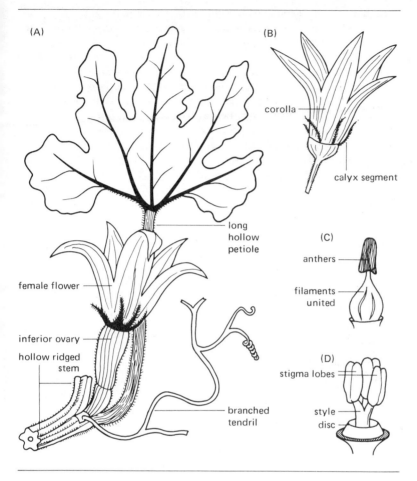

Fig. 6.2 *Cucurbita pepo*: Summer squash or vegetable marrow. (A) A leaf axil ($\times\frac{1}{5}$). (B) A male flower. (C) The androecium. (D) The style and stigma.

(not pointed as in *C. pepo*). The calyx-lobes of the flowers are broader and more leaf-like than those of *C. pepo*. The large, variable fruits are borne on stiff, angular peduncles which are markedly expanded where they join the fruit.

Cucurbita mixta is less widely cultivated than other pumpkins and squashes because its fruits tend to be fibrous and watery without distinctive flavour. It originated in Mexico, where it is grown more for its edible seeds than for its fruits, though named cultivars of pumpkins of this species such as 'White Cushaw' and 'Japanese Pie' are recognized in the United States.

Sweet melons and cucumbers: *Cucumis* spp.

The genus *Cucumis* includes about 40 species, but only two are commonly cultivated, *C. melo*, the sweet melon ($2n = 24$), which originated somewhere in tropical Africa, and *C. sativus*, the cucumber and gherkin ($2n = 14$), which was probably domesticated in northern India. Both species are now cultivated throughout the world, even in cold temperate countries where they are produced commercially in heated glasshouses. The two cultivated species can be distinguished from those of *Cucurbita* by their corolla in which the five petals are fused only at the base, spreading widely above (petals fused at least half their length in *Cucurbita*), and by the clusters of two to three male flowers in a single leaf axil (male flowers solitary in *Cucurbita*).

Cucumis sativus is a trailing monoecious annual with rough, bristly stems and large, alternate light-green leaves which are roughly triangular, but variously lobed and divided in different cultivars. Its tendrils are simple and unbranched. Male flowers are carried in clusters of two to three in the leaf axils, whereas the female flowers, which open later than the males and are fewer in number, are borne singly in the leaf axils. After fertilization the inferior ovary develops into a fleshy pepo which varies greatly in size and shape when, though still immature, it is ready for harvest. The cucumbers are cultivars with fruits 20—100 cm long (commonly 30—50 cm) and 3—5 cm in diameter which are used mainly as a salad vegetable and eaten raw. The gherkins are cultivars with much smaller, darker-green fruits a few centimeters long which are pickled in vinegar, spices and sugar. The fruits of both types may be smooth or warty and with or without spines. Cucumbers which produce parthenocarpic, seedless fruits are commonly grown under glass in Europe, and F_1 hybrid seed is available to commercial growers who can benefit from the increased fruit yields obtained from the progeny of crosses between carefully selected inbred lines.

Cucumis melo, cantaloupe, honeydew or musk-melon is a 'fruit' rather than a vegetable; the sweet, delicately flavoured, juicy flesh of the pepo is eaten raw, often as a dessert. The species is variable with a considerable range of fruit types, many of them highly esteemed for their delicious flavour. Some authorities classify the various forms as distinct botanical varieties, but there is no general agreement about their true status. The common melon of Europe is the cantaloupe which has large, deeply grooved, oblong fruits with a hard, warty and scaly rind. The musk-melon or netted-melon is the predominant type in the United States; it has relatively small globular fruits with very shallow grooves and a network of white or grey ridges on the surface of the rind. Several other forms are known, some with long, narrow fruits like cucumbers (the snake melons) which are eaten as vegetables in parts of Asia. The flesh of the dessert types may be almost white, or varying shades of yellow to orange, or pale green. Melons are annual trailing vines with fairly short stems which are often angled and softly hairy. The alternate, simple leaves on long petioles

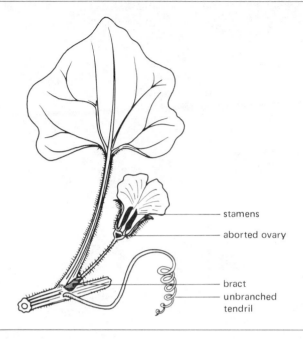

Fig. 6.3 *Cucumis melo*: Melon.   A leaf, a tendril and a male flower in longitudinal section.

are cordate with variously serrated margins; they are very hairy on the under surface. Unbranched tendrils and flowers occur in the leaf axils. Male flowers are produced in clusters, and appear before the female flowers. They have a densely woolly calyx of five lobes, a yellow corolla of five petals fused at their base, and three free stamens with the connective of the anthers prolonged. The female flowers have an inferior ovary densely covered with fine, silky hairs, and containing three to five placentas with numerous ovules. The calyx and corolla are similar to those of male flowers. Occasional hermaphrodite flowers are produced. The ovary is surmounted by a style divided into three to five stigmas at the tip.

Water melon: *Citrullus lanatus*

The water melon originated in Africa and has been in cultivation for more than 4,000 years in the drier parts of the continent, and throughout India and parts of Asia. It is used as a dessert fruit and a thirst quencher, and in the very dry parts of Africa it is relished by both man and his animals as a source of water. The roasted seeds are popular as food in some areas,

Fig. 6.4 *Citrullus lanatus*: Water melon. Water melons in Saudi Arabia. (By courtesy of the FAO.)

especially in West Africa and southern China, and they contain a semi-drying oil. Cultivars with very small fruits and hard, white flesh are used to make preserves and pickles. The genus *Citrullus* is a small one of four species found in Africa, and can be distinguished from *Cucurbita* and *Cucumis* by its deeply pinnately lobed leaves. The water melon is the only species cultivated on any scale, and is a trailing hairy annual with rough, angular stems, and dark-green alternate leaves carried on fairly long petioles. The lamina is deeply divided into several rounded, pinnately arranged lobes, which may in turn themselves be more or less divided. The tendrils are branched, usually into two. Water melons are monoecious with pale yellow flowers which are smaller than those of pumpkins and squashes. They occur singly in the leaf axils, the male flowers being produced before the females. The calyx tube expands into five pointed lobes, and five petals are fused at their base. Male flowers have three free

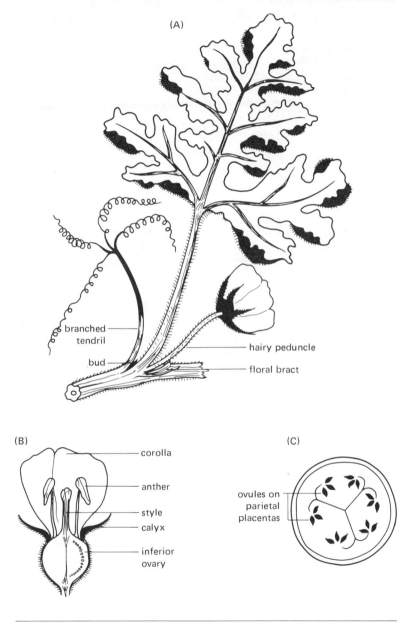

Fig. 6.5 *Citrullus lanatus*: Water melon. (A) The leaf axil – leaf, flower, bud, tendril and bract. (B) A longitudinal section of an hermaphrodite flower. (C) A transverse section of a young fruit.

stamens, the females three non-functional stamens and an inferior ovary with a short style bearing a three-lobed stigma. Nectaries are present in flowers of both sexes, and perfect, hermaphrodite flowers are occasionally produced. The fruit is large and rounded or oblong with a hard, smooth rind which is often mottled light and dark green. It contains a red, pink or yellowish-white flesh in which numerous flattened, black, brown or white seeds are embedded. There is little food value in the flesh, which contains 95 per cent or more of water, and it is considered by many to have very little flavour.

The *Solanaceae*

The *Solanaceae* is one of the most important economic families of flowering plants with about 75 genera and 2,000 species widely distributed throughout the temperate and tropical regions of the world. They are mostly herbaceous plants, though some are soft wooded shrubs or small trees, with alternate, simple leaves which are often softly hairy; there are no stipules. The hermaphrodite flowers are borne in cymose inflorescences. They have a persistent calyx-tube with five lobes, a corolla tube of five, or rarely seven, fused petals, and five to seven epipetalous stamens which alternate with the lobes of the corolla and have large, often coloured anthers. The superior ovary of two united carpels has two locules, though there may appear to be more because false septa occur in the fruits of some species, and there are numerous small ovules in axile placentation. The style is simple with a bilobed or flattened stigma. The fruit of the *Solanaceae* is a berry or capsule, and the seeds have a large endosperm and a curved embryo. The family includes several important crop plants, e.g. *Solanum tuberosum*, the potato, which though commonly grown in the cooler tropics is a temperate crop; *Solanum melongena*, the egg-plant or aubergine; *Lycopersicon esculentum* the tomato; *Capsicum* spp., the chilli and bird peppers; and *Nicotiana* spp., tobacco. There are a few other tropical crop plants of minor importance in the family, and several wild, or occasionally cultivated species, especially of the genus *Solanum*, which are used as pot herbs or whose fruits are eaten as vegetables. The berries of *Physalis peruviana*, cape gooseberry, are a popular dessert fruit in many parts of the world; in South America *Solanum muricatum*, the pepino, and *Solanum guitoense*, the naranjilla, are grown as dessert fruits. *Cyphomandra cetacea*, the tree tomato, is a small tree whose fruits are eaten raw or cooked.

Egg-plant, aubergine: *Solanum melongena*

Solanum melongena is thought to have been domesticated in India where wild plants now grow, but it has spread throughout the warm tropics and is a favourite vegetable, especially in the West Indies and southern United States. It is a much branched perennial, commonly cultivated as an annual, with a strong tap root and a deep, extensively branched, but not widespread root system. Its stems are thick and covered with woolly hairs, and are often more or less spiny. They bear large, alternate, simple leaves up to 20 cm long with wavy margins and a dense covering of hairs. The flowers are solitary or in small groups opposite the leaves. They have a calyx with five large, spiny and woolly lobes which enlarges and persists with the fruit, and a purplish-violet corolla tube divided above into five rotate, incurved lobes. There are five large free stamens which dehisce by apical pores, and a bilocular ovary derived from two united carpels which contains numerous ovules. The aubergine is a large, smooth oblong or ovoid berry up to 15 cm long; it is variously coloured in different cultivars, in some dark-purple, in others green, red or white. Numerous kidney-shaped seeds are embedded in the flesh of the berry. The fruits are usually eaten cooked, and are sometimes stuffed with minced meat.

Tomato: *Lycopersicon esculentum*

The tomato is native to tropical Central and South America where it was cultivated in pre-Columbian times. Its wild progenitor is thought to have been the cherry tomato, *L. esculenton* var. *cerasiforme*, which now grows wild in the Peru—Ecuador area though tomatoes were probably domesticated from weedy forms which had spread as far north as Mexico. It seems likely that the modern tomato has a more complex ancestry, and that after it was domesticated diversity accumulated in it partly as a result of hybridization with several other *Lycopersicon* species which grow wild in the New World tropics. Though its wild ancestors are outbreeding plants, the tomato has become predominantly inbreeding in cultivation. In relatively recent times a great many cultivars have been selected, or have been bred to suit different environments, with fruits suitable for different uses (as a salad vegetable, for purée, paste, ketchup and juice manufacture, or for canning and for pickling), so that the crop is now very diverse. Tomatoes grow well in the tropics, though in humid lowland regions they are often seriously affected by fungus and bacterial diseases, and everywhere susceptible cultivars suffer from virus diseases and often massive infection by root-knot nematodes. A great deal of work is being done, especially in the United States, in which breeders hope to transfer disease resistance from wild South American species to the cultivars.

Tomatoes are annuals or short lived perennials cultivated as annuals with weak, trailing much branched stems covered with glistening reddish-yellow glandular hairs as well as pointed non-glandular hairs. The crop is propagated from seed which is commonly sown in nursery beds or seed boxes. When the seedlings are transplanted their tap root is commonly damaged, and in mature plants is replaced by an extensive, copiously branched adventitious root system. The adventitious roots arise from the base of the stem, and stem cuttings root easily. Different cultivars may have different growth habits; erect, determinate types have fairly thick, solid stems and numerous long lateral branches, while trailing or semi-climbing, indeterminate cultivars have weaker, thinner stems and usually require support to produce a good crop. The stem bears numerous spirally arranged leaves which are unevenly pinnate with variously indented and lobed margins; the main pinnae lie more or less opposite each other along the petiole, with smaller leaflets which vary in number and size between them. Branching at the base of the stem is often monopodial, but tends to be sympodial higher up with inflorescences developed from the apical bud and vegetative growth continued by the development of a bud in the axil of a leaf just below the apex. Consequently the small cymose inflorescences appear to be opposite the leaves, or commonly opposite but slightly above or below them. The flowers are on short pedicels, and are often shed 2—3 days after they have opened; their parts are usually in sixes, but some cultivars have pentamerous (fives) flowers. The persistent calyx enlarges with the fruit; it is covered with glandular and other hairs, and has six united sepals with pointed lobes. The six bright yellow petals are also hairy on the outside, and are fused to form a rotate corolla tube. Six epipetalous stamens with short filaments alternate with the corolla lobes, and have long, coloured, more or less conivant anthers with longitudinal, introrse dehiscence. The tips of the anthers are prolonged into sterile beaks. In some cultivars pollen is shed as the style grows up through the tube formed by the conivant anthers resulting in regular self-fertilization; in others cross-fertilization may occur because the anthers do not dehisce until after the style has grown beyond them. The simple style surmounts a superior ovary of two united carpels. False septa may produce six to twenty locules in the ovary which contains numerous ovules in axile placentation on a large, fleshy placenta. The fruit is a red or yellowish globose berry with a smooth skin, but there is a lot of variation between cultivars in the size and shape of the fruits, in the thickness of the fleshy mesocarp and in the development of the placenta. The coloration of the fruit is due to the pigments carotene and lycopersicin which occur in different concentrations in fruits of different colours; a predominance of carotene produces the least common yellow fruits. Tomato seeds are flat, kidney-shaped and hairy, light-brown in colour and contain a curved embryo. A semi-drying oil is extracted from the seeds which remain with the pulp residues after tomatoes have been processed to make juice, paste, or purée. The press cake is used as a fertilizer or stock feed.

In the tropics tomatoes are grown mostly in home gardens for local consumption, though they are important in commercial vegetable growing enterprises near large towns. In the United States special purpose cultivars have been bred for mechanical harvest; over large areas seed is drilled by machine directly in the field and the irrigated crops yield 20 tonnes per hectare or more of fruit. In the cooler climates of northern Europe the crop is grown on a large scale in heated glasshouses.

The *Cruciferae*

Several important vegetable crops are members of the large and natural family of dicotyledons, the *Cruciferae*. The cultivated species of the genus *Brassica* are important in agriculture throughout the temperate regions of the world, and some of them are grown in the tropics in areas where some months of the year are fairly cool, or at high altitude. *Brassica oleracea* is the most important species. It is an ancient crop, domesticated at least 4,000 years ago in the Mediterranean region. Within it several distinct botanical varieties have evolved in cultivation, the most important of them so far as tropical agriculture is concerned being var. *capitata*, the cabbage, and var. *botrytis*, the cauliflower. The cabbage is a biennial which produces a large, more or less dense 'head' of leaves from a condensed stem in the first year, then a large open racemose inflorescence in the second. Sometimes, in hot tropical environments, the whole development cycle is condensed into a single growing season. Cabbages are harvested after the head of leaves has been formed; they are eaten cooked or raw and are sometimes pickled in brine. Cauliflowers are grown for their immature inflorescences of mostly sterile flowers which are clustered on thick, fleshy condensed branches. A variant called broccoli with a more open inflorescence is predominant in North America. The inflorescence is usually cooked, but cauliflower is sometimes pickled. *Brassica rapa*, the turnip, and *Brassica juncea*, Indian mustard, are both cultivated in tropical countries, the latter most extensively as an oil-seed in India, where several distinct cultivars are recognized. Several other species of *Brassica* are grown for the oil in their seeds, and chief reference will be made to them in the chapter on oil-seeds. The genus *Raphanus*, another member of the family, includes about ten species, one of which, *R. sativus*, the radish, is known the world over for its pungent storage roots which are commonly eaten raw as a salad vegetable. *Eruca sativa*, rocket cress, is of minor importance in Africa and Asia as a salad vegetable. Its young leaves have a sharp flavour and are used to give piquancy to salads.

Purslane: *Portulaca oleracea*

The genus *Portulaca* includes 100 or more species in the small and interesting family of herbaceous dicotyledons, the *Portulacaceae*. The species occur throughout the warmer parts of the world, some as weeds of cultivation, some as small, succulent, drought-resistant wild plants in the driest tropics, and *P. oleracea*, which is cultivated extensively in arid regions for its succulent leaves used as pot herbs. Purslane will grow in the hottest of climates, and often provides the only green vegetable when others have succumbed to the heat. The plant grows wild in most parts of the tropics, sometimes as a troublesome weed, but the cultivated types are in general larger than the weeds, and improved French types differ greatly from their wild relatives in flavour as well as size. Purslane is a low growing fleshy annual with thick, spatulate, opposite leaves which appear to be dull silvery-green owing to the presence in them of water storage tissue. The leaf is made up of three layers of tissue of approximately equal thickness, the middle layer including vascular tissue, palisade and parenchyma cells, while the upper and lower layers are composed of large water storage cells without chloroplasts, and are bounded by a fairly thick epidermis. Calcium oxalate crystals are commonly laid down in the leaves and stems. The thick erect stems are often tinged with red and are very succulent; they carry terminal inflorescences of clusters of yellow flowers (or flowers are solitary in the leaf axils). The flowers, which remain open for a very short time, consist of two sepals, five yellow petals and an indefinite number of up to twelve stamens. The ovary is semi-inferior, and consists of five united carpels with a single loculus in which numerous ovules are carried in basal placentation on funicles (stalks) of varying length. The fruit is a small, almost spherical capsule which splits transversely when mature, then the 'lid' falls off. The small black or dark brown seeds are coiled, and the embryo is curved around the endosperm.

Jew's mallow: *Corchorus olitorius*

Corchorus olitorius of the family *Tiliaceae* is grown throughout the Middle East and in parts of Africa and India as a pot herb, but it is chiefly important in India as the source of the bast fibre 'tossa' jute. The botanical features of the plant are described in the section on jute in Chapter 10. The cultivars grown as pot herbs are rarely more than 30 cm tall, and they branch much more freely than those grown for fibre. Their leaves and young shoots are cooked with a main meat dish, and are one of the commonest of the many pot herbs used in Africa. In parts of West Africa *C. tridens, C. aestuans* and *C. trilocularis* are also grown as minor vegetable crops, and in India the leaves of both *C. olitorius* and *C. capsularis* are eaten, the latter most often as a medicine, though both these species are chiefly important as bast fibre crops. Apart from small amounts of protein (up to 5–6 per cent in old leaves), the leaves of Jew's mallow contain calcium, iron, phosphorous and vitamin C.

Okra, lady's finger: *Hibiscus esculentus*

Okra was domesticated in West or Central Africa, but it is now cultivated widely in the tropics for its immature fruits which are used as a vegetable when they are fresh and green, or after they have been sliced, dried and stored. Like two other *Hibiscus* species, *H. cannabinus* and *H. sabdariffa* in

Okra, lady's finger: Hibiscus esculentus 147

Fig. 6.6 *Hibiscus esculentus*: Okra. A crop of okra with flowers and young fruits.

the economically important family *Malvaceae*, *H. esculentus* produces bast fibres which have many local uses, though they are of poor quality and are not extracted on a commercial scale. The plant is an erect annual up to 2 m tall, with slightly hairy stems, and large, alternate, cordate leaves which are divided into three or five pointed lobes with notched or toothed margins; the upper leaves tend to be more deeply lobed than lower ones.

They are carried on long petioles which are often red, the pigmentation extending into the base of the lamina near the petiole. The leaves are subtended by narrow, pointed stipules. Bright yellow flowers are carried singly in the leaf axils on peduncles rarely more than 2.5 cm long. They have the typical Malvaceous floral structure, with eight to ten very narrow, hairy bracteoles forming an epicalyx; five hairy sepals are fused to form a complete protective covering for the flower bud, then they split into five lobes when the bud opens; and there are five large, showy, bright yellow petals which have a deep reddish-purple spot on the claw. The rounded petals are contorted in the bud and united by their thickened bases. They have raised veins on their under surface. The filaments of numerous stamens are united to form a central staminal column which is fused to the petals at the base. Through the centre of this staminal column rises the style which bears five to nine deep red, apical stigmas. The superior ovary is derived from five united carpels, and has five locules with the ovules in axile placentation. The fruit is a long (10—30 cm), pointed or more or less oblong, ridged capsule. When it is still immature, but ready for harvest, it is dark green and covered with upward pointing hairs. The fruits are very mucilaginous and contain numerous round seeds which vary in colour from dark-green or brown to black. If they are not harvested the capsules become dry, and they dehisce. The fruits are a very popular vegetable and the leaves are eaten as pot herbs throughout the tropics. Numerous cultivars are recognized, some of them the product of breeding and selection; they differ chiefly in the form and slimyness of the fruits, which may be smooth or spiny.

Onion: *Allium cepa*

The common onion, *Allium cepa* var. *cepa*, is the most important and widespread of several cultivated forms of the species, which has been referred by some authorities to the family *Liliaceae*, by others to the *Amaryllidaceae*, but which is now considered to belong to a family intermediate between these two, the *Alliaceae*. The onion is an ancient crop thought to have been domesticated in Central Asia, though its wild ancestor is unknown, nor do onions occur as wild plants. It was cultivated in ancient Egypt, India and China, and in Europe during the Middle Ages, and as recently as the late sixteenth or early seventeenth centuries spread to the New World with the Spanish. The crop is now widespread in both temperate and dry tropical regions, with Europe the centre of greatest production, and a total world crop of around 15 m. tonnes of bulbs annually. In the tropics most onions are grown with irrigation during the relatively cool, dry season in semi-arid regions; they are not a good crop for the hot, wet tropics. Onions are herbaceous biennials with a shallow fibrous, adventitious root system produced from the base of a very short, condensed stem. A succession of alternate leaves is produced from the

Fig. 6.7 *Allium cepa*: Onion. Onion cultivation near Khartoum, Sudan.

apical meristem, each with a sheathing base separated by a membranous ligule from a hollow, cylindrical, green lamina. Successive leaves grow up inside, then burst through the sheath of the leaf which preceded them. Toward the end of the first season's growth, when days become long enough and temperatures are warm, the leaf bases around and immediately above the condensed stem become swollen with food reserves, and form the bulb for which the crop is grown. The outer leaf bases remain as a thin, membranous protective covering around the bulb. Cultivars vary in their long-day requirement for bulb production; those suitable for the tropics produce bulbs in relatively 'short' days, but cultivars adapted to the higher latitudes of temperate regions have relatively long (up to 16 hours) day-length requirements for bulb production. They do not produce a crop in tropical latitudes near the equator where the longest days in the year are shorter than their critical requirement. Provided that there has been a sufficiently long period of growth and that days have become long enough, bulb production is stimulated by warm temperatures and retarded by cool ones. Eventually, after the bulb has been produced, the apical meristem stops producing leaves and becomes reproductive. Cultivars adapted to temperate climates have a marked cold requirement for the induction of flowering, and though some which produce bulbs in the tropics will also flower there, others do not because they are not exposed to cold enough temperatures. The inflorescence is a terminal umbel carried on a hollow,

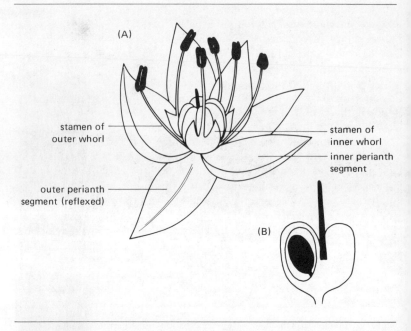

Fig. 6.8 *Allium cepa*: Onion. (A) A general view of the flower (perianth segments separated). (B) A diagram of a section of the ovary.

erect, cylindrical axis, the scape, which is commonly 30–60 cm long. When it is young the inflorescence is enclosed in a membranous spathe, but this splits into two to three persistent, papery bracts as the inflorescence develops. The numerous flowers are carried on narrow pedicels; each has six free, acuminate, greenish-white perianth segments, six stamens inserted on the perianth, and a superior ovary derived from three carpels. The filaments of the three stamens inserted on the inner whorl of perianth segments tend to be thickened at the base. The anthers are oblong with introrse dehiscence. There are usually three locules in the ovary, each containing two ovules, and a filiform style which is at first shorter than the stamens, but after they have dehisced it elongates and becomes receptive. The flowers are thus protandrous, and are pollinated by insects attracted by nectaries at the base of the inner whorl of stamens. The onion fruit is a globular capsule which dehisces by three longitudinal splits exposing the small black seeds. Cultivars of *A. cepa* var. *proliferum* are viviparous and produce small bulbils instead of flowers in the inflorescences; these are used to propagate the plants vegetatively. Common onions are propagated from seed, which is either sown directly in the field, or in seed beds from which young plants are transplanted to the field; or the small bulbs obtained by growing crowded plants are harvested, dried and stored for later use as 'sets' to establish a new main crop vegetatively. F_1 hybrid onion seed is produced commercially utilizing cytoplasmic male sterility to ensure natural cross-fertilization between parents carefully selected for their ability to combine well and give high-yielding progeny. Male sterile plants contain the cytoplasmic factor S, and are homozygous recessive at the male sterility locus (ms ms); they produce no pollen.

The onion bulb is eaten as a cooked or raw vegetable and for flavouring soups and stews, or small mature bulbs are pickled in vinegar; the leaves and leaf bases of young plants are eaten as a salad vegetable. The mature bulbs contain little starch, but appreciable quantities of sugars, a little protein (around 1–2 per cent), and vitamins A, B and C. There is a great deal of variation between cultivars in the size, colour (commonly reddish-purple or white), pungency and storage quality of mature bulbs. Good storage qualities are important because the dependence of the crop on day-length and temperature for bulb development limits production to one crop each year which must be stored after harvest, and eaten or released to market over a protracted period. The pungency of onions develops only when their tissues are cut or bruised, and is due to the presence of allyl sulphides in their tissues.

Several other species of *Allium* are cultivated in parts of the tropics. *Allium cepa* var. *aggregatum* has several forms, one of which, the shallot, is grown in parts of India, Malaysia, Sri Lanka, Africa and some of the West Indian Islands. It is a smaller plant than the onion, and produces clusters of greenish-white or red bulbils at the base of its leaves. *Allium ampeloprasum* var. *porrum*, the leek, has solid leaves and a pithy scape, and an elongated bulb which has a milder flavour than the onion. *Allium sativum*, garlic, is used mainly as a strong flavouring, and resembles the onion except that it has flattened solid leaves, and produces a composite bulb consisting of a number of thin, sheathing, membranous leaf bases enclosing numerous small bulbs, or cloves, derived from axillary buds in the leaf axils.

Further reading

General

King, K. W. (1971). The place of vegetables in meeting the food needs of developing nations, *Econ. Bot.*, 25, 6–11.

Knott, J. E. and Deanon, J. R. (1967). *Vegetable Production in South-east Asia*, University of Philippines Press.

Martin, F. W. and Ruberte, R. M. (1975). *Edible Leaves of the Tropics* Mayagüey: Antillian College Press.

Norman, J. C. (1972). Tropical leafy vegetables in Ghana, *World Crops*, 24, 217–19.

Purseglove, J. W. (1968 and 1972). *Tropical Crops*, 2 vols, London: Longman.

Terra, G. J. A. (1966). *Tropical Vegetables*, Amsterdam: Royal Tropical Institute.

Tindall, H. D. (1968). *Commercial Vegetable Growing*, Oxford University Press.

Cucurbits

Porterfield, W. M. (1955). Loofah – the sponge gourd, *Econ. Bot.*, 9, 211–23.
Whitaker, T. W. and Bohn, G. W. (1950). The taxonomy, genetics, production and uses of cultivated species of *Cucurbita, Econ. Bot.*, 4, 52–81.
Whitaker, T. W. and Davies, G. N. (1962). *Cucurbits*, London: Leonard Hill.

Tomatoes

Jenkins, J. A. (1948). The origin of the cultivated tomato, *Econ. Bot.*, 2, 379–92.
Chapman, T. and Acland, A. D. (1965). Investigations on out-of-season tomato production in Trinidad, *Trop. Agriculture, Trin.*, 42, 153–62.

Onions

Jones, H. A. and Mann, L. K. (1963). *Onions and their Allies*, London: Leonard Hill.

Okra

Joshi, A. B., Gadwal, V. R. and Hardas, M. W. (1974). Okra, in: Hutchinson, Sir Joseph (ed.), *Evolutionary Studies in World Crops. Diversity and Change on the Indian Subcontinent*, Cambridge University Press.

Chapter 7

The Cultivated Tropical Fruits (and Nuts)

In the heading for this chapter the term 'fruit' is used in the popular sense normally understood to mean swollen ovaries which have soft, often juicy and sweet flesh which is commonly eaten raw. Exceptions to this general definition are the fruits of some date palms which are firm and dry, those of some bananas which are mealy and not sweet, and the avocadoes which are soft and oily. Of course, the cereal grains, and the produce of vegetables like tomatoes, chilli peppers, pumpkins, melons, squashes and 'snap' beans are botanically fruits. Large numbers of tropical plants produce edible fruits which fit the restricted description used for this chapter, but very few of them are cultivated, and only five to six have been exploited as commercial crops or have been improved by plant breeders. Though bananas, dates and avocadoes are important food crops in some parts of the world, fruits are most often eaten as a refreshing delicacy and are nutritionally valuable for the minerals and vitamins they contribute to diets.

There is no botanical reason for including 'nuts' in the same chapter as 'fruits'; but they are nowhere important as a major food, and like most fruits are eaten as a delicacy. 'Nuts' in the commonly used sense are the edible, often oily seeds of several species of trees, very few of them important in the tropics. They are not to be confused with other important species whose seeds are also called nuts, such as groundnuts or coconuts. The nuts described here are mostly exported from the tropics for consumption in temperate countries.

Bananas: *Musa* spp. and cultivars

Bananas are large, rhizomatous perennial herbs, cultivated throughout the tropics for their parthenocarpic, seedless berries. They are the most important of the tropical fruits, and a major tropical food crop with annual world production estimated to be around 35 m. tonnes. The fruits are an important export from South and Central America and the West Indies to North America and Europe, with world trade amounting to more than 2.5 m. tonnes annually. The crop thrives in a wide range of environments between latitude 30° N. and S. of the equator, predominantly in the hot, humid lowland forest zones with well distributed annual rainfall greater than 2,000 mm and warm temperatures around 25°–30°C, but such is the range of cultivars within the crop that it extends to high altitudes, and into the semi-arid tropics wherever sufficient water is available. Though the terms tend to be used indiscriminately, some cultivars produce relatively sweet fruits called bananas which are eaten ripe but raw, while others produce starchier fruits called plantains which are eaten unripe but cooked. Until Simmonds's recent studies (see Further reading) of the relationships of the cultivated bananas and plantains and their wild progenitors, they were referred to several species and subspecies, the most important species being the diploids *M. acuminata* and *M. balbisiana* and their triploid derivatives *M. sapientum* and *M. paradisiaca*. Simmonds rejects the classification of cultivated forms into species and sub-species and, in summary, interprets the origin and relationships of bananas and plantains as follows.

Musa acuminata (diploid, $2n = 22$. Genome A) has its centre of diversity in and near Malaysia where fertile forms with small, inedible fruits grow wild in the rain forests. In this area sterile, parthenocarpic, seedless diploids (genomic formula AA) and hardier autotriploids with larger berries (AAA) were selected and propagated vegetatively by man because they had edible fruits. These clones, which were the first cultivated bananas, were dispersed from their centre of origin, and some of them reached the centre of diversity of wild *M. balbisiana* (diploid, $2n = 22$. Genome B) in India. *Musa balbisiana* has inedible fruits, but natural diploid (AB) and allotriploid (AAB and ABB) hybrids with edible fruits arose following natural crossing with *M. acuminata*, and were selected, propagated vegetatively and eventually dispersed world-wide. Hybrid clones with the B genome (there are no cultivated BB diploids or BBB autotriploids) are hardier and more tolerant of drier climates than AA or AAA clones of *M. acuminata*, they are less susceptible to diseases, and have starchier fruits; these are the plantains of the present day. Some diploid AA and AB clones, and one ABBB tetraploid clone are cultivated, but all the important bananas and plantains in the world today are triploids. The triploids were dispersed from their centres of origin into Asia and the Pacific islands in ancient times, but not until the fifth or sixth centuries A.D. did they reach East Africa, from whence they spread across the continent south of the Sahara Desert to Central Africa and eventually to West Africa. The Portuguese introduced the crop to the Canary Islands from West Africa, and though it was unknown in the New World in pre-Columbian times, it was soon introduced after the European arrival there.

Although all bananas and plantains are sterile clones, they have become diverse through the accumulation of somatic mutations. The former rationalization of this diversity by raising species and sub-species is rejected by Simmonds because such taxa cannot legitimately be used to describe sterile, vegetatively propagated hybrids. Instead, he suggests that

Fig. 7.1 *Musa* spp.: Banana.   A banana plantation in Jamaica.

the cultivated clones are grouped according to their genomic constitution, and on this system most of the world's crop falls into two such groups.

1. The AAA triploids occur throughout the tropics, in some areas as a major food crop (for example, Uganda), and they are the only bananas in international trade. Two main groups of AAA clones are recognized.
(*a*) The 'Gros Michel' type is a tall, robust plant with large, highly esteemed fruits which are bright yellow when ripe; it became the mainstay of the South and Central American banana industry, though its susceptibility to Panama disease (banana wilt), caused by the fungus *Fusarium oxysporum*, to some extent limits production in South America.
(*b*) On the other hand, the 'Cavendish' type is resistant to Panama disease and has replaced 'Gros Michel' in the West Indies and parts of South America. It includes several clones which vary in height and leaf size, most of them smaller plants with smaller fruits than 'Gros Michel'. One clone, 'Dwarf Cavendish' or 'Canary' banana is more tolerant of cold than other AAA triploids, and is the most important banana where the crop is grown in the sub-tropics.
2. The other major group of cultivated clones are the AAB triploids. They are the plantains of India (where they originated), and of Africa and tropical America. They have slender fruits which are starchy, but acid and unpalatable if eaten raw, so they are usually picked before they are ripe and are cooked. ABB triploids are also plantains, but are less widespread

than AAB clones, even though they are more vigorous and are resistant to drought and diseases because they have two *M. balbisiana* genomes.

Bananas and plantains have short basal underground stems (corms) from which short rhizomes grow to produce a clump of aerial shoots (suckers) close to the base of the parent plant. The crops are propagated vegetatively, either by planting pieces of old corms with a bud, or using suckers which are lifted for planting with their underground stem. The adventitious root system is not extensively developed either laterally or in depth, and, considering that some clones are 8 m tall, the plants are poorly anchored so that protection from wind is usually essential on estates and plantations. The aerial shoots are cylindrical pseudostems of overlapping leaf bases which are tightly rolled round each other to form a rigid bundle about 30 cm across. New leaves continuously grow up through the centre of the pseudostem with their laminas tightly rolled; they expand at the top into a large, oblong blade with a pronounced supporting mid-rib, and well-marked, pinnately arranged parallel veins. Each plant carries a handsome crown of ten to fifteen leaves, with a new one appearing every 1 or 2 weeks to replace an old one which has died. The lamina of mature leaves is commonly torn by strong winds, but then offers less resistance to them because it hangs in ribbons from the mid-rib. It is light-green, smooth and glossy, and may be very large, as long as 4 m and as wide as 1 m. Bananas and plantains provide good shade, and are often used as nurse crops for young cocoa or coffee.

When plants grown from suckers are about 8–10 months old and have produced 40–50 leaves the apical growing point of the corm becomes reproductive, and instead of leaves it produces an inflorescence which grows up inside the pseudostem on a long unbranched axis. Once exserted from the pseudostem the axis bends down under the weight of the inflorescence, or in some cultivars (such as Gros Michel) because it is positively geotropic, so that the inflorescence is pendent. It is a compound spike of flowers which are arranged in several groups; each group is enclosed in a large, reddish-coloured subtending bract and consists of two rows of flowers, one row morphologically above and closely appressed to the other. The bracts are oval and pointed, and with their axillary groups of flowers are arranged spirally around the axis. They overlap each other closely so that the young inflorescence is compact and conical. As the flowers develop they are exposed because the bract subtending them becomes reflexed, then when the fruits start to develop it is shed, leaving a scar on the axis. The bracts at the base of the inflorescence have female flowers in their axils, a few in the middle may subtend neuter flowers, while at the top of the inflorescence all the flowers are male. Female

---

Fig. 7.2 *Musa* spp.: Banana. (A) Hermaphrodite flower – perianth separated (x1). (B) Diagrammatic longitudinal section of an hermaphrodite flower (x1). (C) Diagrammatic longitudinal section of a female flower (x1). (D) Transverse section of a young fruit (x2). (E) Transverse section of a mature fruit (x1).

The Cultivated Tropical Fruits (and Nuts) 157

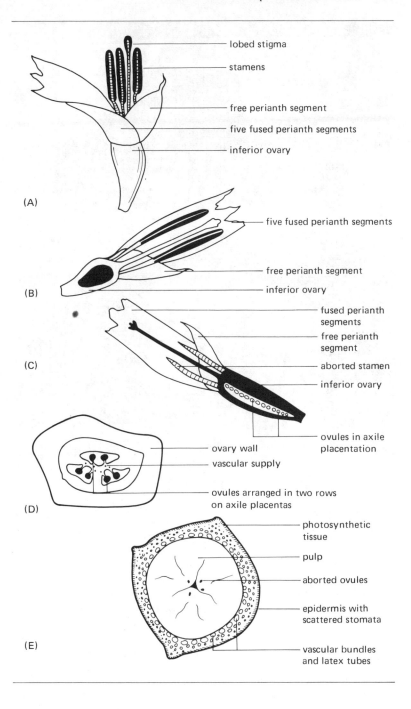

Fig. 7.3 *Musa* spp.: Banana. A good commercial bunch of banana fruits.

flowers have an inferior ovary of three united carpels, roughly triangular in section, surmounted by a short perianth of five fused segments and one free segment which together form a tube around the style and the sterile androecium. The male flowers are usually deciduous; they contain five stamens, but in cultivars do not produce functional pollen. The fruits develop parthenocarpically from the ovaries of the female flowers, and though the numerous ovules on the axile placentas abort, they remain visible as black dots in the centre of mature berries. The fruits mature about 4 months after flowering, but they are usually (always for export) picked before they are mature when about 3 months old. The ripe fruit is elongated, commonly 15—25 cm long, curved, more or less round in cross-section but with the triangular form of the ovary still visible. At the tip the perianth persists for a short time, separated from the fruit by a brown corky layer. Beneath the epidermis a layer of parenchyma contains vascular bundles and latex tubes, but the bulk of the fruit within the 'skin' consists of a pulp of large cells filled with starch which is partially converted to sugars during ripening.

In the best commercial export bananas like Gros Michel, the fruits are negatively geotropic and curve upwards as they develop until they come to lie more or less against the fruits morphologically beneath them (but in clones where the inflorescence is pendent, spatially above them). Developed in this way the fruits do not stick out, and whole inflorescences can be exported without protective packing. The inflorescence is harvested as a 'bunch' of fruits; the groups of fruits which develop from flowers in the axil of the same bract are called 'hands', so many hands go to make up one bunch. The bunch is harvested before the fruit is fully ripe, whether destined for a foreign market or for home consumption. Bananas grown in the New World for the European market are cut when about three-quarters ripe, and are then shipped at low temperatures in an atmosphere where humidity and the oxygen/carbon dioxide ratio are carefully controlled. By this means fruit ripening is slowed down, and the fruit arrives at its destination still green. It is placed in warm, humid rooms, sometimes with ethylene in the atmosphere, to promote ripening before sale.

It has been possible to breed improved clones of bananas because a few seeds are produced in each inflorescence when flowers of male sterile triploid (AAA) cultivars are pollinated with pollen from male fertile diploid (AA) cultivars of *M. acuminata*. The progeny are tetraploids which arise from the fusion of a triploid (AAA) egg in the female parent (meiosis fails or produces sterile gametes in triploids) with a haploid (A) male gamete. Breeding has so far been confined to bananas (none with plantains) with the objectives of achieving resistance to important diseases in dwarf cultivars with good quality fruit.

Citrus fruits: *Citrus* spp.

The original home of the genus *Citrus* is not known with certainty, but the history of the cultivated species suggests that they may have been domesticated in the drier tropics of South-east Asia. Some species have been cultivated there and in China for thousands of years for their juicy fruits, but it is only during the past century that a large citrus industry and international trade in citrus fruits has developed in other parts of the world. Though the crop originated in the tropics it is now cultivated most extensively in the sub-tropics with a Mediterranean climate; it is not productive

Fig. 7.4 *Citrus sinensis*: Sweet orange. An orchard of orange trees in north Nigeria.

in the humid tropics, but thrives best in warm, sunny climates with 900 mm or more annual rainfall, or with irrigation. The leading producers of citrus fruits are the states of Florida, California, Texas and Arizona in the United States, the Mediterranean countries of Spain, Italy and Israel, and Mexico, Brazil, South Africa and Australia. Total world production of all kinds of citrus is now around 40 m. tonnes annually, with only one-sixth of this coming from Africa and Asia. Oranges and grapefruits are the chief citrus fruits in international trade, with European countries, especially West Germany, France and Great Britain, the largest importers.

Citrus is a member of the large dicotyledonous family *Rutaceae* with 130 genera around 1,500 species, most of them evergreen aromatic trees and shrubs of the tropics. There are sixteen species in the genus; those which are cultivated for their fruits are in sub-genus *Eucitrus* with ten species characterized among other features by their large juicy fruits. The fruits are berries, but because of their unusual structure they are called hesperidiums. All of the cultivated species are diploids ($2n = 18$), and though they are described as distinct species their taxonomy has been confused because they hybridize readily. They are all evergreen trees or small shrubs with very hard wood, and often with spines which arise beside the buds in the axils of the simple alternate leaves. Though the leaves are unifoliate with a simple entire margin they have been derived by reduction from an ancestral pinnate leaf; in some species the petiole is winged, and there are joints or articulations between the petiole and lamina and between the petiole and the stem. The trees are aromatic because the leaves and the 'rind' of the fruits are dotted with glands which produce

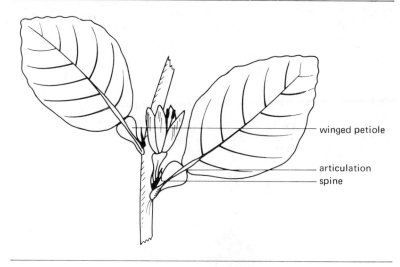

Fig. 7.5 *Citrus paradisi*: Grapefruit.   Leaves and the flower of grapefruit.

essential oils; those from lemon (mostly citral) and sour orange (mostly limonene) are extracted for use in perfumes, while in France the sour orange is grown for its flowers which yield 'oil of neroli' or 'orange blossom oil'. Seedlings produce a tap root, but the main feeding root system is a shallow mass of lateral fibrous roots, usually with few if any root hairs, but with mycorrhizal fungi associated with the finer roots. The sweet smelling hermaphrodite flowers are single or in small groups in the leaf axils on young twigs. They have a variable number of white petals, most often five, and commonly five small, greenish-coloured sepals. The stamens are more or less numerous, always more than four times the number of petals, and collected in five whorls, the filaments of each whorl being partially united at the base. The superior ovary has eight to fifteen united carpels, each locule containing several ovules in axile placentation, and a short deciduous style in which there are as many stylar canals as there are locules in the ovary.

In most species of *Citrus* the androecium and gynaecium mature at about the same time, though some species are slightly protandrous. The pollen grains are sticky, and flowers are self- and cross-pollinated by insects attracted by their colour and strong scent and by the abundant nectar they produce. Some cultivars have predominantly defective, non-viable pollen but produce parthenocarpic fruits even if the flowers are not pollinated; in others, though parthenocarpy does occur, parthenocarpic fruit development does require the stimulus of pollination.

The phenomenon of polyembryony is common in *Citrus*. Following normal fertilization and the growth of an embryo from the zygote, other

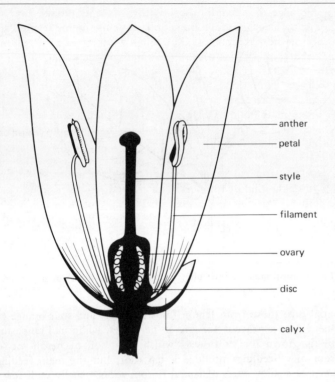

Fig. 7.6 *Citrus* spp.   A longitudinal section of a flower (×4).

asexual embryos may develop in the same ovule from the diploid somatic cells of the nucellus, a phenomenon called adventitious embryony (which, like parthenocarpy, may require the stimulus of pollination and the mediation of hormones produced as a result of it). Asexual embryos compete with the sexual embryo for food and space within the seed, and may eventually suppress it entirely, in which case the seeds can be used to propagate a clone with a genotype identical to that of the parent tree. On the other hand, the sexual embryo may not be suppressed, in which case there is no certainty that the seeds from one tree will not produce genetically heterogeneous seedlings. Consequently, *Citrus* species are not propagated from seed, but by grafting buds from desirable clones on to vigorous root stocks, commonly of a different species — the commonest root stocks used are rough lemon and sour orange.

In the typical hesperidium the exocarp and mesocarp are leathery, and protect the juicy inner tissues derived from the endocarp from damage and desiccation. The epidermis of the fruit has a thick cuticle and varying numbers of stomata; the exocarp, or 'flavedo', is a layer of irregular, photosynthetically active parenchyma cells which is green in young fruits, becoming orange or greenish-yellow as they mature, though this colour change may not occur in crops grown in the hot and humid tropics. Within the tissues of the flavedo there are numerous ductless oil glands which contain essential oils under considerable pressure due to the turgor of the cells surrounding the glands. Different species produce different essential oils which are extracted as an important subsidiary industry of citrus production. The mesocarp is thicker than the exocarp, and consists of a spongy white mass of irregularly shaped parenchyma cells and extensive intercellular spaces. The mesocarp varies in thickness depending upon differences between species, and between cultivars of the same species, and is called the 'albedo'. It is rich in vitamin C, sugars, cellulose and in pectin which may be extracted for use in industry. The exocarp and mesocarp together form the rind of the fruit. The centre of the fruit is occupied by the developed carpels of the ovary which are disposed around the pithy axis in the form of several closely packed segments. Each segment develops from a single carpel and is surrounded by the thin, transparent endocarp or 'rag' from which multicellular hairs grow to fill each segment. Each huge cell or pulp vesicle of these hairs fills with juice, and they form the edible part of the fruit for which the crops are grown. The seeds lie on axile placenta close to the central axis, and in the mature fruit are surrounded by the pulp vesicles. In the fruits called 'navel' oranges a second row of carpels develops at the top of the fruit and becomes embedded in the mesocarp, though they are visible through a circular hole in the flavedo.

The change in colour which occurs as the fruits ripen is due to the gradual breakdown of chlorophyll in the plastids of the exocarp, which continues until the predominant pigments are xanthophyll and carotene giving the fruit its yellow or orange colour. In commerce this change is promoted in various ways, but especially by exposing the fruits to small concentrations of ethylene, though sometimes oranges are coloured artificially with pigments sprayed on to their surface. The composition of mature fruits varies between species and cultivars, but is about 40–45 per cent juice, 30 per cent rind and 30 per cent pulp (rag) and seeds, which taken together consist of about 90 per cent water, 5–8 per cent sugars, 1–2 per cent pectins, various acids, essential oils, protein and minerals. On a dry matter basis the fruit contains 80–90 per cent of sugars and acids, with the relative proportions varying between species; limes and lemons contain more acid than sugar, but in most other species sugars predominate in the ripe fruits. Citric acid is the most abundant acid in the sap, though small quantities of others occur in all species. As the fruits ripen the amount of sap increases, so there is an apparent decrease in acid content. Pectins in the juice give it a cloudy, colloidal appearance, and it also contains mineral salts, glucosides, small amounts of protein and vitamins; the citrus fruits are good sources of vitamin C.

The difficulty of classifying cultivated species of citrus has already been referred to, but Swingle's classification (see Further reading, Weber and Bachelor (1948)) is perhaps the most acceptable and is adopted here, though it does not take into account the complete range of types found in Asia. The cultivated species are: *Citrus paradisi*, Grapefruit; *C. sinensis*, Sweet orange; *C. limon*, Lemon; *C. aurantifolia*, Lime; *C. aurantium*, Sour orange; *C. reticulata*, Tangerine and Mandarin oranges: *C. grandis*, Pummelo; *C. medica*, Citron.

Citrus paradisi, Grapefruit. The fresh fruit of the grapefruit is a popular breakfast food in many parts of the world, and large quantities of the juicy flesh or of the juice alone are canned. The rind is a commercial source of pectin, and the pulp remaining after the juice has been expressed may be used as a cattle feed. Annual world production of grapefruits is greater than 3.5 m. tonnes, some 70 per cent of this amount coming from the United States. The crop probably originated in the West Indies, perhaps as recently as the beginning of the eighteenth century as a somatic mutation (a 'bud sport') in the pummelo, *C. grandis*, or as a hybrid between pummelo and sweet orange, *C. sinensis*. It is a vigorous tree, large compared with other cultivated species, growing 10—15 m tall with strong, spreading branches. Its angular, glabrous shoots bear leaves which are pale-green when young, becoming darker-green and leathery with age. They are smaller than the leaves of pummelo, and have petioles with smaller wings. The large, sweetly scented flowers are single, or in axillary groups, and conform in structure to the usual *Citrus* type. Large, globose fruits, commonly 10—12 cm in diameter, are borne in clusters which weigh down the branches bearing them; they are pale-yellow or greenish-yellow when ripe with a rind which varies in thickness, and flesh which varies in colour from very pale-yellow to pink, depending upon differences between cultivars. Grapefruit juice is acid, and sometimes bitter; it is rich in vitamin C and contains appreciable amounts of thiamine (vitamin B_1). The seeds are white, and they are not ridged like those of pummelo; they are polyembryonic.

Citrus sinensis; Sweet orange. Oranges originated in southern China thousands of years ago but did not reach Europe or the New World until the end of the fifteenth century. Now they are the most popular and widespread of the citrus fruits, with world production greater than 29 m. tonnes each year, and numerous diverse cultivars which vary in growth, habit, plant size, and fruit colour, size, juiciness and sugar content. The United States is the leading producer with around 9 m. tonnes of oranges each year, followed by Brazil (4 m.), Spain (1.9 m.), Mexico (1.9 m.), Italy (1.5 m.) and Israel (1.1 m.). The fruits are eaten fresh or they are used to make marmalade, the juice (but not the flesh) is canned, and essential oils are obtained from the flowers, leaves and rind. *Citrus sinensis* is a spreading evergreen sometimes spiny tree up to 12 m tall with ovate-elliptic leaves which are commonly 7—10 cm long, dark-green and rounded at the base. They are carried on short, articulated petioles with very narrow wings. The leaves are much darker on the lower surface than those of

The Cultivated Tropical Fruits (and Nuts) 165

Fig. 7.7 *Citrus paradisi*: Grapefruit.   Mature grapefruits ready for picking.

grapefruit, and they are not strongly scented; the white, sweet-smelling flowers are smaller than those of the grapefruit. The rounded fruits are up to 12 cm in diameter, though most commonly 6–8 cm in diameter, deep yellow to orange, or in the humid tropics remaining green when ripe, with

Fig. 7.8 *Citrus sinensis*: Sweet orange. An orange tree bearing fruit in Indonesia. (By courtesy of the FAO.)

a rind which varies in thickness and in the ease with which it can be removed. The flesh is deep-orange or yellow-coloured. The juice is rich in vitamin C, and contains 5—10 per cent of sugar and 1—2 per cent citric acid.

Citrus limon; Lemon. Though its centre of origin is not known with certainty, it seems likely that lemons originated in South-east Asia and that they have been cultivated there since ancient times. They reached Europe in the twelfth and thirteenth centuries, and were spread to the New World at about the same time as oranges. The lemon tree is small, though larger and more spreading than the lime, commonly 4—5 m tall with leaves 5—10 cm long on almost wingless, short, articulated petioles. The tree is spiny, though the spines are short, and the dark-green leaves are not strongly scented as they are in some other species. The flower buds are characteristically pink whereas the open flowers are white, but the flowers are otherwise typical of the genus. The oval, yellow fruit has a thick rough skin and a characteristic nipple-like tip. The flesh is pale yellow, and the juice is strongly acid with about 5 per cent citric acid, 5 per cent sugar and appreciable quantities of vitamin C, carotene and thiamine. Lemons are not eaten fresh because they are too sour, but the juice is extracted to make drinks, and the rind is used as a flavouring and garnish for foods.

Citrus aurantifolia; Lime. The lime is less hardy than most other cultivated citrus fruits and is confined mainly to the tropics. It occurs wild in northern India where it was probably domesticated, and is now cultivated there on a large scale as well as in Italy, the West Indies, Florida and California. Elsewhere in the tropics limes are a common feature of home gardens grown for their fruits which are used as a flavouring or to make a refreshing drink, or when densely sown and regularly clipped, as an impenetrable spiny hedge. The lime is a small spreading tree or low shrub with spiny branches and small leaves, 5–8 cm long, which are dark-green above but much paler below. The leaves are on narrowly winged petioles. Flowers are produced in large numbers all year round, often singly in the leaf axils; they are small, white and strongly scented. The oval fruit is only 4–6 cm in diameter, more or less pointed and green or greenish-yellow when ripe. It has a thin rind and pale-yellow flesh with very acid juice. In India sliced fruits are often used in pickles. Essential oils used for flavouring drinks and confectionery are distilled from the juice and the flavedo. Though *C. aurantifolia* is a variable species, two distinct groups of cultivars are recognized, the sweet lime, which probably arose as a hybrid between *C. aurantifolia* and *C. medica* (citron), and acid limes, which are the most widespread and important of the two groups. There are two kinds of acid limes, the West Indian (or Mexican or 'key' limes), which have small fruits and are widespread in the tropics, and the Persian limes (or Tahitiian limes), which are triploid with seedless fruits which are larger and have a smoother skin than West Indian limes. The common West Indian type is propagated from seed unlike other species of *Citrus*, which are propagated vegetatively.

Citrus reticulata; Mandarin, tangerine.   The mandarin has been cultivated in China and Japan for centuries, but did not become widespread and popular as a dessert fruit until the nineteenth century. Now it is also cultivated extensively in Southern Europe, the Gulf States of the United States and throughout the tropics, with annual world production around 6 m. tonnes; more than half of this still grown in Japan. It is a spreading shrub or small tree with drooping, thin stemmed, thornless branches. The elliptical leaves are rather narrow, and are carried on short, almost wingless petioles. They are dark green on both surfaces and have a strong, very characteristic scent when crushed. The small flowers have thin petals, and are produced singly in the leaf axils. The small fruits are rounded or more or less flattened, bright orange (mandarins) or yellow (tangerines) with a very loose, easily removed rind. The flesh is juicy and sweet, and has a very distinctive flavour.

Citrus grandis; Pummelo or shaddock.   The pummelo is the largest of the citrus fruits, but it is coarse, thick-skinned and bitter with a tough, solid pulp. It originated in Malaysia, and is popular throughout Asia and in India as a dessert fruit, but though it has spread throughout the tropics it is nowhere cultivated on a large scale. Pummelo is a large tree reaching a height of 10 m or more, with dark green, broad leaves on broadly winged petioles. The fruits are carried singly, and may measure up to 30 cm in diameter.

Citrus aurantium; Sour or Seville orange. The sour orange is native to South-east Asia, and was introduced to Europe in the eleventh century. It was cultivated in Spain long before the introduction of the sweet orange. The fruit is used largely in the manufacture of marmalade, the juice is a source of vitamins A and B, an essential oil used in perfumes is obtained from the stems, leaves and flowers and the wood is used to make furniture in some parts of the world. The sour orange is a spiny tree up to 10 m tall with fairly large, ovate leaves which are up to 10 cm long, wedge-shaped at the base, and pale-coloured on the under surface. The leaves are strongly scented and are carried on broadly winged petioles. The large, sweet-scented flowers are in axillary racemes, and produce large round fruits with rough warty skins. The fruit is very acid and bitter and often develops a hollow core as it ripens. The seeds are polyembryonic with predominantly adventitious, nucellar embryos.

Sour orange is chiefly important as a root-stock for the propagation of oranges, grapefruits, lemons and tangerines on heavy soils. It has a deep, vigorous root system and is tolerant of drought; it is resistant to gummosis disease caused by the fungus *Phytophthora* spp., but susceptible to two other important *Citrus* diseases, tristeza caused by a virus, and scab caused by the fungus *Elsinoe fawcettii*. Seeds of sour orange are sown densely in nursery beds from which the least vigorous seedlings, which are usually those which grow from sexually produced, zygotic embryos, are discarded. From the nursery seedlings are transplanted to beds at wider spacing, or into polythene bags or baskets where they grow for several months until their stems are the thickness of a pencil. Then they are budded, using an inverted T incision in their bark about 30 cm from the ground. The budwood is selected from the young branches of disease-free, desirable trees and bound to the root-stock with budding tape to ensure close contact between the tissues of the stock and scion, and to exclude pathogens. When the bud has 'taken' and begins to grow the stem of the root-stock is cut off above it.

Rough lemon (a hybrid of *C. limon* and *C. medica*) is used as a root-stock on light sandy soils. It is resistant to tristeza. Sweet orange is resistant to scab and tristeza, and is used as a root-stock for lemons and oranges on medium but well-drained loams.

Date palm: *Phoenix dactylifera*

For thousands of years the fruits of the date palm have provided the inhabitants of the hot dry regions of Arabia, the Middle East and North Africa with one of their most important foods. It is one of the few food plants which thrives in the harsh, dry desert environments of these areas.

The fruit contains a large proportion of sugar and other carbohydrates, and together with milk forms an important article of food, especially for many Arabs, but every part of the palm is used for a variety of other purposes. The houses of poorer people are constructed largely from the leaves and stems of date palms; door posts, window frames and rafters are made from the trunks; the leaves and the strong sclerenchyma fibres extracted from them are made into crates, mats, baskets, ropes and containers of every description. The coarse fibres are also used to make donkey saddles, and those parts of the palm not put to other good use provide fuel. Even the seeds are fed to cattle and camels, and palm wine is prepared from the sap which is tapped from the bases of inflorescences.

The place of origin of *P. dactylifera* is unknown, though it seems probable that it was in the Persian Gulf area or in Mesopotamia. It became naturalized in Arabia in ancient times, and spread to North Africa, to Spain and up the Nile Valley as far as the Sudan early in the history of those areas. Date palms were introduced to the New World from Spain two or three centuries ago, and are now grown commercially in the drier parts of the American tropics, especially in southern California and in Arizona. World production is estimated to be 2 m. tonnes of fruit annually, with Egypt, Iran, Iraq and Saudi Arabia the leading producers. The crop is restricted to areas with a long, hot and completely dry period which is necessary for flowering, successful natural pollination and for fruit development. It thrives in temperatures of $30°C$ or hotter, but can also withstand brief exposure to very cold night temperatures, even colder than $0°C$. The palm is tolerant of a wide range of soil types, including those which are very alkaline, provided that they are well drained and well aerated. It is cultivated in 'date gardens' with irrigation or where there is an assured supply of ground water. The environmental demands of date palms are summed up by the Arab saying: 'The date palm needs its feet in running water and its head in the fire of the sky.'

It is an unbranched palm, up to 30 m tall, though the trunk never becomes very thick. Numerous offshoots are produced from the base of the trunk, which, if not removed as part of normal cultural practice, or used to propagate the crop vegetatively, develop into subsidiary trunks which eventually grow to form a characteristic clump of palms. Date palms produce a mass of adventitious roots from the base of the stem; they are long and rope-like, dark brown, unbranched and fairly even in thickness for most of their length. They spread widely around the tree, many of them in the surface soil. Mycorrhizal fungi in association with the root tips have been reported. The trunk is protected throughout its life by the persistent fibrous leaf bases. It is surmounted by a handsome crown of about 100 leaves, ten to twenty new ones expanding each year to replace those which are old and which are shed. The pinnate leaves are produced in a spiral from the apical growing point, subtended by cylindrical sheaths of tough, reticulate fibres which form a tight protective envelope for the apical bud, and later, when the leaves have expanded, protective insulation

170 *The Cultivated Tropical Fruits (and Nuts)*

Fig. 7.9 *Phoenix dactylifera*: Date palm.   A date palm in the northern Sudan on which the basal offshoots have been allowed to develop naturally.

for the stem against the severe heat of the sun. When young the leaves ascend vertically in the centre of the crown, but as time passes and they are displaced by younger leaves, they curve away from the centre, and eventually hang down. Each mature leaf is 3—7 m long and grey-green in colour, with a thick, fibrous rigid petiole which in section is convex below,

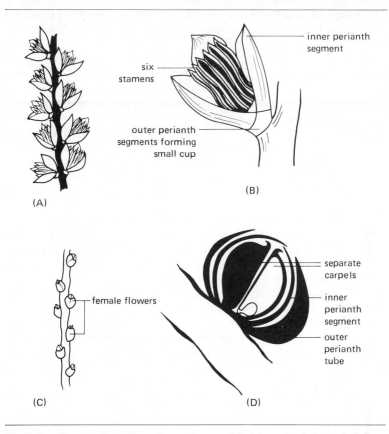

Fig. 7.10 *Phoenix dactylifera*: Date palm.   (a) A branch of the male inflorescence crowded with male flowers.   (B) A single, open male flower (x3).   (C) Part of a branch of a female inflorescence (x$\frac{2}{3}$).   (D) A diagrammatic longitudinal section through a female flower (x12).

but flat on top. The petiole is terminated by a sharply pointed leaflet, while at the base it is expanded and thickened into a strong, broadly triangular sheathing attachment to the stem. The leaflets are 20–40 cm long, those at the base of the petiole being much modified to form short, sharp spines whose number, length and arrangement varies between cultivars. Both spines and leaflets tend to occur in opposite pairs near the base of the petiole, but higher up they do not. Instead, of each pair of pinnae, one lies horizontally at right angles to the petiole while the other points upwards at an acute angle from it. This arrangement is repeated on both sides of the petiole so that the pinnae appear to be carried in four ranks. The lanceolate leaflets are rigid, pointed and covered with a waxy bloom; each is folded up around its mid-rib.

The date palm is dioecious, and though male and female palms look much alike, experienced observers are said to be able to differentiate between them on vegetative characters alone. In both sexes the inflorescence is a much branched axillary panicle, the males with 100—150 branches 12—15 cm long, the females with fewer branches 30—50 cm long which eventually become pendulous on the long inflorescence axis when they bear fruit. Both kinds of inflorescence are protected during development by a sheathing, fibrous spathe which splits longitudinally to release the inflorescence when it is mature. The small yellow or cream-coloured flowers are less than 1 cm long and they are sessile along the inflorescence branches. They have six perianth segments, the outer three forming a cup-shaped calyx-tube which is much smaller in male than in female flowers. The three inner perianth segments of the female flower are small, almost circular, and closely appressed to the three free carpels of the superior ovary, each of which has a short, hooked stigma. On the other hand, the three inner perianth segments of the male flowers are valvate, thick and leathery with thin margins and bluntly pointed tips; they open to expose six stamens with short filaments and long flat anthers with undulate margins. The centre of the male flower is occupied by a small, rudimentary ovary. In cultivation dates are usually pollinated by hand by dusting the light, smooth pollen into the spathe of a young female inflorescence; one male palm provides enough pollen to pollinate about 100 females in this way, and though it is commonly used fresh, the pollen remains viable for several years in store. The pollen from different male palms may have different effects upon the size, quality and time of maturity of the fruit. These effects of the source of pollen on fruit tissues derived from the diploid, somatic tissues of the female parent are rare in plants, and are called 'metaxenia'. They suggest that, as well as good females, male palms should be selected and propagated in date gardens.

After fertilization the branches of the female inflorescence elongate as large, pendulous bunches of fruits are formed. Each female palm may carry 10—30 bunches in different stages of development, so they do not mature together. The fruit is a one-seeded berry 2—8 cm long, bright-yellow to reddish-brown in colour, and usually longer than broad, though in some cultivars the fruits are more or less globular. The branches of the female inflorescences may be pruned as the fruits develop to ensure the best quality of those that remain without straining the food resources of the palm in fruit production. The narrow, cylindrical seeds are commonly around 2.5 cm long with a papery testa, and a longitudinal groove along one side; they have a hard endosperm, and a small circular mark at one end of the seed which marks the position of the embryo.

Date palms are heterozygous, dioecious outbreeders, and because many of the palms of Arabia and North Africa have been grown from seed, the crop in these areas is also very heterogeneous; a large number of vegetatively propagated clones is recognized, but even so they are referred to only three main groups depending upon whether they produce soft, dry or

Fig. 7.11 *Phoenix dactylifera*: Date palm. (A) Male and female inflorescences — spathes just opening. (B) Fully exposed male inflorescence. (C) Fully exposed female inflorescence with a branch of a male inserted for pollination.

semi-dry fruits. Soft dates are harvested before they are fully mature, and are eaten fresh or preserved while still moist. Most of the dates in world commerce are of this kind, the 'Mishrig' and 'Deglet nour' types being among the best known of them. Soft dates are grown in environments where the air does not become very dry when the fruits are ripening, so that they are unlikely to dry out before harvest. They contain less sugar than dry or semi-dry types, commonly around 60 per cent of the fruit weight as glucose or fructose. Semi-dry dates have firmer flesh than soft dates, though they too are usually harvested before they are fully ripe. Dry dates are left on the palm until they are fully ripe and are hard and dry; they contain 65–70 per cent of sugar, most of it as sucrose. Much of the local trade in dates throughout North Africa and Arabia is in the dry type, which stores well for long periods and which is easy to pack and to transport.

Mango: *Mangifera indica*

The mango originated in the India–Bangladesh–Burma region, and had spread into cultivation and common use in the Indian sub-continent by 2000 B.C. Though the tree is now to be found throughout the tropics, and has become naturalized in many areas outside its centre of origin, nowhere is it so important as in India, where annual production is around 8 m. tonnes of fruit compared with a world total of 12 m. tonnes annually. Mangoes spread throughout South-east Asia during the first millenium B.C. but did not reach Africa until about 1,000 years ago. They were introduced to the New World at the beginning of the eighteenth century. There are no defined limits of rainfall necessary for the successful cultivation of mangoes, provided that there is a distinct annual dry season when the crop flowers. Fruit set tends to be least in very wet, humid areas, and irrigation may be necessary in areas with little rainfall. Similarly, the crop will grow in a wide range of temperatures, but thrives when the temperature is around 25°C. A very large number of cultivars is recognized, the best of them being those which have been selected for their delicious, delicately flavoured fruits which have very little fibre in their flesh. All such superior cultivars are propagated vegetatively by budding, but there are unfortunately many inferior mangoes grown from seed which have coarse, fibrous fruits often thought to taste of turpentine.

The genus *Mangifera* belongs to the family *Anacardiaceae* which includes many tropical trees and shrubs with resinous sap and alternate leaves, among them economically important crops such as cashew (*Anacardium occidentale*) and pistachio (*Pistacia vera*), and the poison ivy plant of North America, *Rhus toxicodendron*. There are around 60 species in the genus *Mangifera*, most of them found in South-east Asia, from northern India to New Guinea, where they occur as trees in the savannas and in the lowland, wet forests. Several species have edible fruits, and a few are cultivated in restricted areas in Asia, but only *M. indica* is widespread and important.

The mango tree is a large, spreading evergreen, sometimes having a distinct crown of branches, but often of indeterminate shape with long pendulous branches and no definite crown. Many vegetatively propagated clones are relatively small, compact trees about 10 m tall, whereas trees

Fig. 7.12 *Mangifera indica*: Mango.   Mango trees in Tanzania. (By courtesy of the FAO.)

grown from seed (such as those in Fig. 7.12) may be very large and up to 40 m high. All mangoes are renowned for the dense shade which they cast in hot, sunny climates. The alternate, simple leaves are 12–40 cm long and up to 10 cm broad, and are carried on petioles up to 10 cm long with a pulvinus at their base. The tree grows in flushes in which a number of thin, flaccid tan-red coloured leaves is produced; as they mature the leaves become stiff and dark-green. These periodic growth flushes do not occur at regular intervals, nor do they necessarily occur all over a single tree at the same time. The inflorescence is a terminal panicle 10–50 cm long with three or four orders of branching and a very variable number (1,000–6,000) of reddish, pink or almost white flowers. As few as 1–35 per cent of them are hermaphrodite, the rest are male; both kinds of flowers occur in one inflorescence, but the ratio of males to hermaphrodites varies between cultivars. Hermaphrodite flowers have a deciduous calyx divided into four to five lobes, and four to five free petals about twice as long as the calyx lobes. At the centre of the flower a raised disc carries five stamens, but only one (rarely more) has a fertile anther, while the rest are sterile and represented by staminodes. A small, greenish-yellow, one-celled ovary arises obliquely from the disc; it has a lateral style and contains a single pendulous ovule. The male flowers are similar, but have no gynaecium.

Mango trees come into bearing when 4–6 years old, and become

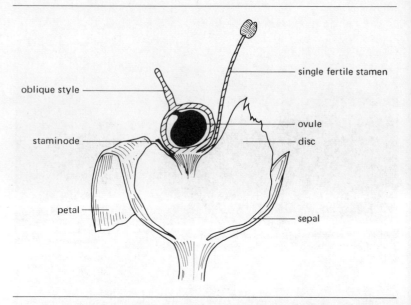

Fig. 7.13 *Mangifera indica*: Mango.   A longitudinal section through a flower (×5).

increasingly productive until they are about 20 years old, then yields decline. They do not flower profusely every year, but tend to produce large crops in alternate years because flowering and fruiting deplete their food reserves which must be replaced before a subsequent large crop is borne. After mature branches have flowered a new flush of growth occurs on them and this in turn produces inflorescences after it has matured and accumulated sufficient reserves of assimilates. Very large crops of fruit may deplete food reserves so much that the subsequent flush of new vegetative growth, and so the next period of flowering, are delayed. In some, but apparently not all, cultivars the onset of the recurrent reproductive phase is correlated with the ratio of carbohydrates (mostly starch) to nitrogen in the tree, and occurs when this ratio is large. A mature tree in full bearing may produce 1,000 inflorescences, and though the total number of flowers per tree is consequently enormous it is unusual for a large tree in full bearing to produce more than 2,500 fruits. As few as one-third of the total numbers of hermaphrodite flowers are ever fertilized, and of these many are subsequently shed; eventually only 0.1–0.25 per cent of the hermaphrodite flowers on a tree produce mature fruit. Flowers are shed most abundantly when wet, humid weather occurs as they open. The number of fertilized ovaries which develop to produce mature fruit is affected by husbandry, and can be increased by the use of fertilizers and organic manures.

Fig. 7.14 *Mangifera indica*: Mango.   'Julie' mangoes in Trinidad.

The flowers open during the night and early morning and they are pollinated by short-tongued insects attracted by their nectar. Though self-incompatibility is rare and many flowers are self-fertilized, cross-fertilization is also common. The fruit is a large ovoid, asymmetrical drupe which varies greatly in size in different cultivars and is from less than 5 cm to as many as 30 cm long. It matures 2—5 months after fertilization and when mature is yellow, green or red with a smooth, thick skin. The fleshy mesocarp is bright orange-yellow; in some cultivars, especially those propagated from seed, it is more or less fibrous and tends to be resinous with an unpleasant flavour, while in better, selected clonal cultivars propagated by budding it has little fibre and a delicious, delicate flavour. The endocarp of the drupe is very hard and fibrous. It encloses a single smooth, light brown seed which is within a papery envelope and which may have a single zygotic embryo, or several apomictic embryos which develop from the tissues of the nucellus and which may or may not suppress the zygotic embryo. The fruits contain approximately 85 per cent water, 10—20 per cent sugar and small amounts of protein; they are a good dietary source of vitamin A, and contain vitamins B and C.

Trees grown from seed may be from zygotic or apomictic embryos; four to six of the latter may occur in one seed. To avoid the risk of multiplying variable trees from zygotes it is necessary to propagate clones of desirable trees vegetatively by approach grafting or by shield-budding.

Pineapple: *Ananas comosus*

Pineapples are among the most popular of tropical fruits, and are grown in most tropical and sub-tropical countries where temperatures are moderate, between 16° and 32°C, and where annual rainfall of around 1,300 mm is evenly distributed throughout the year. They are cultivated not only for their mature, ripe fruits for local consumption, but also for their flesh and juice for canning and export, or for the export of whole, fresh fruits harvested before they are fully ripe. Some cultivars are grown in small amounts for the extraction of leaf fibres. Annual world production of fruit is around 4.5 m. tonnes, with the leading producers Hawaii (0.8 m. tonnes), Brazil (0.4 m.), mainland China (0.37 m.), Mexico (0.34 m.) and Malaysia (0.33 m.), while the Philippine Islands, Thailand and South Africa also produce important amounts. The crop is native to the tropical regions of South America, and was grown in the New World for food, for its medicinal properties and for the production of wine long before the discovery of the New World. The plant spread quickly throughout the tropics largely because of the ease with which it can be propagated vegetatively, and because the parts used for propagation tolerated the desiccation to which they were subjected during the long sea voyages involved in their dispersal.

The genus *Ananas* is a member of the large and curious family of monocotyledons, the *Bromeliaceae*. Most of the 1,300 or so species in the family are natives of tropical America where many of them are epiphytes on trees or other supports in the rain forests. Among these species roots are often poorly developed, but mycorrhizal associations are common. *Ananas* is a terrestrial genus, but it also has mycorrhizal fungi associated with the smaller rootlets of the rather poorly developed root system. All of the cultivated forms are referred to one variable species, *A. comosus*, which is not known to occur truly wild though it has become naturalized in some areas. It differs in several ways from three wild species *A. ananassoides, A. bracteatus* and *A. erectifolius*. These three species are dis-

The Cultivated Tropical Fruits (and Nuts)   179

Fig. 7.15   *Ananas comosus*: Pineapple.   A pineapple plantation.

tributed throughout the northern part of South America in habitats which range from hot, wet localities in the valley of the Amazon River to the much drier upland regions of Brazil. They differ from the cultivated types in their smaller and less fleshy fruits which contain numerous seeds, their conspicuous floral bracts and their thinner peduncles. There are numerous cultivars of pineapple which differ in plant and fruit size, in the colour and flavour of the flesh of the fruit, and in the form of the leaf margin which is usually smooth, though recessive mutants with spiny leaves do occur.

Pineapples are short-lived perennials or biennials with a very short main axis and a rosette of leaves at first, but later the axis elongates slightly, and carries numerous leaves arranged in a close spiral. The axis may eventually reach a length of 60—100 cm. The long, fleshy, fibrous leaves are sword-shaped, dark-green and taper to a fine point; they often have spiny margins. They reach a length of 1 m or more and are 5—8 cm wide, grooved on the upper surface and clasping the main axis closely at their base. During the first year of growth the axis thickens as starch is laid down, then after a variable period of 12—20 months the terminal bud gives rise to the peduncle which carries the inflorescence. The numerous reddish-purple sessile flowers are arranged spirally on the axis, each subtended by a pointed bract. They have two perianth whorls, an outer one of three short, free calyx segments and an inner one of three larger petals which are also free though they form a tube enclosing six stamens and a narrow style with its three-branched stigma. The inferior ovary has three united carpels with ten to fifteen ovules in each of the three loculi. Flowers open in the early morning and are faded by evening; the lower flowers on the inflorescence open first, then over a period of about 3 weeks flowering proceeds gradually up the inflorescence axis. Pineapples are self-incompatible and do not produce seed after self-fertilization. Cross-pollination between cultivars, or between cultivars and wild forms, results in normal fertilization with the production of seeds in abundance. Humming birds are the main pollinating agents in South America, but they do not occur in other parts of the world where pineapples are grown so that pineapple fruits are usually parthenocarpic and seedless.

The multiple fruit matures 5—6 months after flowering. It is formed by an extensive thickening of the axis of the inflorescence and by the fusion of the small, berry-like fruits produced by each flower. The sepals and the pointed floral bract subtending each flower persist and form the hard rind of the fruit, those of adjacent flowers becoming more or less fused in the process. There are 100—200 individual fruits arranged spirally around the thick central axis and the whole forms a broad, almost cylindrical multiple fruit or syncarp. The 'average' fruit is around 20 cm long and 14 cm broad in the middle; but fruit size varies from those of the cultivar 'Queens' which are relatively small and weigh about 1 kg to the large fruits of 'Cayenne' weighing 2.5 kg. The fruit tapers towards the top where it is surmounted by a rosette of short, stiff spirally arranged leaves called the 'crown'. During the maturation of the fruit starch stored in the main stem is converted to sugars which are translocated to the fruit and stored there; little or no starch is ever found in the fruit. The full flavour is attained only when fruits are allowed to mature on the plant, but whole fruits are harvested for export before they are fully mature. Mature fruits contain citric and malic acids, about 14 per cent of sugars, a protein digesting enzyme called bromelain and vitamins A and B. The yellow-orange colour of the ripe fruit is due to the presence of carotene and xanthophyll. In Hawaii the time of flowering and fruit production is sometimes controlled by spraying crops with hormones.

Pineapples are propagated vegetatively by planting the crown of leaves cut from the top of the fruit, or from shoots called 'slips' which grow on the peduncle just below the fruit, or from 'suckers' produced in leaf axils lower down the stem. The crown is removed only from fruits destined for the cannery, fresh pineapples always being sold or shipped with the crown attached. On pineapple estates the crops are planted through a paper or plastic sheeting mulch at populations of around 45,000 plants per hectare. The mulch helps to conserve water and to control weeds and its use gives substantial increases in yield of fruits. In commercial production very large amounts of fertilizers are applied to crops, but the frequency of application and the amount and kind of fertilizer used vary between countries. Plants grown from crowns develop slowly and produce fruit about 2 years after planting; those propagated from slips develop more quickly, but suckers produce fruit in the shortest time, about 15 months after planting.

In the Philippines, southern China and Taiwan pineapple leaves are the source of a very strong, durable fibre. The fibres are shining white when extracted, flexible and of good quality. They are used to make pinacloth, which is rather like linen but with the sheen of silk; such cloth is expensive to make and is little used outside areas where it is woven.

Pawpaw: *Carica papaya*

Carica papaya belongs to the family *Caricaceae*, a small group of four genera of trees or shrubs with their leaves in terminal clusters and latex vessels throughout their tissues. There are about 40 species of the genus *Carica* in the American tropics and sub-tropics, and the pawpaw or papaya probably originated in Central America, perhaps as a hybrid between other species. It is not known to occur wild, but is easy to propagate from seed, and soon spread throughout the tropics after the discovery of the New World. It is a popular fruit wherever it is grown, and though it is not cultivated on a large scale, it is common in home gardens everywhere in the tropics. The fruits are eaten fresh, or they may be used in drinks, or for flavouring desserts. The ripe flesh is canned, and immature fruits can be cooked and eaten as a vegetable; when tapped they yield a latex containing the proteolytic enzyme papain. Pawpaws thrive on a variety of soils but will not tolerate waterlogging, and though they are not demanding in their environmental requirements the best fruits are produced only in warm climates.

The pawpaw is a large, quick growing, herbaceous plant or soft-stemmed tree, rather like a palm in appearance with an unbranched main stem up to 10 m tall (commonly 4—5 m tall) and a crown of large leaves. Sometimes branches arise from the base of old plants to form a cluster of erect stems, or the main stem itself may branch if the apical meristem is damaged. The stems are hollow with soft, fleshy tissues, and are covered with smooth, grey bark marked externally by numerous large, orbicular leaf scars. The whole plant contains vessels which exude a thin white latex when they are cut. Pawpaw leaves are very large, up to 75 cm across, and deeply divided into about seven palmate lobes, each lobe being again pinnately lobed. The leaves are dark-green, often drooping from the ends of hollow petioles up to 1 m long, and are eventually shed, leaving a large scar on the stem.

Pawpaws are usually dioecious, though hermaphrodite plants and even an hermaphrodite cultivar ('Solo'), do occur. Sex is determined by five closely linked genes on the sixth chromosome whose phenotypic effects can be described as though they were determined by three alleles of a single gene; M_1 is dominant male, M_2 dominant hermaphrodite and m recessive female. The only viable combinations can be represented by the symbols M_1m (male), M_2m (hermaphrodite) and mm (female). The inflorescence of a male plant is a long, pendulous panicle of sessile male flowers produced from the axil of a leaf high on the plant. Male flowers

Fig. 7.16 *Carica papaya*: Pawpaw.   A pawpaw tree in full fruit.

are small, about 2–3 cm long, and have a minute calyx of five small united sepals, and a long corolla tube divided about one-third of the way from its mouth into five pointed lobes. The pale yellow corolla is often fragrant; at the mouth of the tube ten epipetalous stamens are arranged in two rows of five, one row with the longest filaments alternating with the petal lobes, the other with shorter filaments lying opposite the lobes. Small insects visit the flowers and may be pollinating agents, though it has been suggested that pawpaws are anemophilous, or that they are pollinated by moths at night.

Female flowers are much larger than the males, around 4–5 cm long,

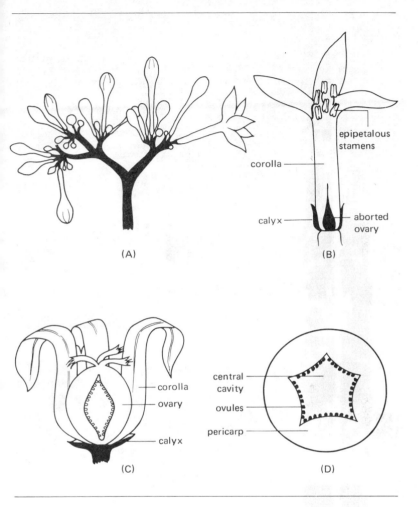

Fig. 7.17 *Carica papaya*: Pawpaw. (A) The end of a branch of the male inflorescence. (B) A longitudinal section of a male flower with its tubular corolla (×1). (C) A longitudinal section of a female flower with its free petals (×½). (D) A transverse section through a fruit (×¼).

and are more or less sessile on the main axis, though they also occur in few-flowered cymes in the axils of leaves. They have a short calyx-tube with five short lobes, and five waxy, yellow petals united at the base but free for most of their length and with twisted, pointed tips. The large superior ovary is sessile, globular and green with five much branched sessile stigmas. It is composed of five united carpels with a single locule which contains many ovules on parietal placentas.

Fig. 7.18 *Carica papaya*: Pawpaw. (A) Female flowers. (B) Male inflorescences.

There are two kinds of hermaphrodite flowers; one with a long corolla tube and ten stamens produces an elongated fruit, the other with a short corolla, and only five functional stamens, produces a globular fruit with five distinct furrows. Hermaphrodite flowers sometimes occur on male plants and they may produce fruits which do not develop. Complete sex reversal has been known to occur, from male to female, but never from female to male.

The fruit is a berry up to 50–60 cm long, elongated or globular and with a general resemblance to a melon. In ripe fruits the smooth skin is yellow or bright orange, and the thick yellow-orange flesh encloses a central cavity containing many seeds which are embedded in a mass of mucilage derived from their arils. The seeds are small, rounded, dark green or brown, and about the size of a pea. Though dry seeds remain viable for a considerable time, it is best to use fresh seeds to propagate the crop because they germinate sooner than old ones, some 3–4 weeks after sowing. The seeds sometimes germinate while they are in the fruit.

No satisfactory method of vegetative propagation has been devised to overcome the variation in fruit quality which is a consequence of propagation by seed. The best plants have fruits with deeply coloured, soft, delicious flesh, but in inferior types the fruits may have pale, insipid flesh. The seeds contain about 25 per cent of a pale-yellow, non-drying oil, but it has not become commercially important.

Small latex vessels ramify through all the tissues of the plant and exude latex containing papain when they are cut. Papain is a proteolitic enzyme with action similar to that of the enzyme pepsin in the animal stomach. The latex is obtained from unripe fruits by tapping them with a series of shallow longitudinal cuts made with glass or sharp bone. Metal knives are never used because chemical changes occur in papain when it is in contact with metals. Latex oozes from the cuts to be gathered beneath the fruit in porcelain or earthenware containers. It is dried in the sun or in ovens, and then sold. The largest producer of papain is Tanzania, and most is imported by the United States. It is used in brewing to 'chill-proof' beer because papain prevents the cloudiness caused by the precipitation of proteins during chilling; it is used in canned meats and in meat-tenderizing preparations, in medicine and pharmacy, and in the leather tanning industry. Shrink-resistance processes applied to wool, and to textiles made from wool or silk, also involve the use of this enzyme. Papain is also used in the manufacture of easily digestible children's foods and in chewing gum. Wherever pawpaws are grown the properties of the latex seem to be known, and tough meat is often wrapped in pawpaw leaves to make it tender.

Guava: *Psidium guajava*

The guava tree is native to tropical America where the fruits were eaten and the trees were domesticated more than 2,000 years ago. It was spread rapidly throughout the world's tropics by the Spanish and Portuguese soon after the discovery of the New World, helped no doubt by its ability to grow well in a wide range of environments and in a variety of soils. In some areas it has spread so extensively as to become almost a weed. Guava is a member of the large family *Myrtaceae* which includes about 3,000 species of woody plants characterized by the presence of essential oils in some of their tissues. The clove, cinnamon, allspice and eucalyptus all belong to the *Myrtaceae*, and many members of the family have edible fruits, though few have been exploited. The 150 species of the genus *Psidium* are all native to tropical America, but only the guava, *P. guajava*, has become important. The guava is a large shrub or small spreading tree up to 10 m tall, with a fairly thin trunk which has a scaly, often multi-coloured bark. Several more or less erect branches arise from the base of the trunk and carry spreading lateral branches, but when guavas are cultivated in orchards they may be pruned to a more regular, roughly obconical shape. The smaller branches are square in cross-section and carry almost sessile, opposite, light-green, simple, oval leaves. The largest leaves are up to 15 cm long with prominent veins on the soft, downy under surface, but with the veins markedly depressed on the upper surface. Attractive white flowers, about 2.5 cm in diameter, are carried singly on short, slender peduncles in the axils of the leaves on the younger branches, or they are sometimes in two to three flowered cymes. The calyx-tube completely encloses the flower bud, but bursts open as four to six irregular lobes when the flower is mature; four free white petals spread widely as the flower opens. Numerous stamens have long filaments and short, rounded anthers; they are in groups around the top of an inferior ovary of four united carpels, each carpel containing many ovules in axile placentation. The long style extends beyond the stamens and has a knob-like, terminal stigma. The flowers are self- and cross-pollinated by insects. After fertilization the ovary develops into a globular or pear shaped berry up to 10 cm long and 5 cm or more in diameter with a thin greenish-yellow or bright yellow skin and the persistent remains of the calyx-tube at the top. Within the fruit the small yellowish seeds are embedded in the white or pink flesh of the mesocarp which usually contains sclerids (stone cells). Guavas are commonly propagated from seed and are consequently very variable with diverse kinds of fruits; the best cultivars have fruits with a mild, pleasant flavour and smell, whereas others have very acid and acrid fruits. Vegetative propagation by patch-budding, root cuttings, layering and approach-grafting is possible.

The fruit is an excellent source of vitamin C, though some cultivars contain much more than others, and they are also fairly rich in vitamin A, iron, calcium and phosphorus. The fruit is eaten fresh or made into jelly, its juice is used to make a drink, while guava paste or 'guava cheese' are popular dishes in Florida, the West Indies and parts of South America.

Psidium littorale is a smaller, bushier tree than *P. guajava* with smaller, thicker leaves, a smoother grey-brown bark, and small dark-red fruits about 3 cm in diameter which are known as strawberry guavas. They are cultivated on a small scale in the West Indies.

Mangosteen: *Garcinia mangostana*

The mangosteen is perhaps one of the finest tropical fruits with a flavour and quality much admired by visitors to Malaysia and Indonesia. The fruit is obtained from a small, slow-growing, dioecious tree *Garcinia mangostana*, a member of the family *Guttiferae*. The genus *Garcinia* includes 150—200 species of small trees indigenous to Eastern Asia and parts of Africa. The yellow dye gamboge is obtained from the latex of *G. hanburyi*. Other species yield a gum which is soluble in turpentine, but not in water, which has been used as a varnish; several others have edible fruits, though none have the quality or flavour of the mangosteen.

Garcinia mangostana is a native of Malaysia, and is cultivated there, in Indonesia, Sri Lanka and the Philippine Islands. Attempts to introduce mangosteens to other parts of the tropics have been unsuccessful largely because the tree is difficult to establish. The seeds remain viable only a short time after they have been removed from the fruit; the root system requires several years to become well established and the trees do not bear fruit until they are 10 or more years old. Mangosteens are usually grown from seed because it is not easy to propagate them vegetatively, though hardy, quick-growing wild species of *Garcinia* are compatible root stocks for scions of the cultivars. Mangosteen is a crop of the lowland humid tropics; it requires a hot, wet climate, a good well-drained soil, and shade in the young stages of growth.

The tree is up to 15 m tall with a compact, conical shape, and has opposite simple leaves on short petioles 12—25 cm long. The leaves are around 20 cm long, roughly elliptical in shape, with an acuminate tip; they are thick and leathery in texture, and dark shining-green in colour with prominent mid-ribs and numerous parallel lateral veins. Though male and hermaphrodite flowers do occur, most cultivated trees have only female flowers with sterile stamens; they produce parthenocarpic fruits. The flowers are terminal on the small branches; they have four thick leathery sepals; four broad, spreading yellow petals; and a globular ovary with four to eight locules, each of which may contain a single apomictic seed. The ovary is surmounted by a sessile four-lobed stigma. The fruit is a globular, indehiscent berry, flattened at the base where it is subtended by the persistent calyx-lobes, and at the top where the remains of the stigma also persist. The fruit is the size of a small orange, 4—7 cm in diameter, and a beautiful reddish-purple colour. It has a smooth, thick pericarp containing yellow resin, and though the rind is tough the fruits must be handled with care at harvest to avoid bruising. The soft tissues of the berry consist of five to eight ivory-white segments, resembling those of the mandarin or tangerine, which are easy to separate. Each segment is covered by a delicate fibrous network and consists of a delicious, soft, melting flesh. The rind contains tannin which is used for tanning leather and also in medicine.

Breadfruit: *Artocarpus altilis*

The genus *Artocarpus* is a member of the family *Moraceae*, and is thus related to the figs and mulberries. It includes about 50 species of trees which grow in the hot moist regions of the South-east Asian tropics and the Polynesian Islands. Several of them are cultivated in the tropics for their large and curious fruits, and of these the breadfruit is perhaps the most important, though it has never become very popular outside its original home, despite its wide dissemination in the tropics. *Artocarpus altilis* is a handsome tree up to 20 m tall which grows on well-drained soils, especially on coastal plains where there are up to 2,500 mm of rainfall annually, and where temperatures are hot. The tree has a smooth light-coloured bark, and all its tissues contain latex vessels. The very large, striking leaves are arranged spirally on its many branches; they have thick, fairly long petioles, and laminas 30–60 cm long and up to 40 cm broad. The shining, dark-green lamina is thick, stiff and leathery, and broadly ovate in general shape, but with several deep, pointed lobes. It is slightly hairy on the lower surface and along well-marked veins. The leaves are subtended by two large, deciduous stipules. The trees are monoecious with male and female flowers in different inflorescences on the same tree, the males in dense yellow catkins, the females crowded together around a fleshy receptacle. The male inflorescence may reach a length of 15–25 cm; it is a pendulous, club-shaped mass of numerous small flowers made up of a two- or four-lobed perianth and a single stamen. The female inflorescences are more globular in shape with the numerous flowers embedded in the receptacle, each having a tubular perianth and a small, two-celled ovary topped by a narrow style and a lobed stigma. The whole female inflorescence develops parthenocarpically into an aggregate seedless fruit which may be 10–30 cm in diameter, and which changes from dark green to brownish-yellow as it matures. Breadfruits have a thick warty rind enclosing a pale yellow, starchy, fibrous flesh which is a rich source of food energy; when fresh it contains around 20 per cent of carbohydrate as well as small amounts of protein, minerals and vitamin A. The flesh has a distinct smell and is cooked and eaten as a vegetable. Cultivars with seeded fruits are called 'breadnuts'; they have little edible flesh, but their seeds are cooked and eaten. Seedless breadfruits are propagated by root cuttings, or by removing and planting the offshoots which grow from exposed surface roots when they are damaged. On the other hand, breadnuts are propagated from seed. In its original home in Polynesia numerous cultivars of the breadfruit exist, but little work has been done on the crop except for the selection of apparently superior trees for testing, despite the fact that the fruit is such good food.

Of various other species of *Artocarpus* cultivated for their edible fruits,

Fig. 7.19 *Artocarpus altilis*. Breadfruits.

A. heterophyllus, the jackfruit tree, is the most common. This is a large handsome tree up to 20 m tall with a smooth, round trunk supporting a dense crown of leafy branches. It is a native of India and Malaysia, and is found growing throughout South-east Asia where its large fruits are valued as an article of food. The plant has been introduced to most parts of the tropics, but has not attained any general popularity. It differs from the breadfruit in having stiff, usually entire elliptical leaves which are rarely longer than 20 cm, glabrous and terminated by a short blunt point. The male inflorescences are somewhat shorter than those of breadfruit, and the female inflorescences are cauliflorous arising directly from the main trunk or branches. They develop into enormous aggregate fruits, often 60 cm or more in length and up to 50 cm in diameter. The fruits have a thick, warty and spiny rind which is green when young, but which turns brown as the fruit ripens. The flesh consists of the swollen receptacle of the inflorescence together with the numerous flowers; it surrounds the seeds, each one embedded in a yellow fleshy sheath which some people consider to be the best flavoured part of the fruit. The fleshy receptacle is eaten as a cooked vegetable, or when ripe as a dessert fruit; the flesh is also preserved or dried, and the seeds are boiled and roasted for eating. There are numerous cultivars, but unlike the breadfruit there are no seedless jackfruits. The plants are propagated by seed, which retains its viability for a short time and must be sown soon after collection.

Fig. 7.20 *Persea americana*: Avocado. The foliage of the avocado tree. (By courtesy of the FAO.)

Avocado pear: *Persea americana*

The avocado pear originated in Central America and there is archaeological evidence to suggest that is was introduced to Mexico as long as 9,000 years ago. It was cultivated in these areas when Europeans discovered the New World, but did not spread into the West Indies until the seventeenth century, and it is relatively recently that it has become widespread in the world's tropics. Though avocados are now cultivated throughout the tropics and are an important food crop in Central America, they are economically important as a cash crop only in the southern United States, Brazil, Argentina, South Africa, Hawaii and tropical Australia. They have a unique and delicate flavour and are among the most nutritious of all tropical fruits. Their smooth flesh contains up to 4 per cent protein and 30 per cent of a readily digestible oil as well as appreciable quantities of vitamins A, B and E, and of iron. They are the richest source of energy of all fruits.

Persea americana is an evergreen tree native to the rain forests of Central America, and is a member of the family *Lauraceae*. It grows to a height of 20 m or more, but the shape and form of the tree are variable; it may be erect with a stout trunk or small, spreading and bushy. The tree grows in flushes and bears spirally arranged simple leaves on short petioles. Young leaves have a reddish tinge, but they become bright green and broadly ovate to almost lanceolate as they mature. Both the trunk and roots have a thick fleshy bark. Flowers are produced in large numbers in compact axillary panicles near the ends of branches. They are greenish-yellow in colour and have six calyx segments which are united at the base, and arranged in two whorls, those of the inner whorl being slightly larger than the outer segments. There is no corolla. The calyx is softly hairy in some cultivars, but almost glabrous in others. Nine stamens are arranged in three whorls; those of the inner whorl have the longest filaments and there is a pair of orange nectaries at the base of each of them. The anthers have four pollen sacs which dehisce through terminal valves. The innermost whorl of the androecium consists of three staminodes surrounding the superior, unilocular ovary which contains a single ovule and which has a short, often bent, hairy style and a simple stigma.

Avocado flowers open twice, and all the flowers on a tree open and close together. When the flowers open for the first time the ovary is mature and the stigma is receptive, but the anthers are immature and shed no pollen. The flowers close after 1 day and open again 12 or 36 hours later by which time the anthers have matured and the stigma has withered. In some groups of cultivars the flowers open first in the morning while in others they do so in the afternoon. The periodicity of flower opening appears to be influenced by weather conditions, and is most marked on warm, sunny days. This seems to be an extremely effective physiological mechanism to ensure cross-fertilization, but avocados are self-fertile and the two flowering periods may overlap sufficiently for self-pollination to occur. The flowers are pollinated by bees which visit them to collect nectar.

The fruits of avocados are single seeded berries, which vary greatly in size (up to 20 cm long), shape and colour between cultivars. They may be rounded, egg or pear-shaped, with a thin membranous skin or a thick, tough rind, which is pale-yellow, green, reddish or dark-purple in colour. The cream-coloured or yellowish-green flesh encloses a very large, oval seed. Most cultivars of avocado show the same tendency to alternate bearing of good and poor crops as the mango, presumably for the same reason, although there is some evidence which suggests that seasonal variation in yield is genetically controlled.

The numerous cultivars now grown have been developed fairly recently, and since the practice of propagating the crop vegetatively by budding or grafting to establish selected clones has become established. Cultivars are grouped into three races, the Mexican race (considered by some to be a botanical variety *P. americana* var. *drymifolia*), the West Indian race and the Guatemalan race. Mexican types have strongly scented leaves, very hairy flowers and fruits with a thin membranous skin which are usually smaller than the fruits of other types. Mexican cultivars are used mostly as hardy root stocks for the propagation of other, better cultivars. West Indian cultivars have fruits with a leathery skin and light-coloured leaves without any strong scent; because they are less hardy than other races, they are best suited to cultivation in warm climates. Guatemalan cultivars were derived from ancestors in the upland regions of Central America. They have large rugose fruits with a hard, brittle skin. The different races of avocados have been hybridized, but the crop has not been improved by breeding. Instead superior seedlings have been selected within each race for clonal propagation.

Avocado fruits perish quickly once they are ripe, they are bulky and easily bruised. Consequently they are difficult and expensive to transport, and though they are highly esteemed in temperate countries where they do not grow there is little international trade in them. They are eaten when fresh, either alone or mixed in salads, but because they are so rich in oil and are not sweet or juicy they are used as a savoury, not as a dessert fruit.

Nuts — Cashew: *Anacardium occidentale*

Anacardium occidentale belongs to the family *Anacardiaceae* which also includes *Mangifera indica*, the mango, and a few other less important economic species. The cashew is a small tree native to Mexico, Central and South America with its centre of diversity in the arid regions of eastern Brazil. It grows wild in these areas, often on poor, barren soils, and was widely distributed throughout the wet and semi-arid tropics during the sixteenth and seventeenth centuries. Few economically important tropical plants tolerate such a wide range of soil and climatic conditions as the cashew, and its distribution is limited chiefly by its intolerance of cold. It is commonly grown on soils too poor for other crops and may receive little attention once established; indeed cashews have become naturalized in many parts of the Old World to which they have been introduced. It is cultivated chiefly for its seeds, which are the cashew 'nuts' of commerce, but the pericarps of its achenes yield cashew shell oil which is extracted with solvents or steam in some areas for industrial use; its seeds contain around 50 per cent by weight of an edible non-drying oil, and the pedicel and receptacle of the flower grow to produce a large fleshy 'cashew apple' or 'pear' which is eaten as a dessert fruit. Annual world production of cashew 'nuts' is around 0.2 m. tonnes, some from wild and cultivated trees in tropical America, and some from crops in India, East Africa (especially Tanzania) and the Mediterranean area.

Cashew is a straggling, untidy evergreen tree around 10 m tall with long branches, some of which droop to touch the ground. It has large, alternate pale-green leaves borne on short, swollen petioles. The leaves are narrow at the base, but broaden gradually to a smoothly notched apex, and they have prominent veins. Male and hermaphrodite flowers occur together in loose terminal panicles at the ends of branches, with about six times as many males as there are hermaphrodites. The flowers are regular with five green sepals, five petals which are red when mature and ten free stamens. Commonly the filaments of nine of the stamens are short, while the tenth is about twice as long as the rest, but in male flowers filament length sometimes varies more than this. The ovary is superior with a long, simple style and one locule containing a single ovule. The pollen is heavy and sticky, and the flowers are pollinated by insects, but only six or fewer pollinated flowers produce fruit in each inflorescence. The fruit is a kidney-shaped achene about 3 cm long with a hard grey-green pericarp, which contains, as well as oil, a skin irritant (a feature of the sap of several members of the *Anacardiaceae*). Before the seeds are removed from the pericarp, the fruits are roasted, a process which burns off the shell oil and cooks the seeds so that they lose the astringent flavour they have when raw and become palatable. Cashew seed oil is rarely extracted because the 'nuts' command a high price as a food delicacy.

Further reading

General

Purseglove, J. W. (1968 and 1972). *Tropical Crops*, 2 vols, London: Longman.
Singh, S., Krishnamurthi, S. and Katyal, S. L. (1963). *Fruit Culture in India*, New Delhi: Indian Council of Agricultural Research.

Banana

Champion, J. (1963). *Le Bananier*, Paris: Maisonneuve and Larose.
Haarer, A. E. (1964). *Modern Banana Production*, London: Leonard Hill.
Simmonds, N. W. (1966). *Bananas*, 2nd edition, London: Longman.

Citrus

Hume, H. H. (1957). *Citrus Fruits*, 2nd edition, New York: Macmillan.
Weber, H. J. and Bachelor, L. D. (1948). *The Citrus Industry*, 2 vols, Los Angeles: University of California Press.

Dates

Corner, E. J. H. (1966). *The Natural History of Palms*, London: Weidenfeld and Nicolson.
El Baradi, T. A. (1968). Date Growing, *Trop. Abstracts*, 23, 473—9.
Goor, A. (1967). The history of the date through the ages in the Holy Land, *Econ. Bot.*, 21, 320—40.

Mangoes

Gangolly, S. R. *et al.* (1957). *The Mango*, New Delhi: Indian Council of Agricultural Research.
Mukherjee, S. K. (1953). The mango. Its botany, cultivation, uses, and future improvement, especially as observed in India, *Econ. Bot.*, 7, 130—62.
Mukherjee, S. K. (1972). Origin of mango (*Mangifera indica*), *Econ. Bot.*, 26, 260—4.
Singh, L. B. (1960). *The Mango. Botany, Cultivation and Utilization*, London: Leonard Hill.

Pineapples

Collins, J. L. (1968). *The Pineapple. Botany, Cultivation and Utilization*, London: Leonard Hill.
Py, C. and Tisseau, M. A. (1965). *L'Ananas*, Paris: Maisonneuve and Larose.

Pawpaws

Becker, S. (1958). The production of papain — an agricultural industry for tropical America, *Econ. Bot.*, 12, 62—79.
Storey, W. B. (1958). Modification of sex expression in papaya, *Hort. Adv.*, 2, 49—60.

Guavas

Ruehle, G. D. (1948). The common guava — a neglected fruit with a promising future, *Econ. Bot.*, 2, 306—25.

Avocados

Hodgson, R. W. (1950). The avocado — a gift from the Middle Americas, *Econ. Bot.*, 4, 253—93.

Nuts

Howes, F. N. (1948). *Nuts*, London: Faber and Faber.
Morton, J. F. (1961). The cashew's bright future, *Econ. Bot.*, 15, 57—78.
Morton, J. F. (1972). Avoid failures and losses in the cultivation of cashew, *Econ. Bot.*, 26, 245—54.

Chapter 8

Beverage, Masticatory and Drug Plants

The major non-alcoholic beverages are tea, coffee and cocoa, all made from the produce of tropical crops grown in several countries, either on plantations or by smallholders. They are popular drinks because they have pleasant aromas and flavours derived largely from essential oils in the plant parts from which they are made, and because of the refreshing and stimulating effects of the alkaloid caffeine or related substances which they contain. In small amounts caffeine acts on the central nervous system to increase mental activity and decrease fatigue, but if taken in excess it may have very harmful effects. Caffeine also aids digestion by stimulating increased production of digestive juices, and it has a diuretic effect increasing the excretion of uric acid. The plant parts from which tea and coffee are made contain 1—3 per cent caffeine; cocoa seeds contain 1—2 per cent of theobromine, which is an alkaloid similar to caffeine. The stimulating properties of the popular West African masticatory kola are due to caffeine, theobromine and a heart-stimulating glucoside, while betel nut, which is chewed in parts of India and Asia in a 'quid' with betel pepper leaf, contains the alkaloid arecoline.

Medicinal plants have been used by man for centuries, and they are still important, especially in many developing communities in the tropics where traditional remedies are commonly used in the treatment of numerous ailments. Though we may not be able to ascribe any scientific justification for the supposed effects of many of them, others like opium and quinine are very important drugs in modern medicine. The science of botany began with the study of medicinal plants, first by the ancient Greeks (especially Theophrastus and Dioscorides), then with renewed vigour by the herbalists of Europe in the fifteenth to seventeenth centuries. Several important drugs first discovered in plants are now synthesized more cheaply than they can be extracted from them, but the search for new ones continues, especially in the medicinal plants of the tropics. Very few tropical drug plants are discussed here, and nothing is said about the important fungi and bacteria which are the source of antibiotics like penicillin and streptomycin. None the less, among the crops which are discussed here some are the mainstay of the economies of the countries which grow them.

The beverages

Tea: *Camellia sinensis*

Tea is a beverage prepared by making an infusion of the processed, dried leaves of *Camellia sinensis*. It is perhaps the most popular of all the non-alcoholic drinks; about half the people in the world drink it regularly, and it is especially popular in the very densely populated parts of the Indian sub-continent, South-east Asia and the Far East. Tea has been used as a medicine in South-east China since very ancient times, and the beverage has been drunk there for the past 2,000–3,000 years. The tea-drinking habit began to spread from China around the fifth century A.D., but it did not reach Europe until the sixteenth century and China remained the only exporter of tea leaves. Britain became the chief importer, and Britons came to be among the world's leading tea drinkers; they were largely responsible for spreading the use of the beverage during the nineteenth century, and encouraged the production of the crop in India and Sri Lanka. Now it is also grown in Africa, Russia, Turkey, South America and Australia as well as in China and South-east Asia. Annual world production of 'made' tea (the processed, dried leaf) is around 1.5 m. tonnes, 0.46 m. from India, 0.2 m. from Sri Lanka and 0.3 m. from mainland China; the processed leaf is exported and is available throughout the world. The beverage is made by pouring boiling water on to the 'made' tea (usually a

Fig. 8.1 *Camellia sinensis*: Tea. Tea planted on the contour at the Tea Research Institute, Sri Lanka.

blend of leaf from different sources and of various quality), and is commonly drunk with sugar and milk. Along the coast of North Africa and in Egypt a very sweet, strong brew which includes herbs and spices is popular, while in Asia, as well as drinking infusions of tea leaves, they are pounded and mixed with cereal flour and fat to be eaten.

Tea probably originated in tropical rain forests near the source of the Irrawaddy River, but there is doubt whether wild plants in that area are descendants of the wild ancestors of the crop, or whether they are relicts of ancient cultivation. The crop spread very early into south-west China and the rest of South-east Asia, and is now grown in the wet highlands of many tropical countries. It does not thrive in the hot, lowland tropics, nor is it tolerant of frost, though tea is grown in parts of China and Russia where frosts do occur while the crop is dormant. It grows best on deep, well-drained, acid soils (pH less than 6) in relatively cool, humid climates where annual rainfall is greater than 1,500 mm, with an average of at least 50 mm each month in the driest part of the year. There seems to be no upper limit to the amount of annual rainfall in which tea can be grown, but it does demand moderate temperatures in the range $13°-30°C$; such climatic conditions are to be found only in the high-altitude tropics and the wet sub-tropics.

Tea is outbreeding and until recently was propagated by seed, so that old crops are mixtures of heterozygous plants which vary in productivity and quality. The benefits to be derived from propagating clones of productive, high-quality tea bushes have led to decreased propagation from seed and the increasing use of vegetative propagation (stem cuttings) on a commercial scale. Very large numbers of cultivars have been described; those which have distinct characteristics and which are associated with particular areas are called 'jats'. Despite great variation in the crop, only two main groups of cultivars are recognized, and these have the status of botanical varieties. *Camellia sinensis* var. *sinensis* is 'China tea', which is a hardy, more or less dwarf type with small, narrow coarsely serrate leaves, and solitary flowers. *Camellia sinensis* var. *assamica* is 'Assam tea', which predominates in commercial tea estates outside China. It has larger leaves than var. *sinensis*, with fewer serrations, and its flowers are borne in clusters of two or three in the leaf axils.

Camellia sinensis belongs to a small family of dicotyledons, the *Theaceae*, which includes about 200 species of trees and shrubs in the wetter tropics. The genus *Camellia* has 45 species of evergreen trees and shrubs, all native to tropical Asia. One of them, *C. japonica* (Camellia), is cultivated for its decorative flowers. Tea is a shrub or straggling tree which grows wild to a height of 10 m or more; but in cultivation the growth habit of the plant is carefully controlled by pruning, and by the more or less continuous removal of young shoots which are harvested to be processed. The plant has a large tap root with strong laterals in the upper soil. Feeding roots have no root hairs, but they are associated with endotrophic mycorrhiza, though the role of the mycorrhizal fungus in tea nutrition is not well understood. The roots store starch which is very important for the successful cultivation of the crop because it is the chief food reserve available for regrowth after periodic, severe pruning. Indeed, it is most important to ensure that crops are pruned at the end of or soon after periods of natural dormancy, when starch reserves in the roots are most plentiful. Tea leaves are alternate on the thin ultimate branches of the bush. They have very short, slightly thickened petioles, but no stipules, and obovate—lanceolate, acuminate laminas 5—30 cm long which taper towards the base and towards the tip; they are slightly hairy below, especially when young. The mature leaves are bright green, stiff and leathery with more or less dentate margins; they have no stomata on their upper surface, and the upper epidermis has a thick, glossy cuticle. The leaf has one or two layers of palisade tissue, and a spongy mesophyll in which lignified, much branched sclerids, or stone cells, occur. The characteristic form of the sclerids is easy to recognize, and is useful when it is necessary to determine the purity of a sample of made tea when purity is judged by the amount of leaf from other plants which adulterate it.

Seeds are obtained from selected plants, sometimes selected clones, grown at a density of 170—250 trees per hectare in areas set aside for the purpose where they are allowed to grow without heavy pruning. The sweet scented white or pink flowers are solitary or in groups of up to four in the leaf axils, each one on a short pedicel. Each flower is subtended by a pair of small bracts, and has five to seven persistent pointed sepals which are free from each other except at the base; five to seven thin, curved petals are also united at the base. Numerous stamens are arranged in two series, the outer ones more or less united by their filaments and fused to the base of the corolla, the inner ones free. They carry small, yellow, two-celled, versatile anthers. The superior, slightly hairy ovary has three to four fused carpels, each with four to six ovules in axile placentation. There are as many styles as there are carpels, fused below but divided above into separate stigmas. The flowers open in the afternoon, and remain open for 2 days. They are self and cross-pollinated by insects. The fruit is a woody, dehiscent, more or less globular capsule, 1.5 cm or more in diameter. It matures after 9—12 months and splits from the apex into valves, each with one or two flat or spherical brown seeds which fall to the ground, where they are gathered and must be sown soon after because they do not remain viable for long.

Tea seeds can be sown 'at stake' in the field where the crop is to grow, but it is better practice to raise them in nurseries where shade and water are carefully controlled. When they are 3—4 years old the main stems of seedling plants are pruned back to within about 10 cm of the soil before the plants are transferred from the nursery to the field. At this age they have a long tap root and are not easy to transplant. When tea is propagated vegetatively short stem cuttings of a single leaf with its axillary bud and the internode below it are taken from the young stems of plants selected for their growth vigour and leaf quality. They are planted in propagating

Fig. 8.2 *Camellia sinensis*: Tea.

beds where there is careful control of water and shade. Rooted cuttings are transferred to a nursery or to the field after about 1 year. Tea is commonly planted in the field at populations of 5,000–7,000 plants per hectare (spacing of 1.2 × 1.5 m). In the past various leguminous trees were planted with tea to provide shade for the mature crop, but the benefits derived from this practice are uncertain and it is becoming uncommon.

Fig. 8.3 *Camellia sinensis*: Tea.   A tea bush grown to a convenient plucking height.

Young tea bushes are pruned to stimulate the growth of lateral branches and a large root system. The object of early pruning is to establish a substantial, low-branching 'frame' bearing many leafy shoots at a uniform level, the 'plucking table', at least 60 cm above the ground. The leafy shoots grow in flushes during which four leaves are produced, followed by a dormant bud in the axil of the terminal leaf. This cycle of flush and dormancy lasts 70—90 days, which is also the interval between successive harvests from each shoot. However, a single bush has leafy shoots in various stages of this physiological growth cycle, and is plucked every 7—14 days, depending upon variation in the environmental determinants of growth rate, but chiefly on temperature. At each plucking two to four leaves and a bud are taken, most commonly by hand though mechanical pluckers have been developed. For the best quality tea only the bud and the first two leaves are taken; poorer quality tea is made when three or four leaves are taken with the bud because these lower leaves are older. The constant removal of the tips of young shoots stimulates the growth of axillary buds lower on the shoot which develop into leafy shoots to be plucked in their turn. After a varying number of flushes over a period of 2—5 years, the shorter intervals at lower, warmer altitudes, it becomes necessary to prune severely because the growth vigour of the plants decreases as a consequence of the continuous removal of young leaves, and because the plucking table gradually increases in height until it is difficult

Fig. 8.4 *Camellia sinensis*: Tea. Pruning the tea bush.

to reach. Several different methods of 'production' pruning are used which vary in severity, but they are all timed to coincide with a period of natural dormancy in the crop which is induced by cool temperatures or dry weather. Severe pruning may remove all the foliage from the crop. Fertilizer is applied after pruning, but regrowth is largely dependent upon the carbohydrate reserves stored in the roots.

The harvested shoots are taken to a factory on the tea estate soon after they are plucked and there they are subjected to a series of complex processes during which the flavour and aroma of the 'made' tea are developed. Most tea marketed in the world is black tea which is prepared by first withering the leaves, then twisting and breaking them on rolling drums which break the tissues and release the cell sap, and finally by allowing them to ferment under very carefully controlled conditions before they are dried quickly in hot air to stop the fermentation. During this process caffeine produced as a result of protein breakdown is released from its union with polyphenols in the leaf. The final caffeine content of made tea is 1.5–5 per cent. The polyphenols give tea its 'body' or 'strength'; they are most concentrated in young leaves which is one reason why the bud and first two leaves on a shoot make the best tea. Good tea has a high caffeine content, and is rich in the essential oils which impart aroma and flavour. The dry tea is graded and packed in plywood tea chests at the factory. Different grades, and black tea from different sources, are blended in the importing country before the tea is sold. Green tea produced in China and Japan is derived from unfermented leaves which are dried quickly after they are plucked. Oolong tea is produced from partially fermented leaves in Taiwan.

Tea seeds contain about 20 per cent of a golden-yellow non-drying oil, but it is not extracted commercially. Two other species of *Camellia, C. sasangua* and *C. japonica*, produce seeds with 60—70 per cent oil which is extracted in India, China and Japan in small amounts to make soap and for illumination.

Coffee: *Coffea* spp.

Coffea is a member of the family *Rubiaceae*, which includes around 400 genera of trees and shrubs, most of them tropical plants, but the only economic genera are *Coffea*, coffee, and *Cinchona*, quinine. The family is characterized by its simple opposite leaves, by the fusion of the sepals to form a calyx tube and of the petals to form a corolla tube, and by the inferior ovary with two locules. The genus *Coffea* includes about 60 species, though not all of them are clearly defined. They are handsome evergreen trees, shrubs or rarely climbers native to the tropical rain forests of parts of Africa, the Malagasy Republic, Mauritius and South-east Asia. Most of the world's coffee comes from the tetraploid ($2n = 4x = 44$) species *C. arabica*, Arabian or Arabica coffee, which is a native of the highland, wet forests of Ethiopia where it grows wild as a constituent of the understory of shrubby vegetation, more or less densely shaded by taller forest trees. The diploid ($2n = 22$) species *C. canephora*, robusta coffee, is native to the rain forests of equatorial Africa; it accounts for as much as one-quarter of world production, used mostly to make powdered 'instant' coffee. A third species, the diploid *C. liberica*, liberica coffee, is indigenous in Liberia; the beverage made from its seeds is of poor quality, and liberica coffee is unimportant compared with the other two species.

Arabica coffee was domesticated recently compared with most other crops. It was cultivated, and a drink was first made from its roasted seeds, in the Yemen during the fifteenth century A.D., and though we do not know when it reached Arabia from Ethiopia, it seems certain that it arrived there later than the fifth century, and perhaps not until the fourteenth. The coffee-drinking habit had become established in Europe by the end of the seventeenth century. Arabica coffee was introduced to India and Sri Lanka from the Yemen at the end of the seventeenth century, and very soon afterwards to Java. The subsequent history of this coffee, which is the botanical variety *C. arabica* var. *arabica*, is truly remarkable. A single tree was taken by the Dutch from Java to Amsterdam in 1706; there it flourished in the Botanic Garden, and one of its progeny raised from seed was given to Louis XIV of France in Paris. During the first half of the eighteenth century the descendants of these trees in Europe were spread throughout Dutch and French possessions in the New World, Africa and Asia. Today they account for a large part of world coffee production, though in Brazil, where some 30 per cent of the world crop is grown, an increasing proportion of coffee is from the cultivar 'Mundo Novo' which is a hybrid between var. *arabica* and a mutant form of it, the botanical variety *C. arabica* var. *bourbon*. The first records of the cultivation of var. *bourbon* (which is also 'arabica' coffee) are from the island of Réunion, where it was introduced by the French at the beginning of the eighteenth century. Much later it became important in East Africa.

Annual world production of coffee is around 4 m. tonnes of which 2.5 m. are from South and Central America and Mexico (Brazil 1.0 m., Colombia 0.5 m.), 1.3 m. from Africa, where most of the world's robusta coffee is grown, and 0.4 m. from Asia. Coffee production was once the chief industry, and coffee was the leading export from Sri Lanka until the crop was entirely destroyed at the end of the nineteenth century by coffee leaf rust, a disease caused by the fungus *Hemileia vastatrix*. Arabica coffee requires around 2,000 mm annual rainfall; where rainfall is less than 1,500 mm annually the crop can be grown only with supplementary irrigation and the regular use of deep mulches to conserve water. It grows best on deep, well-drained loams, in climates with moderate temperatures in the range 16°–25°C where seasonal, but short periods with little rainfall and cool temperature restrict vegetative growth. Under such conditions 'overbearing' is avoided, and the crop continues from season to season to give moderate, but economically worthwhile yields. On the other hand, when the crop is grown at low altitude where the climate is warm and wet all year and there is no seasonal check to growth, excessive fruit production may so deplete food reserves that young shoots die, luxuriant foliage may favour pest and disease organisms, and eventually production may cease to be worthwhile. Robusta coffee can be grown in a wider range of environments than arabica, it thrives at lower, generally warmer altitudes, and it is resistant to coffee leaf rust.

Coffea arabica is a small tree which grows up to 5 m tall if it is not pruned. In cultivation the architecture and form of the tree are commonly controlled by careful pruning to produce a more or less densely branched bush 2–3 m tall, though much coffee is grown without pruning. The branches of *C. arabica* are dimorphic. Orthotropic, vegetative main stems and branches, which grow more or less vertically, have two buds in the axil of each of their leaves; from the upper bud of this pair a plagiotropic, reproductive branch is produced which grows more or less horizontally in var. *arabica*, and at a more acute angle of about 50° from the main stem in var. *bourbon*. As they age and bear fruit the plagiotropic branches droop towards the ground. If the apical growing point of the main, orthotropic stem is damaged, or removed as it may be during pruning, the lower bud of the pair in a leaf axil near the base of the stem grows to produce an orthotropic branch. Consequently, by selective pruning the coffee bush can be made to have numerous vegetative, orthotropic branches, but usually one (single stem pruning), two or four (multiple stem pruning) are maintained. Alternatively, by bending the main stem over and fixing its tip near the ground, several buds can be induced to grow into orthotropic branches. The thin, dark-green leaves are oval or elliptical, 10—15 cm long and 5 cm or more broad with a long, pointed tip when they are fully expanded; a pair of small stipules occurs between the short petioles of each pair of opposite leaves. There are six buds in the axil of each leaf on a plagiotropic branch; one of them may grow to produce a secondary plagiotropic branch, or three or four of them may grow into inflorescences. Each inflorescence consists of four flowers on a short pedicel with a pair of bracts at its base. Buds in the axils of these bracts usually remain dormant a long time before they too produce inflorescences, often on relatively old, leafless parts of the plagiotropic branch. The white, heavily scented flowers have five united sepals which form a short cup at the base of the corolla. The five, or sometimes six, petals are fused to form a tube about 1 cm long at their base, but expanded above into lobes about as long as the tube. Five stamens inserted on the corolla tube alternate with the corolla lobes; they have long anthers on slender, upright filaments. The inferior ovary of two united carpels contains two ovules attached to the base of the ovary wall, and has a long, slender style with two short, pointed stigmas. Flower initiation is influenced by day-length and temperature (at least in var. *bourbon*), and flowering is correlated with rainfall; mature, but unopened flower buds remain dormant until they are wetted, then they open together about 2 weeks later. The flowers open in the morning, and wither a few hours later. Arabica coffee is predominantly self-pollinated, though some cross-pollination by insects does occur. (*Coffea canephora* and *C. liberica* are self-incompatible and out-breeding.) The fruit is a spherical drupe which develops slowly over a period of 9 months, during which time it changes gradually from dull green, through yellow to bright crimson when it is mature. The ripe drupe is called a 'berry'. It has a thin red exocarp, a pulpy yellow mesocarp, and a thin, fibrous endocarp called the 'parchment'. The two seeds are grey-green, about 12 mm long, elipsoidal but flat where they are pressed together within the fruit; their thin testas as called 'silver skins'. The bulk of each seed is composed of a curiously folded endosperm which encloses a very small embryo. Coffee seeds are commonly called 'beans'.

Coffee is propagated from seed, though it can be propagated vegetatively by budding, grafting or from stem cuttings. Seeds from selected superior bushes are sown in a nursery where seedlings are raised, at first under shade, for periods of 6—24 months before they are transplanted to the field. Arabica coffee is predominantly inbreeding, and much of it originated from the narrowest possible genetic base — the single tree taken from Java to Amsterdam in 1706. Consequently the crop is so homogeneous that little improvement can be expected from single plant selection to provide planting seed. None the less, useful mutations have occurred in the crop, some of them now recognized as distinct cultivars.

Even though the natural environment of coffee is within the shade cast by forest trees, most coffee is grown on plantations without interplanted

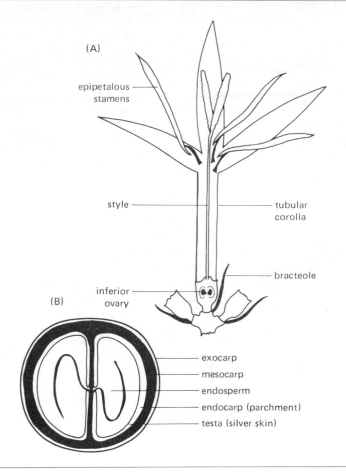

Fig. 8.5 *Coffea arabica*: Arabica coffee. (A) A longitudinal section of the flower (×4). (B) A section through the drupe.

shade trees. Coffee bushes begin to bear fruit when they are 3–4 years old, and may continue to do so for 50 or 60 years. The fruits are collected by hand when they are mature, either directly from the branches or after they have been shaken to the ground. The best quality arabica coffee is prepared from berries subjected to wet processing in which the mesocarp is removed in a pulping machine after which the washed seeds, still within the parchment, are allowed to ferment while wet for 12–24 hours. During fermentation the remains of the mesocarp and the mucilage which surrounds the parchment are removed and the desirable bright, greyish-blue colour of the seeds is developed. The seeds are then dried on racks in

Fig. 8.6 *Coffea arabica*: Arabica coffee. (A) Flowers. (B) Fruits.

the sun for a week or more, during which they are regularly turned and are protected from rain. Finally, when they are dry, the beans are 'hulled' by a machine which removes the parchment, and polished to remove the silver skin. Fermentation and drying require careful, skilled supervision to ensure that the processed coffee is of the best quality. Alternatively the mature berries may be dried in the sun without being pulped, then hulled without prior fermentation, but the coffee beans processed in this way are inferior to those subjected to wet processing.

Processed, dried coffee beans contain approximately 12 per cent water, 35 per cent cellulose, 14 per cent protein, 10—13 per cent oil including the volatile oil caffeol which is largely responsible for the distinctive aroma and flavour of the beverage, 7—10 per cent sugar, and 1.5 per cent caffeine. The flavour and aroma of coffee are accentuated when the beans are roasted; they are ground and used soon after roasting because the grounds loose their flavour, and the oils in them become rancid when they are kept.

The flavour and quality of coffee vary greatly depending upon the maturity of the berries at harvest, upon the environment in which the crop was grown, especially altitude and temperature, upon differences between cultivars, upon the skill and care with which the berries were processed and the method used, and upon the temperature and length of time the beans were roasted.

Coffea canephora is a larger tree than *C. arabica*, and grows to 10 m tall, with larger, thicker leaves 15—30 cm long, rounded at the base and with wavy margins. It was discovered in Zaire in 1895, and is now grown chiefly in the lowland tropics of Central Africa and Indonesia. Its seeds contain more caffeine than those of the other two important cultivated species. *Coffea liberica* is also a large tree, up to 17 m tall. It is hardy and disease resistant, but produces beans of poor quality which make a bitter drink. The mature berries of liberica coffee are larger than those of arabica or robusta, up to 1.5 cm long.

Cocoa: *Theobroma cacao*

Cocoa and chocolate are prepared from the seeds of the cocoa (often and equally correctly spelt cacao) tree, which presumably originated in the equatorial rain forests to the east of the South American Andes Mountains where a wide range of types now grow in the dense shade cast by tall forest trees. They extend into Central America and as far north as Mexico, though in these areas it is difficult to distinguish between truly wild indigenous types and those which are escapes from cultivation. Cocoa has been cultivated for centuries in the northern part of South America and in the Central American countries, and was an early introduction into Mexico where the potentialities of the crop were fully realized, and where its value as a food beverage and its stimulating properties made it popular. Cocoa seeds were introduced into Europe to make the beverage in the sixteenth century, but the crop was not cultivated outside the New World until two centuries later. At the present time annual production of cocoa 'beans' (the seeds) is about 1.3 m. tonnes, about 1 m. from Africa (Ghana 0.35 m., Nigeria 0.22 m.), 0.3 m. from South America, 0.1 m. from Central America and Mexico, and less than 0.02 m. from Asia.

Cocoa is a strictly tropical crop, restricted in cultivation to lowland areas where annual rainfall is around 2,000 mm, though it can be grown with as little as 1,200 mm or as much as 7,000 mm of rain each year. To grow well the trees need a well-drained, deep soil of good crumb structure, containing adequate supplies of phosphorous and potassium. Cocoa will tolerate periods of little rainfall provided that it is grown on soils with a large water-holding capacity and is protected from drying winds; it does not tolerate sudden fluctuations of humidity. The shade trees under which the crop is commonly grown help to provide this protection and, in fact, create conditions somewhat similar to the natural habit of the cocoa tree, which is in the lower storey of rain forest. The high humidity inseparable from such conditions, at least at some seasons of the year, is unfortunately ideal for the growth and spread of numerous pests and diseases, and wherever cocoa is grown these are one of the main factors which limit production. Among the most important pathogens are *Phytophthora palmivora*, the fungus responsible for the disease called 'black pod' and *Crinipellis* (*Marasmius*) *perniciosus*, a fungus which causes witches' broom disease.

The genus *Theobroma* is a member of the family *Sterculiaceae*, which includes 50 genera and 700 or more species of tropical trees and shrubs. The family is closely related to the *Theaceae* and has some affinities to the *Malvaceae*. As well as cocoa, the important West African masticatory *Cola* spp. is in the family. Its members have alternate, often simple leaves usually subtended by deciduous stipules. The flowers are hermaphrodite or unisexual, and are actinomorphic with partially fused sepals, usually five petals (though some species have none) and five or more stamens, some of which may be represented by staminodes. The stamens have two-celled anthers, unlike those of the *Malvaceae*, which are one-celled. The superior ovary is made up of several united carpels and the numerous ovules are carried in axile placentation.

There are about 20 species in the genus *Theobroma*, but only *T. cacao* is of major economic importance; all the wild, semi-wild and cultivated cocoas are referred to it and (provided that they are compatible at the 'S' locus — see later), they all intercross readily, though the diversity of types within the species is considerable. The prevalence of cross-fertilization, the long period during which the crop has been cultivated in different environments, the ease with which different types hybridize and the fact that until recently propagation has been entirely by seed are all factors which have contributed to the present heterogeneity and heterozygosity in the crop. Consequently the classification of cultivated and wild cocoas is by no means easy. Although there is a general similarity in vegetative characters between all types of cocoa, considerable variation exists between them in fruit and seed characters, and in hardiness and vigour of growth. Depending largely upon differences between their fruits and seeds the cocoas can be divided conveniently into two main groups which are fairly easy to distinguish by both farmers and by those who buy the seeds. These two groups, the 'Criollo' and 'Forastero' cocoas, probably had a common origin somewhere in the region of the upper waters of several tributaries of the Amazon River, near the equator.

The Criollo group was carried across the Andes Mountains and eventually spread throughout Central America and Mexico, providing the original cultivated cocoas of those regions. The presence of red pigment in the pericarp of their fruits is common in Criollo cocoas which have yellow or red mature pods. In addition to this distinctive colour, the pods are characteristically deeply furrowed and tend to be rough and warty, with a distinctly pointed, not a rounded, end. The seeds of Criollo cocoas are large and rounded with white or pale-violet cotyledons; they produce a better quality beverage which has better flavour and aroma than that made from the seeds of the more commonly cultivated Forastero types.

The Criollo cocoas can be separated into two sub-groups:

1. The Central American Criollos probably arose by human selection for good quality white seeds within the heterogeneous stock that was carried north from the centre of diversity of the crop in South America. They have been cultivated in Central America and Mexico for centuries.
2. The South American Criollos include types which are apparently indigenous in Colombia and Venezuela with less sharply pointed, redder pods than the Central American group.

Forastero cocoas were probably domesticated more recently than the Criollos, and spread east and west from their centre of origin (not north). Their pods are much less deeply furrowed than those of the Criollos, or they may even be smooth. Red pigmentation in the pericarp is rare, and the mature pods are green or yellow with a much thicker, woodier wall than the Criollos; the seeds are flat with dark-red or deep-purple cotyledons. Forastero cocoas are generally much hardier and more vigorous, and they give larger yields than Criollos, but the beverage made from their seeds is of inferior quality.

The Forastero cocoas are also divided into two sub-groups:

1. The Amazonian Forasteros represent the original Forastero population with yellow pods containing flat seeds with deep-purple cotyledons. This group seems to have evolved without the influence of human selection. Their chief importance lies in the fact that they gave rise to the Brazilian Forastero population from which the West African 'Amelonado' cocoas were derived, and to the *Cacao Nacional* (National cocoa) of Ecuador.

2. The Trinitario Complex arose when Forastero and Criollo stocks merged in the Orinoco River region to form a single very diverse and variable population from which material was selected for cultivation in the West Indies and parts of South America. The Trinitarios are thus a heterogeneous collection of hybrids of fairly recent origin and exhibit a wide range of morphological and physiological characters. Superior types have been selected and are maintained in the West Indies as clones by propagation with stem cuttings. More recently improved seed yields have been achieved by hybridizing selected cross-compatible clones.

Some 80—90 per cent of the world's cocoa crop comes from cultivars of the Amazonian Forastero type, and most of the rest from Trinitario. The large West African Amelonado crop was derived mainly from introductions of the Amazonian type from Brazil, first made in 1879. Most of the West Indian crop, and that of several of the less important producers, derives mainly from mixed populations which can be classified as Trinitario. The Criollos are not important in world trade; in both Central and South America they have lost their identity in commercial plantations through mixture and hybridization with introduced Trinitario and Amazonian types.

The chief morphological features of these heterogeneous groups of cocoas are sufficiently constant to warrant one general description of the plant. The cocoa tree has a strong tap root with lateral branches which produce a mat of fibrous feeding roots extending several metres around the tree in the surface soil. There is evidence of mycorrhizal associations in the finer rootlets. Branching in cocoa is dimorphic. The short main stem which develops from the seedling is orthotropic, and grows vertically to about 1.5 m; at its apex three to six more or less horizontal plagiotropic branches are produced which form a group called the 'fan'. As the fan develops a new orthotropic leading shoot called the 'chupon' grows from an axillary bud just below the apex of the main stem, and in its turn produces a fan of plagiotropic branches above the first. This dimorphic branching pattern is repeated time and again until the spreading form of the mature tree develops to a height of 6—8 m. The orthotropic stems bear spirally arranged leaves, whereas on plagiotropic branches the leaves are alternate in two ranks; both kinds of branch bear flowers. Growth proceeds by a series of flushes, each followed by a period of apical bud dormancy during which the new growth hardens. The young leaves are delicate, flaccid and variously coloured from yellow to red; the leaves of any flush are shed after two subsequent flushes have been produced. When mature the simple ovate leaves are up to 30 cm long, thin but leathery, with a sharply pointed tip and prominent veins. The short petioles are subtended by deciduous stipules which leave small stipular scars on the branches when they fall. Cocoa is commonly propagated by seed, sown in the field or in a nursery where the seedlings are raised under heavy shade; or the crop can be propagated vegetatively by stem cuttings from plagiotropic branches by which means clones of high-yielding, good quality trees can be established. The crop is normally cultivated under shade trees, though under optimum climatic and soil conditions it is productive without shade once a continuous canopy has developed between trees.

The most curious feature of the cocoa plant is the way its flowers are carried in groups arising directly from the bark of the main trunk and older, leafless branches. They are said to be 'cauliflorous'. The clusters of flowers grow from small cushions which are greatly modified branches growing from the axils of very small, sessile leaves. Each flower is hermaphrodite and regular with five conspicuous pink sepals which are united at the base, but expand above into five narrow, pointed lobes. The five yellowish petals are smaller than the sepals; they are very narrow at the base, then expand to form a pouch (they are saccate) from which there extends a long, narrow reflexed lobe with a broad, flattened end or ligule. The androecium consists of an outer whorl of five long, narrow, pointed sterile staminodes fused at their base to the filaments of an inner whorl of five stamens. The filaments of the stamens are curved so that the anthers, each with four pollen sacs, develop within the petal pouches. The ovary consists of five united carpels with a single style which is shorter than the 'fence' of staminodes which surround it. At the top the style is divided into five stigmatic lobes, and there is a stylar canal which penetrates to the ovary where the ovules are in parietal placentation at the top (around the canal) and in axile placentation below.

Cocoa pollen grains are fairly large and sticky. Pollination is effected by insects, though the flowers have no nectar or scent, nor is their structure adapted to any of the more usual kinds of insect pollination. The anthers are more or less hidden within the petal pouches, and the styles are enclosed by the 'fence' of staminodes. None the less, some flowers are pollinated (though as few as 5 per cent), usually in the very early morning, sometimes by crawling insects like ants or aphids, but pollen is also carried between trees over distances as great as 45 m by midges of the 'sand fly' family, the *Ceratopogonidae*. They appear to be attracted by, and to feed on, the purple guide lines on the staminodes and within the petal pouches. Though insects do effect cross-pollination, cross-fertilization is restricted to the union of compatible gametes because an 'S' allele sporophytic incompatibility system operates in some populations of cocoa trees, especially those at the centre of diversity of the crop in South America. On the other hand the economically important West African populations of Amelonado trees, though they were derived from Amazonian Forasteros, consist entirely of self-compatible individuals. Whether gametes fuse to form a zygote is determined by the diploid constitution of the parents at the 'S' locus, where there are multiple alleles among which some exhibit dominance and others are independent. The incompatibility system does not depend upon the failure of pollen germination, nor upon differential rates of pollen tube growth, but operates by the failure of gametes carrying incompatible alleles to fuse in the embryo sac.

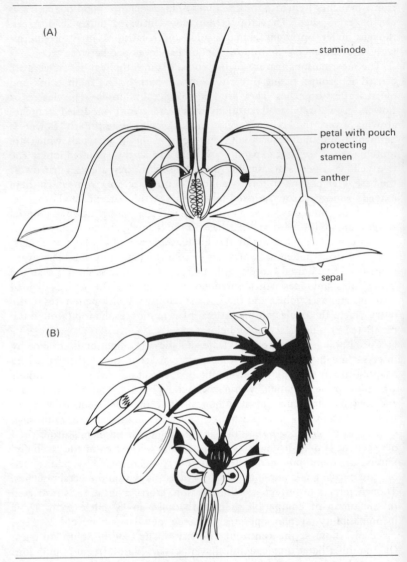

Fig. 8.7 *Theobroma cacao*: Cocoa. (A) A diagrammatic longitudinal section through a flower (×8). (B) A flower and flower buds arising from a flower cushion (×1½).

Cocoa trees begin to bear fruit when they are 4–5 years old, and continue to do so for 50 years or more. The fruits mature after about 6 months, though it is not easy for the unpractised observer to tell when they are ripe, partly because they attain their maximum size about 5 months after fertilization, and before they are mature. A large proportion of flowers never produce fruit even if they are fertilized, and large numbers of fertilized ovaries are shed prematurely. It has been estimated that only 1 in 500 flowers produces a mature fruit, largely of course because so few are pollinated, but any nutritional or physiological disturbance causes some young fruits to shrivel and to be shed. This occurs mostly when the fruits are 7—8 weeks old; after 3 months of growth there is little risk of fruit failing to mature. Each fruit contains 20—60 seeds embedded in a mass of white or pink pulp developed from the outer layers of the testa. Seed development follows a normal pattern, except that growth of the embryo is very slow until the third month, after which it grows quickly and consumes all the endosperm in the process so that the mature seed, which is about 3 cm long and up to 2 cm broad, consists of two wrinkled cotyledons without endosperm. The indehiscent fruits are pendulous on short stalks growing from the trunk and branches. They are called pods, though they are thought to be drupes. They vary in shape from cylindrical to spherical, and may be red, yellow, orange or purple depending on differences between cultivars. The pods are up to 30 cm long and 10 cm in diameter with a pericarp which may be thick and woody, or thinner and soft, and with a surface which may be corrugated and warty, or smoother without ridges.

The seeds constitute about 25 per cent of the weight of the mature fruit. They are removed after the pericarp has been split open by hand, then the mass of seeds with the mucilaginous pulp in which they are embedded is fermented. The seeds are piled in perforated boxes, or, in West Africa, in heaps covered by banana leaves, and allowed to ferment for up to 1 week during which time they may be stirred to allow good aeration and to prevent the temperature rising above about 50°C. Sugars in the pulp are broken down by yeasts during fermentation so that the pulp and mucilage disintegrate into a vinegary fluid which escapes through slits in the base of fermentation boxes. During the process the embryo is killed, various chemical changes occur in the seeds and the flavour and aroma of cocoa develop. The colour of the purple cotyledons changes gradually until, in well-fermented seeds, the tissues are pale brown. When the mass has cooled to about 35°C the fermentation process is considered to be complete. The Criollo cocoas take less time to ferment than the Forasteros with their thicker, tougher testas. After fermentation the seeds are dried, usually in the sun on large platforms with movable roofs to protect them from rain, dew or excessive heating by the sun. Their moisture content falls from about 50—60 per cent to about 6 per cent over a week of drying which is at first slow, though later the seeds may be exposed to the sun all day. The dried seeds are exported. Apart from sugars, starch, theobromine and essential oils, 'cocoa beans' (the dried seeds) contain 50—57 per cent of a pale yellow, non-drying fat known as cocoa butter. In the manufacture of chocolate the beans are shelled, roasted and ground to a mass to which extra cocoa butter, sugar and other ingredients are added. The residue of beans from which the cocoa butter has been expressed provides, after processing, the cocoa powder used for making the beverage cocoa, and for flavouring.

Maté: *Ilex paraguariensis*

Maté is the favourite drink of much of the population of South America where the leaves of wild species of *Ilex* have been used to make a stimulating caffeine drink since long before the arrival of Europeans. The leaves of numerous species of *Ilex* contain caffeine, and wild plants occur in southern Brazil, Paraguay, Uruguay and Argentina, but *Ilex paraguariensis*, which was first cultivated by Jesuit priests, is the most important cultivated species and provides most of the maté drunk in South America. The beverage is made in much the same way as tea, as an infusion of the processed, dried leaves to which sugar and lemon, but not milk, are added. Very little maté is exported from South America, nor has the plant been introduced in cultivation to other tropical countries; in other parts of the world the beverage does not compete with tea or coffee.

The genus *Ilex* is one of three in the predominantly Central and South American family *Aquifoliaceae*, better known to Europeans and Americans as the Holly family. Its members are small evergreen trees or shrubs with alternate leaves and small, unisexual flowers carried in axillary cymose inflorescences. The flowers have a calyx-tube of four sepals which persist with the fruit, four petals and four stamens, and a superior ovary which matures to form a drupe containing several one-seeded stones. *Ilex paraguariensis* is a small tree about 6 m tall, though in cultivation it is pruned to form a low-growing, many-stemmed bush. The leaves are collected from the bushes every 3 or 4 years. Whole young shoots are cut or pulled off and 'toasted' over a fire to dry them thoroughly. The leaves are separated from the twigs and ground to small pieces, and then may be kept to age and develop flavour before they are used. The drink made from them contains about 2 per cent of caffeine, and is a mild stimulant with slight laxative properties; it has a pleasant smell, but a slightly bitter taste.

The masticatories

Kola: *Cola* spp.

Kola 'nuts' are the cotyledons of some species of *Cola*, a member of the *Sterculiaceae*, and have been a popular masticatory in West Africa for at least 800 years. About 50 species of the genus grow wild as trees in the coastal rain forests of West Africa, and among several sub-genera recognized by taxonomists the sub-genus *Cola* includes seven species which have

Fig. 8.8 *Cola* sp. A pendent cluster of mature kola follicles, each containing several large seeds. (By courtesy of the Cocoa Research Institute of Nigeria.)

edible seeds. By far the most important of these is *Cola nitida*, 'gbanja' kola, which originated in Ghana and the Ivory Coast. It was taken to the West Indies in the seventeenth century, and has more recently spread to other parts of the humid tropics, but it has not become important outside West Africa. A second less important cultivated species is *C. acuminata*, 'abata' kola, which has a natural distribution from Nigeria to Gabon. A great expansion of the number of kola trees cultivated in West Africa took place at the beginning of this century, and south-western Nigeria, especially those areas where cocoa is the major crop, became the centre of production. Annual production of kola nuts in Nigeria is now more than 110,000 tonnes from some 18 m. trees, 90 per cent of them *C. nitida*, and smaller amounts are produced in several other West African countries. An important feature of kola nut production in West Africa is the large trade in 'nuts' between centres of production in the forest zone near the coast and the peoples of the savanna zone hundreds of miles inland. Though tobacco, coffee and tea are becoming increasingly important among urban communities, kola nuts remain the principal stimulant used by the Muslim population of West Africa, and are important in social and religious customs among them.

Cola nitida is an evergreen tree, commonly around 12 m tall, though exceptionally taller than 20 m, with a smooth trunk which sometimes has short, narrow buttresses at its base, and spreading branches which carry dense foliage. The leaves are around 20 cm long, and often oblong or elliptical, though their shape is very variable; they have an acuminate tip, prominent lateral veins, and are carried on long petioles which have pulvini at their base and tip. The tree grows in flushes which occur two to three times each year. Male and hermaphrodite flowers occur together in axillary inflorescences on recent growth, but not until it is more than 5 months old. The flowers have no corolla, but the five perianth segments, which are fused for about one-third of their length, are creamy white with red markings inside their base. In male flowers, which are about 2 cm in diameter, the filaments are fused to form a column so short that the ten red anthers appear to be sessile at the base of the perianth. In hermaphrodite flowers, which are around 5 cm in diameter, the anthers surround the base of the ovary, and though their pollen is viable they do not dehisce so that hermaphrodite flowers function as females. The ovary consists of five to seven free carpels, each of which develops into a smooth, green dehiscent follicle up to 13 cm long. The groups of follicles are pendent on a short stalk; each follicle is curved upward with a prominent keel on its lower, convex surface, a short, curved beak at its tip, and contains four to twelve large seeds which are up to 5 cm long with red, pink or white cotyledons. In *C. nitida* each seed has two cotyledons, whereas in other species with edible seeds there are commonly three or more. Their bitter taste and stimulating properties are due to their content of tannins, about 2 per cent of caffeine and small amounts of the alkaloids theobromine and betain.

The natural environment of the kola tree is in rain forests where annual rainfall is 1,200–1,500 mm unevenly distributed with a marked dry season, and temperatures around 25°C. When they are subjected to severe water stress, as they may sometimes be when the annual dry season is unusually long, kola trees shed some or even all of their leaves. Well-established, mature trees may survive such conditions, but young trees are killed by drought. In West Africa the cultivation of kola has extended from the forest zone along the coast into regions with less rainfall and a longer dry season only along rivers and in swamps where there is an assured supply of soil water. Young kola trees, whether seedlings or rooted stem cuttings, require shade for several years until they are well established, after which they can withstand direct exposure to the sun. The annual yield of nuts from old kola trees in West Africa is very variable because all of the long established trees were propagated from seed. This practice produced heterogeneous populations because kola is predominantly out-breeding; both self- and cross-incompatibility systems occur within populations of *C. nitida* (though some trees are self-fertile), but the incompatibility mechanism is not understood. Variation of yield between trees has made it possible to select superior individuals for the vegetative propagation of high-yielding clonal cultivars, but only after single tree yields have been recorded for many years, because the yield even of good trees varies greatly from year to year. Of course, in plantations of clonal material, self-compatible, or mixtures of cross-compatible clones must be used.

Betel-palm and betel-pepper: *Areca catechu* and *Piper betle*

The seeds of the betel-palm, *Areca catechu*, sliced and mixed with spices (cardamoms or cloves) and sometimes with tobacco, then wrapped in a leaf of betel-pepper, *Piper betle*, which has been smeared with slaked lime form a 'quid' which has been a very popular masticatory in the Indian sub-continent, Sri Lanka, South-east Asia and the central Pacific for more than 2,000 years. Chewing the betel quid stimulates salivation due to the hot, astringent nature of the betel-pepper leaf, and has a mildly narcotic and stimulating effect because betel-palm seeds contain small amounts (less than 0.5 per cent) of the alkaloid arecoline as well as smaller quantities of other alkaloids.

Areca catechu is a slender, erect palm up to 30 m tall with a smooth trunk and a crown of large pinnate leaves 1—2 m long. The palm is monoecious with the number of male flowers greatly exceeding the females on the inflorescence, which is a much branched spadix typical of the family *Palmae*. The fruit is an ovoid drupe up to 5 cm long, orange coloured when ripe, with a hard fibrous endocarp and mesocarp and a single seed, commonly called a 'nut'. The seeds are used at all stages of development, and are chewed fresh or after curing which decreases their content of tannins and improves their colour, palatability and storage properties. *Piper betle* is a perennial creeping or climbing vine native to the tropical rain forests of the Malaysian peninsular. It belongs to the family *Piperaceae* which includes *Piper nigrum*, pepper, grown for its spicy fruits. Betel-pepper is an evergreen with broad, pointed, dark-green simple leaves 3—4 cm long carried alternately and subtended by deciduous stipules.

Drugs and fumitories

Tobacco: *Nicotiana tabacum*

The stimulating and slightly narcotic effects of smoking, snuffing or chewing the dried, cured and fermented leaves of tobacco have led to the almost universal use of this New World plant. The smoking habit, and tobacco is an habituating narcotic, whether of pipes, cigarettes or cigars, or the taking of powdered tobacco as 'snuff', is found in every part of the world in spite of considerable opposition over the years from State rulers and religious leaders, and despite the recent demonstration of an association between cigarette smoking and ill-health, especially heart disease and lung cancer. Tobacco is native to tropical America where it was used for chewing, smoking or as snuff long before it was discovered by Columbus in the West Indies in 1492. The Spanish carried tobacco from Mexico to the court of Spain in 1558, and Sir John Hawkins introduced it to England from Florida in 1565, though it was Sir Walter Raleigh who brought the pipe-smoking habit to England in 1585, where it immediately became popular. One of the last things Raleigh did before he was beheaded was to call for his pipe! Cigarette smoking did not become widely popular until soldiers acquired a taste for them from the Turks during the First World War. The crop was introduced to Turkey in 1610, and to Virginia in 1612; it was spread widely to India, Asia and Africa by the Spanish and Portuguese at the beginning of the seventeenth century. The narcotic and relaxing properties of tobacco are due chiefly to the presence of 1–3 per cent of the alkaloid nicotine in the processed leaf, though the amount varies with differences between *Nicotiana* species and cultivars and the environments in which they are grown. The smell and flavour of tobacco are due to the liberation of various essential oils and other aromatic substances during the curing process, but they vary with the method used. Though the most important component of tobacco smoke is nicotine, it contains numerous other liquids, solids and gases which determine its smell and flavour, as well as its hazard to health. Among these are resins, phenols and pyridine compounds and various gases, some of them poisonous (e.g. carbon monoxide). Flavourings such as rum, apple juice, menthol and sugars may be added to the leaf when it is processed.

Nicotiana is a member of the economically important family Solanaceae, which is described in the chapter on vegetables. Two species are cultivated. *Nicotiana rustica* is no longer important except in India, but it was the common tobacco of the West Indies and Mexico in the fifteenth and sixteenth centuries, and the one first introduced to Europe and Virginia. Its leaves have a higher nicotine content than *N. tabacum*, and in recent times it has been grown mainly as a source of nicotinic acid, and of nicotine for use as an insecticide. *Nicotiana tabacum* soon replaced *N. rustica* as a fumitory when tobacco cultivation and the smoking habit spread in the seventeenth century. Both species are tetraploids

($2n = 4x = 48$), probably allotetraploids, though their diploid ancestors have not been identified with certainty, and neither of them occur as wild plants. *Nicotiana tabacum* may have originated from the natural hybridization of the ancestors of the diploid species *N. sylvestris* and *N. otophora*, perhaps in the region of middle South America where these species now grow wild together. It is a coarse, herbaceous, short-lived perennial cultivated as an annual, with a rosette-like growth habit when young, later producing a stout, erect main axis around 1.5 m tall, though rarely as tall as 3 m. The seedling has a single main root which, if undamaged, becomes a strong tap root with extensive lateral development; but the tap root is often broken when seedlings are transplanted, and an extensive mass of fibrous roots develops instead. The stem is covered in glandular hairs, and tends to become woody at the base. It bears large, simple, ovate leaves arranged spirally, though the phyllotaxy is variable; the leaves vary greatly in size, thickness, texture and in the prominence of their veins, depending upon differences between cultivars. They are commonly pointed with entire but somewhat undulate margins, about 50 cm long and sessile, though in some cultivars the leaves have short, winged petioles. Each cultivar produces a characteristic number of leaves commonly in the range 20—30, though rarely as many as 100. The slightly irregular flowers are in terminal panicles of racemes, which may be short and compact or more or less elongated. Each flower has a short peduncle and is subtended by a bract. Five sepals are fused to form a calyx tube about half as long as the corolla; it is covered with soft white hairs and expanded above into four to five, often unequal, pointed lobes. The corolla tube, which is also densely hairy outside, is up to 5 cm long, pink or white, inflated in the middle and spreading widely at the top where there are a number of indistinct lobes. Five stamens are inserted at the base of the corolla tube, each with a long filament and a small anther which dehisces longitudinally. The filaments are not all the same length; the longest pair carry their anthers to the mouth of the corolla tube, a second pair is slightly shorter, while the fifth filament is shorter still. Two carpels are fused in a superior ovary which has two locules and a massive axile placenta with many ovules. The long slender style carries a capitate, two-lobed stigma to the mouth of the corolla tube. The anthers dehisce and the stigma is receptive when the flower opens so that self-pollination is the rule. Up to 4 per cent crossing does occur because bees and other insects visit the flowers for nectar, so if breeders wish to ensure inbreeding they place a bag over the inflorescence before the flowers open. An 'S' allele gametophytic incompatibility system operates in some tobacco; pollen fails to germinate, or pollen tube growth is slow if the haploid pollen grain contains the same S allele as either of those in the diploid tissues of the style.

The fruit is a two-valved ovoid capsule which is almost completely covered by the calyx. It dehisces longitudinally to release as many as 8,000 minute dark-brown seeds which are elliptical with a reticulate surface and a prominent raphe along one side. Tobacco seeds have been known to remain viable 20 years or longer. When they germinate the testa bursts open near the micropyle, the radicle emerges and the hypocotyl grows rapidly, carrying up the cotyledons and stem apex still enclosed in the testa. Tobacco is propagated by seed which has been sterilized to control pathogens, and which is sown densely on to seed-beds of sterilized, fertile soil with a fine tilth. The seedlings are shaded in the nursery, but the amount of shade is gradually decreased to harden them before they are transplanted to the field 6–9 weeks after sowing. Tobacco is one of the most lucrative of the tropical crops, but it is expensive to grow; it requires skilled management, much hand labour, heavy manuring, careful harvesting and facilities in which the harvested leaf can be subjected to carefully controlled and sometimes prolonged curing and fermentation. Though tobacco is sensitive to minor variations in the soil which not only determine the quality of the leaf produced, but also the kind of tobacco which can be grown (cigarette, cigar filler, wrapper or binder), it is one of the most adaptable of tropical crops, and grows well in the sub-tropics, and even in temperate climates (it is killed by frost). In general tobacco demands a well-drained, acid, loamy soil rich in potassium, calcium and organic matter. The optimum temperature for the crop is around $25°C$, and it requires a moderate rainfall of around 400 mm during the 3–4 month growing season followed by dry weather when the leaves are maturing; the crop needs protection from strong winds. Tobacco is grown in a remarkable variety of localities, from the banks of the Nile and in the Persian Gulf States to the home counties of England. Annual world production of all kinds of tobacco leaf is greater than 4.5 m. tonnes, more than 2 m. from Asia (mainland China about 1.0 m., India 0.4 m.), around 1 m. from North and Central America (most from the United States), 0.6 m. from Europe, 0.35 m. from South America and 0.2 m. from Africa.

The vegetative characters of cultivated tobacco are considerably modified by the practice of 'topping' in which the terminal bud is removed once the desired number of leaves has been produced. As a consequence of the removal of the apical bud lateral shoots develop, first from leaf axils near the top of the stem, then progressively further down. These are also removed as they appear, usually by hand in two operations, though chemical sprays are increasingly used to retard lateral bud development, and so to save expensive labour. The removal of buds ensures that the maximum amounts of nutrients and assimilates are used by the leaves, which become abnormally large and contain more nicotine than the leaves of untopped plants. As a result of topping the leaves also tend to ripen more evenly. Two methods of harvesting the leaves are common, depending upon the kind of tobacco grown and the way in which it is to be cured. Individual leaves are gathered as they become ripe for flue-cured cigarette and cigar wrapper leaf. The time to maturity of the lower leaves on the plant, which are the first to ripen, varies widely between cultivars from 3–4½ months after sowing. As they mature their colour changes gradually from dark-green to greenish-yellow, and they become more brittle and rougher to the touch. Experienced workers know by their colour and feel which leaves are ready for harvest, and confirm their judgement by the way the mid-rib of a mature leaf snaps. For air and fire-curing the whole plant is harvested when most of the leaves near the middle of the stem are ripe. In general the harvested leaf contains 70–80 per cent of water, variable amounts of starch, various nitrogenous compounds, oxalic, citric and malic acids as well as 1–8 per cent of nicotine, with the finer tobaccos having least.

The harvested leaves are cured by subjecting them to varying temperatures and degrees of humidity for some time. The leaves are stacked on racks in barns without heat (air-curing), over the smoke of fires (fire-curing), or over heated air (flue-curing). During the first part of the process the air is kept fairly moist so that the leaf dries out slowly; later the curing barn is ventilated and moisture is removed as quickly as possible while the temperature in heated barns is kept high. During early curing and the slow drying of the leaves respiration and enzyme action proceed with little interruption, water is lost slowly, and as the chlorophyll of the leaf gradually disappears the leaf becomes yellow. In the later stages of sun-, air-, or fire-cutting the leaf becomes much darker, and changes to reddish-brown. Most of the starch and sugars and up to 60 per cent of the protein in the leaf are lost during curing, some of the protein as ammonia. The amount of citric acid increases, some malic acid is broken down, whereas the amount of oxalic acid remains more or less unchanged. Nicotine occurs in the leaf in two forms, volatile and fixed; during curing the volatile nicotine is lost, so that total nicotine decreases. Fire-curing takes 3–4 weeks and air-curing 6–8 weeks, whereas modern methods of flue-curing take only 4–6 days during which time the leaves retain their bright yellow colour, and their sugar content remains practically unchanged.

After curing the leaves are heaped and allowed to ferment for a period of 4–6 weeks, using techniques which vary with the ultimate use to which the tobacco is to be put. Care must be taken that the temperature does not rise too high during fermentation, so the piles of leaves are rearranged and disturbed at intervals during the process. Fermentation brings about further changes in the chemical composition of the leaf and results in a further decrease in nicotine content, while at the same time the characteristic odour and flavour of tobacco develop. Once cured and fermented the leaves are graded according to quality, tied in small bundles, then packed into bales in which they are left to mature in store for up to 2 years before use. Several classes of tobacco are recognized depending upon the way the leaf was cured and upon whether it is to be used for cigarette manufacture, for cigar filler, binder or wrapper, or for snuff or chewing. Certain tobacco-growing countries have a reputation for producing particular kinds of leaf, the United States and Canada for flue-cured cigarette tobacco, Puerto Rico and Cuba for cigar tobacco, Sumatra for cigar wrappers, and Greece and Turkey for air-cured 'Turkish' tobacco.

Bhang, ganja, marijuana, hashish: *Cannabis sativa*

The botany of hemp, *Cannabis sativa*, is described briefly in the chapter on vegetable fibres. It is mentioned again here because some cultivars have been cultivated for thousands of years in China, India and the Middle East for the hallucinogenic drug which occurs in resins produced by glandular hairs on the leaves, stems and inflorescences, especially on the inflorescences of female plants. The active principal in *Cannabis* resin is the alkaloid tetrahydrocannabinol. The drug is consumed in several ways, and has several names depending upon the part of the plant used and the way it is consumed.

Hashish or charas is the crude resin which is obtained mostly from female inflorescences by gathering the glandular hairs on cloth, to which they adhere. The resin is a dark-green viscous substance when fresh, but becomes a dark-brown, brittle solid with age. It is sometimes eaten, but is usually heated until it vapourizes and can be inhaled from a pipe.

Marijuana (or marihuana), bhang and ganja are names for dried preparations of young inflorescences (sometimes including young leaves). Bhang and ganja are commonly drunk as a concoction of the powdered plant material in milk or water, but all these forms of the drug are also smoked.

Pot and grass are recent terms which describe dried leaves, petioles and even pieces of stem which are smoked in pipes, or in cigarettes sometimes called 'reefers' or 'joints'. These forms of the drug are the least potent.

Legislation forbids the growing, possession, sale or use of *Cannabis* in many countries, though it appears to be non-addictive and some authorities believe it to be less injurious to health than either tobacco or alcohol.

Quinine: *Cinchona* spp.

Quinine is one of the four alkaloids, together called totaquina, which are extracted from the bark of the trunks, branches and roots of a few species of trees in the genus *Cinchona*, and used in medicine as a cure for malaria. No other drug has been of such great value in the tropics, and the story of the discovery of quinine and of the spread of *Cinchona* as a cultivated crop is a fascinating one. The medicinal properties of the thick grey bark of *Cinchona* trees were realized by the Spanish when they penetrated into South America after their conquest of Mexico in the sixteenth century. Perhaps they learned of its value from the Andean Indians who had used it long before the arrival of the Spanish. Legend has it, unfortunately wrongly, that the wife of the Spanish Viceroy to Peru, Countess Cinchon, was cured of malaria by the drug.

Cinchona bark was taken to Spain by Jesuit priests in 1639, and as the source of a cure for the 'ague' (malaria) became widely known in Europe as Jesuit's or Peruvian bark. For at least 200 years after its introduction to Europe the only source of *Cinchona* bark was the Andean Mountain rain forests, from Bolivia to Colombia, where wild trees were felled in large numbers, especially during the first half of the nineteenth century. The destruction of wild populations of trees to gather bark stimulated the Dutch and British, despite considerable local opposition, to collect seeds and seedlings for European botanic gardens, and then to establish extensive *Cinchona* plantations in Java and India during the second half of the nineteenth century. Later, at the beginning of this century, the crop was introduced into East Africa. Until the Second World War quinine was the only drug known which effectively controlled the malarial parasite in the human blood stream; so when war interrupted supplies from Java in the 1940s demand for the drug to treat malaria in soldiers fighting in the tropics stimulated renewed exploitation of wild trees in South America, and extensive planting in Central and South America and parts of Africa. Then, not only was expensive synthetic atabrin produced, but other synthetic malarial prophylactics like paludrine and chloroquine became available cheaply enough to limit the expansion of natural quinine production from *Cinchona* trees. Finally, during the 1960s, it became apparent that the synthetic prophylactics were less effective in controlling malaria than they had once been thought to be; sometimes they suppress but do not kill the parasite in human blood. So the production of natural quinine from *Cinchona* bark is once more increasing.

Cinchona is a member of the large family *Rubiaceae* in which the only other economic genus is *Coffea*. The members of the family are mostly tropical trees and shrubs characterized by their opposite or whorled, simple leaves, their gamopetalous tubular corolla and epipetalous stamens, and by their inferior ovary which commonly has two united carpels. The two important genera give their names to two tribes of the family, the *Cinchonoideae* with many ovules in each locule of the ovary, and the *Coffeoideae* with only one ovule per locule. Cinchona is a genus of small, mostly evergreen trees. The economic species have a limited natural distribution in the rain forests of the Andean Mountains near the equator in South America where they grow on the porous, well-drained acid soils of steeply sloping valley sides around 1,000–3,000 m above sea level. They thrive where temperatures are moderate in the range $15°–20°C$, and where annual rainfall is greater than 2,000 mm, conditions which in other parts of the tropics are well suited for the cultivation of tea. The genus includes several species, but their relationships, taxonomy and nomenclature are not well defined because natural hybrids between species occur, as well as mutant forms which appear to have become stabilized. Four cultivated species are commonly recognized, though perhaps two of them are no more than botanical varieties of the same species.

Cinchona calisaya is the source of Yellow, Peruvian or Calisaya bark, and though it has more exacting environmental requirements than other cultivated species, it gives the largest yields of totaquina and of quinine. It is a large tree with light-grey bark and opposite, deciduous leaves which are oblong, thick and smooth with slightly pointed tips. Its flowers are pink, and its ovoid capsules are about 12 mm long. *Cinchona ledgeriana*, though given specific status, is probably no more than a variety of *C. calisaya*; it is a weaker, more straggling tree with much branched stems, fairly thick pale bark and smooth, pointed leaves. Its flowers are yellow. *Cinchona ledgeriana* originated from seed gathered against the wishes of the local Andean Indians by the British Resident in Bolivia, the Australian, Mr Ledger. He was subsequently killed for doing so, but not until he had sent seed to London, where the Dutch bought some which they used to establish *Cinchona* in Java. *Cinchona ledgeriana* is now the common 'species' in Java and in India where continued selection has produced improved populations which yield 'Ledger' bark containing up to 16 per cent of quinine, though the amount varies with the environment in which the crop is grown. It is least in warm climates at low altitudes.

Cinchona officinalis provides the Crown or Loxa bark of commerce and is grown at higher elevations than *C. calisaya* or *C. ledgeriana*. It is a slender tree up to 10 m tall with rough brown bark containing less than 2 per cent of quinine. The opposite leaves are narrower and smaller than those of *C. calisaya*, and are smooth and shining with reddish petioles. This species has dark pink flowers and relatively large oblong capsules up to 2 cm long.

Cinchona succirubra is a much taller tree than the other cultivated species, reaching a height of 30 m. It is hardy and can be grown in a wider range of environments than other species, but it is difficult to extract the quinine from its bark. The bark is dark-brown, occasionally streaked with white and cracked horizontally; it is the red bark of commerce, but the tree is little grown except as a root stock to which scions of *C. ledgeriana* are budded or grafted. The large leaves are light-green, the flowers are pale pink and the fruits are up to 3 cm long.

Apart from differences between species in the form and size of the trees, in the size, colour and shape of their leaves, and in the colour of their flowers, the botanical characteristics of cultivated species of *Cinchona* are much the same. The opposite leaves are carried on short petioles subtended by deciduous stipules, though there is only one pair of stipules between each pair of opposite petioles. The inflorescences are terminal panicles of regular, hermaphrodite flowers, each flower with a tubular, four to five lobed calyx, and a tubular corolla which expands at the top into five lobes. There are five stamens adnate to the corolla tube and alternating with the corolla lobes. *Cinchona* is heterostylous with two kinds of flowers, but only one kind occurs in one tree. Flowers with styles longer than the corolla tube have exserted stigmas, but the filaments of their anthers are shorter than the corolla tube. In the other kind of flower the anthers are exserted on long filaments, but the styles are short. Not only are the flowers thus morphologically adapted to cross-pollination, but they are self-sterile, so *Cinchona* is out-breeding and heterozygous. The inferior ovary has two locules; it matures as an elongated capsule which dehisces from below by two valves to release the small, winged seeds which are dispersed by wind.

In the past the crop was propagated from seed, but in modern plantations it is propagated vegetatively by grafting buds from selected trees on to 6-month-old *C. succirubra* root stocks. Budded root stocks are raised in a shaded nursery for 1 or 2 years before they are transplanted to the field at populations of about 6,000–7,000 plants per hectare. There is considerable variation between trees in a plantation of *Cinchona* which was raised from seed because the crop is outbreeding. On the other hand vegetative propagation produces a homogeneous crop made up of one or few selected superior genotypes. Harvest begins when the trees are 4–6 years old, and continues with progressive thinning of the plantation until the last trees are uprooted when they are 10–12 years old. The trees are uprooted, not cut down, because the bark of the roots contains more alkaloids than the branches, and the roots of some clones yield more bark than the aerial parts. After it has been beaten to loosen it, the bark is stripped from the branches, trunk and roots, then dried quickly in the sun or in ovens where heating is carefully controlled. The alkaloid content of the bark varies with locality, even within a species, and from tree to tree in plantations propagated from seed. The largest yields of totaquina are from trees grown at altitudes higher than 1,000 m, though the crop can be grown at lower elevations. The most important alkaloid in *Cinchona* bark is quinine; the others are quinidine, cinchonine and cinchonidine which occur in different proportions in different species. The bark of *C. ledgeriana* is not only richer in total alkaloids than other species, it also contains the largest proportion of quinine. The alkaloids are extracted from powdered, dry bark in solvents such as amyl alcohol or ether.

Opium: *Papaver somniferum*

The habit of eating or smoking opium for its strongly hallucinogenic and analgesic effects has been widespread in Asia for centuries, but especially after its use was promoted by Europeans for gain in relatively recent times. Crude opium is obtained from the young capsules of the opium poppy, *Papaver somniferum*, of the family *Papaveraceae*. Soon after the flowers open the petals are removed, and a few days later the immature capsules are wounded with a number of shallow incisions from which a white latex oozes. The latex hardens and is scraped from the outside of the capsule as crude opium. Its narcotic properties are due mainly to its content of the alkaloids morphine, codein, narcotine and papaverine (heroin is a derivative of morphine). Both morphine, which has not been synthesized, and codein which has, are widely used in medicine for the relief of pain; indeed, opium was commonly used in Europe for that purpose in the eighteenth and nineteenth centuries.

The opium poppy originated in Asia Minor from the wild species *Papaver setigerum*, and spread throughout Asia more than 1,000 years ago. The plant is a quick growing, more or less glabrous or slightly bristly erect annual herb, up to 1 m tall with slender stems and alternate, coarsely toothed leaves. The lower leaves are large and petioled, but the upper ones are smaller and sessile. Two sub-species are cultivated, subsp. *somniferum* with large white flowers and white seeds is grown for opium, while subsp. *hortense* with black seeds and lilac flowers which have dark spots at the base of the petals is grown for the edible, drying oil which constitutes up to 50 per cent of the weight of its seeds. The oil is much used in India in the preparation of sweetmeats and curry, and the seeds are exported to Europe where the oil is extracted for edible purposes and to make soap and paint.

The regular flowers are solitary in the leaf axils on long peduncles. Two deciduous sepals protect the flower bud, but they are shed before the flower opens. Four, thin rounded petals spread up to 18 cm apart when they open to expose numerous stamens and a superior, globular ovary which has a large sessile stigma. Small ovules are on numerous parietal placentas which extend into the middle of the ovary. The fruit is a globular, flat topped pale-green capsule up to 5 cm long carrying the persistent remains of the stigma; it dehisces by horizontal slits which open around its upper rim, between the placentas. Poppies are propagated by broadcasting their very small seeds which may have to be resown several times to ensure that a sufficient number of plants become established.

Other tropical drug plants

Of the numerous other tropical plants known for their medicinal or stimulating properties, and which are cultivated or which grow wild, few are so important to merit detailed description in this brief survey. On the other hand, we must acknowledge that there is much to learn of hundreds, even thousands, of plants whose medicinal use is part of the folklore of inhabitants of the tropics all over the world.

Erythroxylon spp. coca is a genus of small South American trees or shrubs in the family *Erythroxylaceae*. The leaves of two species, *E. coca* from high altitudes, and *E. novogranatense* from lower, warmer altitudes, are used as a masticatory in South America in much the same way as betel-pepper leaves in Asia; but they have a much greater physical and mental effect than betel, and addiction to chewing coca leads to serious physical deterioration. The leaves of 'coca' contain about 1 per cent of the alkaloid cocain which is extracted with solvents for use in medicine as a local anaesthetic. The crop is propagated with difficulty from seed sown in shaded nurseries; after about 9 months seedlings are transplanted to the field where they grow into much-branched bushes with simple, alternate, elliptical leaves on short petioles with prominent stipules. The small white flowers are in axillary clusters, and have a persistent calyx of five united sepals, five free deciduous petals, ten stamens in two whorls and a superior ovary of three carpels with three styles. The first harvest is taken when the bushes are 2–3 years old and about 2 m tall. The leaves are picked individually by hand, cured and dried in the sun or in ovens, then powdered and packed in boxes for export or, in South America, for local use. Plantations of *E. novogranatense* in Sri Lanka and Java supply much of the coca leaf in international trade.

Rauvolfia serpentina of the *Apocynaceae* is a small perennial Indian shrub grown for its roots, which contain the alkaloid reserpine. It has long been used in India for the treatment of mental illness, and more recently reserpine has been used in medicine elsewhere as a tranquilliser in the treatment of hypertension. *Rauvolfia* is propagated by seed or vegetatively with stem cuttings, and is harvested after 2 or 3 years' growth.

Some species of the genus *Aloe*, a member of the *Liliaceae*, are cultivated for their thick, fleshy leaves from which the drug aloes of commerce are obtained. The genus includes 200 or more species, mostly native to tropical and South Africa, which resemble the Agaves (e.g. sisal) in general form, having a short thick stem and often spiny leaves, and an inflorescence produced on a long erect axis, or pole. Many species have beautiful flowers and are grown as ornamental plants. The resinous latex which exudes from the cut leaves has been used for centuries as a laxative, and more recently to treat X-ray burns. *Aloe ferox*, the Cape aloe, is cultivated in South Africa; *A. perryi*, the Socotrine or Zanzibar aloe, is grown on the island of Socotra, and yields the mildest laxative drug, while *A. barbadensis* is grown in the West Indies.

Strychnos nux-vomica is the only important member in a tropical genus of about 100 species of woody plants in the family *Loganiaceae*. It is a moderate-sized, deciduous tree found in India, Burma and Sri Lanka. It grows to a height of 12 m or more, and is often spiny; its large, opposite leaves are simple and ovate, and its small white flowers are in terminal cymes at the ends of branches. The flat, button-like grey seeds yield two powerful alkaloids, strychnine and brucine, which are used in medicine to treat nervous disorders. Other species of *Strychnos* are used in South America as one important ingredient in secret recipes for the preparation of a paralysing poison called curare which is used on blow-gun darts and arrows.

Further reading

General

Purseglove, J. W. (1968 and 1972). *Tropical Crops*, 2 vols, London: Longman.

Beverages

Tea

Carr, M. K. V. (1972). The climatic requirements of the tea plant: A review, *Exptl. Agric.*, 8, 1–14.

Chenery, E. M. (1966). Factors limiting crop production: 4. Tea, *Span*, 9, 45–8.
Eden, T. (1976). *Tea*, 3rd edition, London: Longman.
Harler, C. R. (1964). *The Culture and Marketing of Tea*, 3rd edition, London: Oxford University Press.
Tea. The journal of the Tea Boards of East Africa.
Tea Quarterly. The journal of the Tea Research Institute of Sri Lanka.

Coffee

Haarer, A. E. (1962). *Modern Coffee Production*, 2nd edition, London: Leonard Hill.
Wellman, F. L. (1961). *Coffee*, London: Leonard Hill.

Cocoa

Toxopeus, H. (1968). Establishment of cacao clones in Nigeria, *Euphytica*, 17, 38–45.
Wood, G. A. R. (1975). *Cocoa*, 3rd edition, London: Longman.
Wills, J. B. (ed.) (1962). *Agriculture and Land Use in Ghana*, London: Oxford University Press.

Maté

Porter, R. H. (1950). Maté – South American or Paraguay tea, *Econ. Bot.*, 4, 37–51.

Drugs, fumitories and masticatories

General

Grover, N. (1965). Man and plants against pain, *Econ. Bot.*, 19, 99–112.
Purseglove, J. W. (1968 and 1972). *Tropical Crops*, 2 vols, London: Longman.

Cinchona

Hodge, W. H. (1948). Wartime cinchona procurement in Latin America, *Econ. Bot.*, 2, 229–57.

Opium

Bengt Loof (1966). Poppy cultivation, *Field Crop Abstracts*, 19, 1–5.
Duke, J. A. (1973). Utilization of *Papaver*, *Econ. Bot.*, 27, 390–400.

Coca

Martin, R. T. (1970). The role of coca in the history, religion, and medicine of the South American Indians, *Econ. Bot.*, 24, 422–38.

Cannabis

Joyce, C. R. B. and Curry, S. H. (eds) (1970). *The Botany and Chemistry of Cannabis*, London: Churchill.
Quinby, M. W. *et al.* (1973). Mississippi grown marihuana – *Cannabis sativa*. Its cultivation and morphological variations, *Econ. Bot.*, 27, 117–27.

Tobacco

Akehurst, B. C. (1966). Tobacco. A review of some recent developments with particular reference to flue-cured tobacco, *Field Crop Abstracts*, 19, 245–52.
Akehurst, B. C. (1968). *Tobacco*, London: Longman.
Goodspeed, T. H. (1954). The genus *Nicotiana*, Waltham, Mass.: *Chronica Botanica*.

Betel

Raghaven, V. and Baruah, H. K. (1958). Arecanut. India's popular masticatory — history, chemistry and utilization, *Econ. Bot.*, 12, 315–45.

Kola

Van Eijnatten, C. L. M. (1969). *Kola: Its Botany and Cultivation*, Amsterdam: Royal Tropical Institute.

Chapter 9

The Spices

The development of the culinary art has been accompanied by an increasing use of various vegetable spices and flavourings, and it is from this group of substances that much of the flavour and taste of modern table dishes is derived. The use of spices to add piquancy to diets in the tropics has long been established, and the everyday food of the inhabitants of the tropics is more highly spiced than that of people in temperate regions. Where the staple diet consists of tropical cereals or roots the use of spices becomes important if palatability is to be maintained, and monotony in the diet avoided. Many spices assist digestion, and some also have medicinal properties. In the tropics the more pungent spices are used liberally to flavour cooked foods, and in many cases to disguise the slightly unpleasant taste of dried meats, thereby increasing their palatability, and at the same time causing increased perspiration with a subsequent fall in body temperature. Spices are still used extensively throughout the temperate regions of the world, but much less than in the past when diets were much less sophisticated and varied. Now the more complex dishes of the true gourmet depend to a large extent for their success on the discriminating use of spices, while everyday meals of Europeans or Americans are often given added piquancy by the use of spicy sauces and ketchups.

The spices owe their peculiar flavouring properties to their content of essential oils, and are obtained from various parts of a variety of plants. The majority of these plants occur in the wetter parts of the tropics, a large proportion in the smaller islands rather than in the great land masses. They provide valuable exports as well as supplying the large and constant demand of the local people. The more important spices are discussed below.

Ginger: *Zingiber officinale*

Ginger is obtained from the rhizomes of *Zingiber officinale*, a herb of the wet tropics which grows best in regions of high rainfall and fairly high temperature, but often in shady places. The plant was domesticated in Asia or India, where it has been cultivated since ancient times; it is also grown now in parts of West Africa and in the West Indies, Jamaica

Fig. 9.1 *Zingiber officinale*: Ginger. Weeding ginger in Liberia. (By courtesy of the FAO.)

producing the finest quality ginger known in world trade. The plant is not known to occur wild. It was referred to by ancient Chinese and Sanskrit writers for its medicinal properties, and was known to the Greeks and Romans. Ginger was in common use in Europe during the ninth and twelfth centuries, and is today used extensively as a flavouring in food and beverages, for medicinal purposes and as a constituent of some curry powders.

Zingiber officinale is a coarse perennial monocotyledon, and a member of the family *Zingiberaceae*. The underground parts consist of numerous short, hard rhizomes, often palmately branched, their form depending partly on the texture of the soil in which they are growing. Straight, undeformed rhizomes, which are the most valuable in trade, are produced only in loose, friable soil which offers little mechanical resistance to their development. The rhizomes are covered with small scales and bear numerous fine, fibrous roots which branch repeatedly in the surface soil. The rhizomes are white or buff, and give rise at intervals along their length to annual, leafy shoots. The aerial stems are slender, up to 1 m tall, and closely enfolded by the sheathing leaf bases. The alternate, light-green leaves are oblong and strongly pointed; they are around 15 cm long and 2 cm broad with a pronounced mid-rib, and tend to be rolled upwards. The inflorescence terminates a leafless, or rarely leafy, reproductive shoot about 30 cm long, which arises directly from the rhizome. In some parts of the world ginger rarely, if ever, flowers, while in others flowering is regular, though seeds are seldom produced. The inflorescence is a cone-like spike about 6 cm long with flowers solitary in the axils of greenish-yellow bracts. The flowers are pale-yellow with a short calyx-tube and a longer (1.5−2.5 cm) corolla tube which flares open at the mouth into three unequal, pointed lobes, the upper one curved down as a hood over the anther. There is only one functional stamen which has a short filament, two distinct pollen-sacs and a broad connective prolonged into a spur. The slender style passes between the pollen-sacs, and is held by them. The lower lip of the flower (the labellum) is formed by a large purple and yellow mottled staminode which is fused to the corolla tube; it is thought to have been derived from three non-functional stamens. The three-celled, inferior ovary contains several ovules in axile placentation, but rarely develops into a fruit. When it does it produces a thin-walled, three-valved capsule containing several small, black, angled seeds. Rather few distinct cultivars are recognized within the crop, and they are carefully preserved clones known in trade to be characteristic of the area where they are produced. For example, there are two types in Jamaica, one of high quality with white or yellow rhizomes, and an inferior type with blue rhizomes. The aerial parts of these two clones are indistinguishable.

Ginger appears on the market in two forms, with several grades of each form. Dried, cured ginger is prepared by removing the aerial parts and fibrous roots from the rhizomes, and then washing and partly boiling them. The washed rhizomes are dried in the sun for a while, shaken or peeled to remove the scales and corky skin, and then dried for a further period before being marketed. The superior types of ginger from Jamaica are carefully peeled before sun drying and boiling. The preserved green ginger of commerce comes mainly from China where the rather fleshy rhizomes are cleaned, washed and boiled briefly in water, and then with half their weight of sugar. This mixture is allowed to stand, and is then boiled again with more sugar before it is packed in jars for export.

The rhizomes contain about 50 per cent of starch, and 1−3 per cent of a volatile oil, mostly zingiberene, as well as varying amounts of fixed oils and resins which together give ginger its aromatic and pungent properties.

Turmeric: *Curcuma domestica*

Turmeric is a spice obtained from the rhizomes of *Curcuma domestica*, a plant much like ginger, and belonging to the same family. Like ginger, turmeric is an ancient crop not known in the wild state. It presumably originated in Southern Asia, but has never become as important as ginger for flavouring food. It has a characteristic musty taste and smell, and in India is used extensively in Hindu religious ritual; it is the common yellow constituent of curry powders, and because it is cheap it is used as an adulterant and colouring matter in other spices, such as mustard. The rhizomes produce a yellow-orange dye which is used in India, China and parts of Europe to dye cotton, silk and wool fibres without the need to use a mordant. It is also used as a cosmetic in parts of Southern Asia.

The genus *Curcuma* includes about 70 species, some ten or twelve of minor economic importance as local medicinal plants, flavourings, perfumes, dyes or sources of starch, but the most important is turmeric. Its scaly, much branched rhizomes are thicker and more rounded, blunter and coarser than those of ginger. They have a closely ringed, corky epidermis and a bright orange flesh with a distinctive smell and taste. The rhizomes branch at right angles and bear fibrous adventitious roots which ramify in the surface layers of the soil. The leaves arise in groups of six to ten on pseudostems up to 1 m tall formed by the sheathing bases of their petioles. The laminas are broadly lanceolate and acuminate, around 40 cm long, bright-green, and with a marked mid-rib. The inflorescence is produced on a short, leafless peduncle arising from the centre of the tuft of leaves. It consists of a cone-shaped collection of numerous ovate bracts, the lower ones greenish-white, the upper ones pink, the whole 10–18 cm long and about 5 cm across. The flowers are in pairs in the axils of the bracts, one opening before the other. They are pale yellow and similar in form to those of ginger, though the lower lip of the flower is not mottled, but has a broad yellow band marking its centre. The flower of turmeric has two lateral, petalloid staminodes whereas those of ginger have only one. The fruits are rarely produced, but they too are capsules like those of ginger.

Turmeric is grown in India, China, Sri Lanka and in the East and West Indies where the climate is hot and moist, and on light soils. It is propagated vegetatively from small pieces of rhizome. The washed, peeled and powdered rhizome forms the spice; the depth of its yellow colour varies with the locality in which the plant is grown.

Curcuma amada, mango ginger, is a closely related species found wild in parts of India where it is cultivated to a small extent for its rhizomes. These have the flavour of raw mangoes, and are used in the preparation of pickles and in local medicine.

Curcuma angustifolia, the Indian arrowroot, occurs wild in the hillier parts of northern India and is also cultivated to a limited extent. The fleshy rhizomes of wild and cultivated plants are used as a source of starch.

Curcuma aromatica, wild turmeric or yellow zedoary, grows wild throughout India and is cultivated in some localities. It produces light-yellow rhizomes with bright orange-red flesh which is used as a local medicine and dye.

Curcurma zedoaria, zedoary, which also occurs wild in India, has been cultivated for centuries in parts of Asia and in Sri Lanka for its rhizomes which are a source of starch once popular in Europe.

Cardamom: *Elettaria cardamomum*

Cardamom is a third spice from the family *Zingiberaceae*, but the plants are grown for their aromatic and highly flavoured seeds, not for their rhizomes. The seeds have been used for centuries throughout the Indian sub-continent and in Sri Lanka as a masticatory and as an important ingredient in curry powders. They have never been popular elsewhere, though used in Europe to a small extent in medicine as a carminative and flavouring, and to flavour cakes, biscuits and some liqueurs. In the Middle East they are used to flavour coffee. Until the nineteenth century most cardamoms were gathered from wild plants growing in the hot rain forests of western India and Sri Lanka. Now the plant is cultivated on the humus-rich soils of forest clearings in light shade, at altitudes around 1,000 m above sea level where annual rainfall is about 3,000 mm, or even where it is as great as 5,000 mm. Production is centred in southern India and Sri Lanka, but the crop is grown for export in Guatemala.

Elettaria cardamomum is a perennial herb much like ginger with a thick, fleshy and somewhat woody, irregularly shaped rhizome. Aerial stems of two kinds arise from the rhizomes. Clumps of leafy vegetative stems grow to 3 m tall with numerous alternate leaves having large, spear-like laminas and sheathing bases. The dark-green laminas are up to 90 cm long and 25 cm broad, sometimes with soft, silky hairs on the lower surface. The inflorescence is a panicle which rises from the rhizome to a height of 1 m or more, or which spreads more or less horizontally. The flowers, which are like those of ginger, arise in groups of two or three from the axils of large, green bracts on the branches of the panicle. Each flower is subtended by a secondary bract, and has a pale-green calyx-tube about 4 cm long with three spreading lobes at its tip. There is a single fertile stamen, and a large, white, spoon-shaped labellum or lip (probably three fused staminodes) streaked with purple and 15–20 mm long. The inferior ovary of three carpels has three loculi with several ovules in axile placentation. The mature fruit is a light-brown to yellow ovoid or oblong capsule, roughly triangular in section with a beaked tip. It dehisces into three papery valves which open to reveal the angular, wrinkled seeds. A mucilage from the arils hold the seeds together in the capsule. They vary in colour with cultivar, but are commonly dark-brown or black.

There are two distinct kinds of cardamoms, the Ceylon (Sri Lanka) and Malabar types which are given the status of distinct botanical varieties. Var. *major* is wild in Sri Lanka and parts of India. It has large leaves, an elongated oblong capsule and dark grey-brown seeds; its stems are streaked with pink at the base. Var. *cardamomum*, the Malabar cardamom, includes most of the cultivated forms. It has a larger panicle with more flowers than var. *major*, narrowed leaves and more nearly spherical, smaller capsules. Its stems are greenish-white at the base, its leaves much hairier below than those of var. *major* and, most important, its seeds are more aromatic.

The capsules are harvested before they are fully ripe, when they are changing from green to yellow. The seeds retain their aroma and flavour best when they are contained in the dry capsule, so immature capsules are cut from the plants with scissors, then dried in the sun to prevent them from dehiscing. They may also be bleached in sulphur fumes before they are marketed. The seeds contain 2—8 per cent of a volatile oil which gives cardamom its distinctive aroma and flavour.

Cinnamon: *Cinnamomum zeylanicum*

The spice cinnamon is the bark of the evergreen tree *Cinnamomum zeylanicum*, a member of the family *Lauraceae* which has nearly 2,000 species, mostly evergreen trees and shrubs, all of them aromatic and widely distributed in the tropics. The chief commercial products of the family are camphor, cinnamon and the avocado pear.

Cinnamomum zeylanicum is native to Sri Lanka and southern India, where it occurs wild. The tree is very sensitive to climatic and soil conditions, and though it grows well enough in other parts of the tropics, attempts to introduce it to new areas have met with little success, and Sri Lanka has always been, and remains, the world's leading producer. Trees growing on the white sandy soils of Sri Lanka produce much better quality cinnamon than those on lateritic soils. The Portuguese occupied Sri Lanka during the sixteenth century largely because of the cinnamon industry they discovered there, and subsequently the island was occupied by the Dutch, then the British, largely for the same reason. Under natural conditions the cinnamon tree is a much-branched evergreen up to 15 m tall; in cultivation careful pruning and training and the continual harvest of shoots for their bark promotes the growth of numerous suckers and produces a thick, dense bush about 2 m tall with many fairly thin, much-branched leafy stems. The large leaves are opposite and carried on short petioles; they are bright green above and a dull grey-green below, rounded at the base and bluntly pointed at the tip with three or five veins which are lighter in colour than the rest of the lamina. The leaves are tough but brittle, broad and around 10—15 cm long, but their shape and size are variable. There are no stipules.

The inflorescences are terminal or axillary, long drooping panicles carried on long, softly hairy peduncles. The small yellow flowers are carried on short branches of the panicle, and have an unpleasant foetid smell. Each flower is subtended by a small hairy bract and has a bell-shaped calyx-tube of six softly hairy, pale yellow-white, pointed segments, but no petals. Nine fertile stamens are in three whorls, and there is a fourth, inner whorl of three staminodes. The four-celled anthers dehisce by means of valves, and are carried on short, hairy filaments with glands at their base. The superior, unilocular ovary tapers to a short style; the single locule contains only one ovule. The fruits are fleshy drupes, about 2 cm long, black, ovoid and pointed with the persistent calyx-tube forming a cup round the base. The crop is propagated by seed which is obtained from inflorescences which have been bagged to protect them from birds.

The crop is harvested during the rains when the bushes are growing actively and the bark is easy to peel from the wood. Shoots are cut when they are about 2 m long and 12—50 mm thick. The bark is removed carefully by cutting into two long strips, which are then heaped, covered with sacking and allowed to ferment for about 24 hours. The outer surface of the bark is then scraped off and the strips are dried slowly. During drying the bark strips become rolled, in which condition they are called 'quills'. The quills are packed one inside another for marketing.

The use of cinnamon as a spice has declined in recent decades, though it is still used for flavouring cakes and sweets, and as a constituent of curry powders. Cinnamon oil, used to some extent in medicine, is distilled from the small chips and waste material from the peeling and scraping processes. Cinnamon leaf oil is distilled from the green leaves of the plant, which yield about 1 per cent of a brown, pungent oil containing 70—95 per cent of eugenol. The oil is used as a substitute for clove oil in the manufacture of perfumes and flavourings.

Cinnamomum camphora is a related species from which camphor is obtained. It is discussed in Chapter 12.

Pepper: *Piper nigrum*

Pepper is one of the oldest and most important of the spices, and is still used universally as a food flavouring. It was used by the ancient Greeks and was highly valued by the Romans. The search for a sea route to the 'Indies' was to some extent instigated by the European demand for spices, especially pepper, which, during the Middle Ages was of great economic importance in Western Europe. It is now used extensively as a condiment and flavouring for all kinds of savoury dishes, in preserving and pickling and in the manufacture of sauces. Both forms of the spice, black pepper and white pepper, are obtained from the drupes of *Piper nigrum*; black pepper is the ground, whole dry fruit, and white pepper is the ground dry endocarp and seed with the mesocarp removed.

Piper nigrum is a member of the *Piperaceae*, a family of about ten genera, and more than 1,000 species which are widespread in the tropics. The members of the family are characterized by their catkin-like, pendent, dense spikes of small flowers, and by the absence of a perianth, but the presence of a subtending floral bract. They have a unilocular ovary containing one ovule, which produces a small, indehiscent dry or fleshy drupe. *Piper nigrum* is a perennial, woody climbing shrub native to the south-western coastal region of India, and cultivated there and in most parts of South-east Asia, in Sri Lanka, Brazil and the Malagasy Republic. It requires a hot, wet tropical climate with a long rainy season and is more productive when grown in shade than in direct sunlight because its surface roots must be protected from the heat of the sun. In the long history of its cultivation many distinct cultivars have evolved, differing in the size of leaves, inflorescences and fruits; the majority of them have hermaphrodite flowers, though in wild plants most flowers are unisexual.

The crop is normally propagated using stem cuttings taken near the top of an actively growing young shoot. The cuttings, about 60 cm long, may be planted in a nursery or in the field near some sort of support, a pole or tree, which the flexible stems climb as they grow, being anchored by short adventitious roots which arise from the swollen nodes of the stem. The stems are dark green, or slightly purple when young, but tend to become woody with age, and covered with grey bark. The climbing stems bear no inflorescences, but at each node produce a leaf, adventitious roots and a fruiting branch. The fruiting branches have no roots. In wild plants the climbing stems may grow to a height of 10 m, but in cultivars they are kept shorter and pruning encourages the production of fruiting branches. The stems bear large, alternate, ovate leaves up to 25 cm long and half as broad. They are rounded at the base and bluntly pointed at the tip, dark-green above, but lighter coloured below, and usually have five to nine veins. The leaves have short petioles, subtended by a pair of deciduous stipules.

The inflorescences are borne opposite the leaves on fruiting branches. They are long, slender pendulous spikes of 50—100 small flowers, each partially enclosed by a fleshy subtending bract. The spikes are yellowish-green, and up to 25 cm long; the whole spike is shed at maturity. The minute flowers have no perianth, and when young are protected by their

Fig. 9.2 *Piper nigrum*: Pepper.   A spike of mature drupes. (By courtesy of the FAO.)

subtending bracts. Hermaphrodite flowers have two to four stamens and an ovate, unilocular ovary with a star-shaped sessile stigma which is at first white, but later brown and eventually, after fertilization, black. The ovary contains a single ovule. Hermaphrodite flowers are protogynous and their stigmas are receptive up to a week before the anthers dehisce. Sometimes hermaphrodite and male flowers occur together in the same inflorescence, with the male flowers at the end of the spike; monoecious and dioecious plants are even less common. Self-pollination is probably the rule in hermaphrodite flowers, but the relative importance of wind, rain and insects in pollinating female flowers is not known.

The fruit is a sessile, indehiscent drupe, more or less spherical, and about 5 mm in diameter. During development it changes from dull green when young to bright red at maturity, when it has a thin exocarp and a thin pulpy mesocarp. The dried fruit is black and wrinkled and is called a 'peppercorn'. Fifty or more fruits may develop on each spike, but they do not all mature at the same time. The spikes are harvested when some drupes are red and the rest are yellow, or still green. White pepper is produced by soaking the spikes for as long as a week in running water until the mesocarp has rotted enough to be removed easily. The endocarp with the seed inside it is then dried and marketed. Black pepper is a more pungent spice prepared by drying whole fruits in the sun, usually after soaking them briefly in boiling water. The pungency of pepper is due to the presence in the fruit of various resins, and a yellow alkaloid, piperine, which constitutes 5–8 per cent of the weight of black pepper.

Vanilla: *Vanilla fragrans*

Vanilla is obtained by fermenting the almost ripe capsules of the climbing orchid *Vanilla fragrans*, a native of Central America and Mexico where it grows wild in lowland forests. It has been known and used there for centuries and was introduced by the Spanish to Europe during the sixteenth century. In 1819 the plant was introduced to Java, and thereafter vanilla cultivation spread to various parts of the tropics wherever suitable environments were found. The present world supply comes from the Malagasy Republic, the Seychelle Islands, Réunion, Fiji, Tahiti, Hawaii and the West Indies, as well as Mexico and Sri Lanka. The Malagasy Republic is now the chief producer. Apart from its original home vanilla thus comes from tropical islands where annual rainfall of 2,000–2,500 mm is evenly distributed, where temperature and humidity are fairly high, and where heavy rainstorms during the fruiting season of the plant are rare.

Vanilla fragrans is a member of the very large and distinctive family of monocotyledons, the *Orchidaceae*. The 20,000 or so species of this family extend throughout the tropics and temperate regions, and include perennial climbing plants, epiphytes and small herbs, some saprophytes and others parasites. They are all characterized by their curious floral structure, and the extreme modification of the essential organs of the flower to effect cross-pollination. The flowers are zygomorphic (extremely so) and usually occur in racemose inflorescences, each flower subtended by a bract and having six perianth segments. The outer three segments are narrow, pointed sepals, but the inner three are large, showy petals, the two lateral ones similar in appearance, while the lower petal is much modified, variously coloured and often rolled, convoluted or greatly enlarged. It is called the labellum or 'lip'. The androecium and gynaecium are combined to form a column, or gynostemium, at the centre of the flower on an extension of the floral axis. One or two anthers are carried at the top of the column; they produce adhesive pollen grains which are collected together in masses called pollinia. In contrast the stigmatic surface is within a hollow on the lower front of the column, and separated from the stamens by a projection from it, the rostellum, so that it is extremely unlikely that pollen ever falls on to the stigma. The column has a central cavity leading to an inferior, unilocular ovary containing vast numbers of ovules on three parietal placentas. The fruit is a dehiscent capsule, and the seeds of many species are dispersed by wind.

Vanilla fragrans is the most important economic member of the family, though many species are greatly admired as ornamental plants. Vanilla is a tall, perennial vine with green, fleshy stems attached to supports by adventitious aerial roots growing from the nodes. The stems grow to considerable heights if allowed to do so, but in cultivation they are trained to remain within reach to facilitate harvest, which must be done carefully not to damage the brittle stems. The alternate sessile leaves are thick, fleshy and oblong to lanceolate, 10—25 cm long and about 5 cm broad. The parallel venation of the leaves is indistinct when they are fresh, but becomes apparent when they dry. The greenish-yellow flowers are in dense axillary racemes 5—8 cm long, each with six to fifteen flowers, sometimes, though rarely, with 30 or more. Each flower is subtended by a small pointed bract and is borne on a very short pedicel. The flower has three narrow, pointed sepals, similar to each other and about 5 cm long; two lateral petals are like the sepals, but the third central and lower one is rolled around and fused to the central column of the flower to form the trumpet-shaped lip which is lobed and hooded at its mouth. The lip is 4—5 cm long, and marked at the end by raised veins. The central column bears at its tip the single stamen with two pollen masses. The viscid stigmatic surface lies below the anther, and is protected from it by the rostellum. The fruit is a faintly three-angled capsule 9—25 cm long containing numerous minute black seeds, which are dispersed when the mature, dry capsule splits longitudinally into two valves.

In Mexico, where vanilla is indigenous, the flowers are cross-pollinated by bees, and perhaps by humming birds. Elsewhere natural cross-pollination does not occur, and flowers are artificially self-pollinated by pushing a small pointed stick into the hood of the column. The pollinia adhere to the end of the stick and are transferred to the stigma which is exposed by pressing on the column to raise the rostellum. The flowers open in the early morning and wither by evening of the same day; they are artificially self-pollinated soon after they open, and trained workers can pollinate as many as 2,000 flowers in one day. Only six to nine flowers are pollinated in each inflorescence in order to obtain a few large fruits, and prevent over-bearing.

Support for cultivated vanilla is provided by wooden trellises or, more commonly, by shade trees which serve the dual purpose of supporting and shading the vines. Trees planted as supports are species selected because they grow quickly to a moderate height, and have an open canopy of spreading branches strong enough to support the weight of the vines. Vanilla is propagated vegetatively using stem cuttings which are planted at the base of the supporting tree. As the vines grow from the cutting they are trained along the branches of the tree, and some lateral stems hang down, but are never allowed to reach the ground. Inflorescences on these pendent shoots are easy to reach for pollination and harvest. The vines begin to flower in the third year after planting. After each harvest old shoots are pruned out to promote the development of new growth. The fruits develop slowly in some localities and quickly in others, and it is 4—9 months before they begin to change from green to yellow and are ready for harvest. The freshly gathered capsules, or 'vanilla beans' as they are called, have none of the aroma or flavour which they develop as they are cured. Various methods are used in different countries to process the fruits; they all involve partial drying in the sun for some days, followed by a 'sweating' process during which the fruits ferment and the characteristic flavour and smell develop. The fruits are spread on blankets in the sun until mid-day; then the blankets are folded over them and the bundles are placed in more or less airtight boxes for the rest of the day, where the fruits sweat. This process is repeated daily for several days until the 'beans' become coffee-coloured and their odour has developed. The fruits may be immersed in very hot water for a few seconds before they are sweated to accelerate the curing process. Vanilla owes its flavour and aroma to the presence of a glycoside 'vanillin', which comprises up to 3 per cent of the weight of the fruits, and minute quantities of other aromatic substances. During sweating vanillin is derived from a glucoside by the action of enzymes, and may accumulate on the surface of the 'beans' as a crystalline deposit.

Vanilla essence is used to flavour chocolate, ice-cream, sweets and cakes, and in the production of various liqueurs and perfumes. There is still a demand for natural vanilla, despite the fact that synthetic vanilla can be made much more cheaply. The substitute is inferior in flavour and aroma, perhaps because it lacks the subsidiary aromatic substances present in the vanilla fruits.

Clove: *Eugenia caryophyllus*

The use of cloves as a spice or perfume was first recorded in China as early as the third century B.C. The spice was known to the Romans and was introduced into Europe during the Middle Ages, where it became an important and very valuable article of commerce. It remains one of the most important of the commercial spices and has a variety of uses. Cloves are presumably native to the Moluccas where they grow in a semi-wild state, and though they have been introduced to many tropical countries clove production is restricted to few places where the environment meets the rather rigorous demands of the tree. Its natural habitat in the Moluccas is in lowland, wet forests, with as much as 3,500 mm of rain each year; but in cultivation it grows best on deep, rich loamy soils where there is a maritime climate with 1,500–2,000 mm rain annually, and is most productive of high-quality produce where there is dry weather at harvest time. Around 90 per cent of the world's clove supply comes from the islands of Zanzibar and Pemba in Tanzania; they are also important in the Malagasy Republic and Indonesia.

The cloves of commerce are the unopened flower buds of the evergreen tree *Eugenia caryophyllus*, a member of the dicotyledon family *Myrtaceae* which includes about 3,000 species of tropical and sub-tropical trees and shrubs. They have opposite, simple leaves on short petioles, no stipules, and regular flowers with an inferior ovary surmounted by a calyx-tube, four to five imbricate, free petals and numerous stamens. *Eugenia caryophyllus* is a small, symmetrically shaped tree with smooth grey bark. The trunk tends to branch near the base into several forks which bear ascending branches. Wild trees grow to 18–20 m, but plantation trees are usually 7–12 m tall. The opposite leaves are lanceolate, pointed at both ends and carried on short reddish petioles. The young leaves are pale greenish-yellow, often tinged pink, but they quickly harden and when fully grown are up to 12 cm long and 2–3 cm broad, dark green above but paler below, stiff and shining with numerous veins and, on their under surface, oil glands. The presence of essential oils make the leaves very aromatic when they are crushed.

The flowers are in groups of three in many flowered terminal cymes, and are borne on short, angular pedicels. The calyx is a more or less fleshy tube with four triangular lobes at the top. The calyx, petals and stamens appear to arise from the hypanthium which surrounds the inferior ovary; but (as in *Arachis*) the hypanthium consists of the fused bases of the perianth and androecium. There are four crimson petals, closely folded around the stamens in the bud; the petals are shed soon after the flower opens. The stamens are numerous with short, slender filaments and small anthers. The inferior ovary has a single style and stigma which falls off with the stamens after fertilization. The ovary develops into an ovoid one-seeded drupe, about 3 cm long, with the remains of the persistent calyx at the top. The large, soft seed is about 15 mm long, and grooved down one side. The crop is harvested when the flower buds are fully grown and green, or beginning to turn red, but before they open. Whole inflorescences are picked by hand and the flower buds are rubbed off them, then

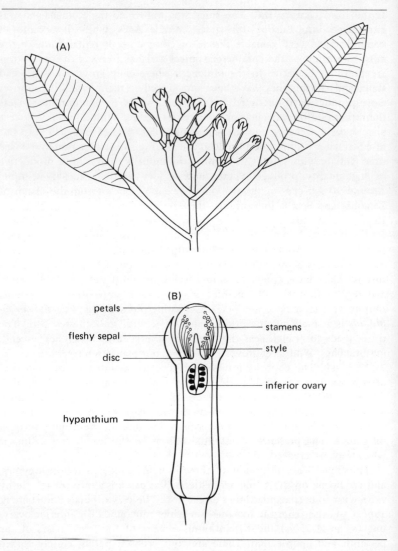

Fig. 9.3 *Eugenia caryophyllus*: Clove. (A) Two simple leaves and part of the cymose inflorescence. (B) A longitudinal section of a flower bud.

dried in the sun as quickly as possible until they become hard and dark brown. The dried flower buds are the cloves of commerce.

Clove trees are propagated from selected, large seeds obtained from new fruits; the proportion of non-viable seed increases quickly as the seed ages. The seeds are sown under shade in a nursery, and the seedlings are transferred to the clove plantation 1 year to 18 months later, after they have been hardened off. The young trees grow slowly and require shade and protection from the wind. They begin bearing after 6—8 years and continue to do so until they are 60 or more years old.

Cloves are used either whole or ground to a powder, and provide flavour for both sweet and savoury foods, and in pickling and the production of sauces and ketchups. In addition to some starch and tannin they contain 14—19 per cent of an essential oil. Clove oil is a colourless or slightly yellow liquid, becoming darker with age and exposure to the air; it is obtained by the distillation of flower buds, inflorescent branches left after the buds have been removed, and the leaves. It contains 80—95 per cent eugenol, and is used in perfumes, in dentistry and as a clearing agent in microscopy.

Allspice, pimento: *Pimenta dioica*

The dried unripe fruits of *Pimenta dioica* of the family *Myrtaceae* provide the spice known as allspice, pimento or Jamaica pepper. It is used extensively to flavour cooked foods, pickles and sauces because it combines the flavours of cinnamon, cloves and to some extent nutmeg. Pimento berry oil is extracted from the dried fruits, and oil is extracted from the leaves with solvents (most pimento oil comes from a different species, *P. racemosa*, the bay tree). The oil is used in toilet preparations, especially hair tonics, and as a scent in soap. *Pimenta dioica* is native to the West Indies and tropical Central America, and most of the world's supply of the spice comes from semi-wild trees in Jamaica. It is closely related to *Eugenia caryophyllus* but differs from the clove tree in the way the flowers are borne (they are not in groups of three in the cymes), and in the form of the embryo which is spirally coiled. Furthermore, even though all the flowers appear to be hermaphrodite, some trees are functionally male and barren while others produce non-viable pollen and are functionally female. Allspice is a medium sized tree up to 10 m tall with smooth grey bark on its slender trunk and many branches. The opposite leaves are 12—15 cm long, bright green and shiny, and are carried on short petioles; they are oval to oblong, pointed and taper towards the petiole. The lower surface of the lamina is covered with oil glands, and is paler green than the upper surface. The inflorescences are corymbs of small cymes in the axils of leaves near the ends of the finer branches. Each flower is less than 1 cm across with four rounded calyx-lobes, four round greenish-white petals, numerous stamens with small white anthers extended beyond the corolla, and an inferior ovary of two united carpels. The fruits are drupes with sweet, fleshy pulp; each contains two seeds. They lose their aroma and flavour as they ripen, so whole inflorescences are plucked while the fruits are still green, then the young fruits are stripped from them and dried in the sun until the seeds rattle inside them. The surface of the dried fruit is rough because it is covered with raised oil glands which contain eugenols.

The dry remains of the calyx-tube surround the shrivelled style at the top of the fruit.

Nutmeg and mace: *Myristica fragrans*

Nutmegs are the dry seeds and mace is the dry aril which surrounds the seed of a dioecious, evergreen tree, *Myristica fragrans*, native to the Spice Islands (the Moluccas north-west of Borneo), and to parts of South-east Asia. The plant has been introduced to many parts of the tropics, and most of the world's supply of nutmegs and mace now comes from Indonesia, the Malagasy Republic and from Granada in the West Indies. The nutmeg tree is a member of the family *Myristicaceae*, a small family of sixteen genera in which *Myristica*, with about 100 species, is the most important. Wild species of the genus occur throughout South-east Asia and in Australia, but tend to be concentrated from New Guinea to Malaysia. Very few wild species produce aromatic or flavoured fruits, but the seeds of those that do are gathered for local use as spices.

In cultivation nutmeg trees require a hot, wet maritime climate with around 2,500 mm of rain annually, and no dry season, a deep loamy soil and some protection from the wind and sun. The crop is propagated from seed, which takes several weeks to germinate in the nursery. As the seeds age the proportion of them that remain viable decreases. Seedlings are transplanted into the grove after about 6 months in the nursery, and eventually grow to produce a handsome tree 8–12 m tall with a stout dark-grey trunk and many branches, some from low on the trunk. The alternate, simple leaves are about 10 cm long and 5 cm broad, ovate-oblong to lanceolate in shape with a pointed tip. They are borne on short petioles, and are darker green above than below, with prominent veins. Young trees begin to flower when they are 4–6 years old; only then can the males be identified, and most of them are cut out, leaving one male tree as a source of pollen for every eight to ten females. The removal of unproductive male trees is wasteful of time and space, and has provided a strong incentive to try various methods of propagating high-yielding female trees vegetatively, though none has been entirely successful. Monoecious trees and hermaphrodite flowers do occur, but they are rare. The flowers are borne singly or in axillary cymose clusters on the finer branches; they are yellow and pleasantly scented, and each one is on a short green pedicel. There is no corolla, but the fused, fleshy sepals form an inflated bell-shaped tube, narrow at the tip and divided into three pointed lobes. The male flowers have numerous stamens with their filaments fused to form a central column with the long, narrow anthers attached to its top. There is no rudimentary ovary in the male flowers. The female flowers are about 1 cm long, larger than the males and with a more oval calyx-tube. The superior ovary is sessile with two short, white, triangular stigmas which rise to the mouth of the calyx-tube. In both sexes there is a nectary at the base of the calyx, and cross-pollination is by small insects. The ovaries are unilocular, and develop into a large fleshy, yellow drupe about 8 cm long; the drupes mature 5—6 months after flowering and split open into two valves to reveal a single dark-brown, shiny seed partly enclosed in a network of the scarlet aril. The seeds are gathered by hand from the trees, or they are allowed to fall to the ground for collection. The aril is removed carefully to avoid damage, then pressed flat by hand or between boards and dried in the sun until is changes colour from red to yellow, and is then sold as mace. It is one of the most delicately flavoured of all the spices used for flavouring savoury foods, and in the preparation of sauces.

After the arils have been removed the seeds are dried in the sun. They are more or less ovoid, about 3 cm long with a thick testa (shell) which is sometimes removed before the nutmegs are marketed. They have been known in Europe since the twelfth century, first as a fumigant as well as a spice, and remain popular especially to add flavour to sweet dishes made from milk. Around 4 per cent of the essential oils in nutmegs and mace is myristicin which is poisonous, so these spices must be used sparingly.

Chilli peppers: *Capsicum* spp.

The *Capsicum* species are natives of the tropics of the New World where their pungent fruits were gathered from wild plants before the development of agriculture, then from cultivars which were among the earliest crops domesticated in South and Central America and Mexico. Cultivated chilli peppers were spread from the New World first by the Spanish and Portuguese in the sixteenth century, and now some species have a pantropical distribution.

Capsicum is a small genus of 20 species in the large dicotyledonous family *Solanaceae* which includes, as well as the chilli peppers, tobacco (*Nicotiana tabacum*), tomato (*Lycopersicon esculentum*) and the 'Irish' potato (*Solanum tuberosum*), all of them domesticated in the New World in pre-Columbian times. Four species of *Capsicum* are cultivated in Central and South America, two of them accompanied by weedy wild relatives which may have been their wild ancestors.

Capsicum annuum is the most widespread cultivated species, both in the New World and throughout the tropics. Its cultivated forms are referred to var. *annuum*, and their wild relatives with dehiscent fruits to var. *minimum*.

Cultivated forms of *C. baccatum* (var. *pendulum*) are found only in southern Peru, Bolivia and southern Brazil where they are accompanied by wild relatives (var. *baccatum*).

Cultivated *C. chinense* has no wild relatives, but it is closely related to, and may have evolved from, *C. frutescens*, which is a weed with a natural distribution in South America, Latin America and the West Indies. *Capsicum frutescens* is not considered to be one of the cultivated species of Central and South America, but both it and *C. chinense* are cultivated

Fig. 9.4 *Capsicum annuum*: Chilli pepper. A chilli pepper bearing a heavy crop of immature fruits.

in Africa and Asia. In the New World *C. frutescens* is cultivated only in the United States where its very pungent fruits are used to make 'Tabasco' sauce.

The fourth cultivated species is *C. pubescens* from the highlands of South America. It has no wild or weedy relatives.

The rest of this account of chilli peppers is confined to *C. annuum*, and to the complex of forms which can be referred to *C. chinense* and *C. frutescens*.

Capsicum annuum includes all the cultivated chilli peppers which are variously called sweet, red, green or bell peppers, or chillies or paprika. Some authorities give these various forms varietal status. *Capsicum annuum* produces both pungent and sweet fruits which vary in length from 1—30 cm, and in colour when ripe from green through yellow, orange and red to brown or purple. The plants are small, bushy annuals 30 cm to 1.5 m tall. They are sometimes woody. The angular stems bear alternate, simple leaves which are very variable in size and shape. They are commonly lanceolate to ovate with a pointed tip, and around 6 cm long, though rarely as long as 12 cm. The petioles are about 2 cm long, and there are no stipules. The regular flowers are borne singly in the axils of the leaves, not in clusters of two or more as they are in *C. frutescens*. Each flower has a cup-shaped calyx of five fused sepals which swell and persist with the fruit. The corolla has five or six white or greenish-white petals which are fused together at their bases, but expanded above into lobes.

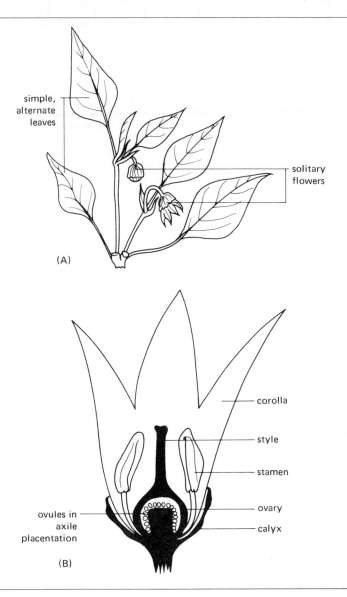

Fig. 9.5 *Capsicum annuum*: Chilli pepper.   (A) Flowers solitary in the leaf axils ($\times \frac{2}{5}$).   (B) A longitudinal section of a flower ($\times 4$).

The five to six stamens are inserted at the base of the corolla and have dark-blue anthers which dehisce longitudinally (not by apical pores, as in many other genera of the *Solanaceae*). The superior ovary has two locules, or sometimes more due to the intrusive growth of false septa, with many ovules in axile placentation.

Capsicum frutescens is a short-lived perennial much like *C. annuum* in morphology, but with its flowers in clusters, and small, very pungent red fruits which are held erect on the pedicel. These are the 'bird peppers' of Africa and Asia, so called because the fruits, despite their pungency, are eaten by birds which play a major role in natural seed dispersal.

The capsicum fruits owe their pungency to the presence of a volatile phenolic compound closely related to vanillin, known as capsaicin, which is distributed throughout the plant, but tends to be concentrated in the placentas. The fruits are a good source of vitamin C, and contain some vitamins A and E. The large non-pungent fruits of *C. annuum* have thick fleshy pericarps and are hollow; they are used as salad vegetables, or may be stuffed with meat and cooked. Red fruits of this kind are used in the preparation of some cheeses, in stuffed olives and various tinned meats; they are often canned to be used in cookery as pimento. Smaller, slightly more pungent fruits of *C. annuum* are dried to make powdered paprika which is used as a condiment, especially in Hungary. The pungent bird peppers from *C. frutescens* and *C. chinense* and pungent fruits from *C. annuum* are commonly dried in the sun, then powdered to make cayenne pepper, or used in the manufacture of hot sauces, pickles and curry powders.

The umbelliferous spices

The *Umbelliferae* is a large family of mostly temperate herbaceous plants, though a few species of minor importance are cultivated in the cool tropics for their aromatic, flavoured fruits. Plants of the *Umbelliferae* usually have thick, hollow, erect, coarse stems with large, alternate much divided leaves on long petioles which have swollen, clasping bases, The characteristic feature of the family is the inflorescence, a simple or compound umbel of small, mostly regular flowers. A simple umbel consists of numerous peduncles arising together from the apex of the stem, where they are subtended by an involucre of bracts; each peduncle carries a single flower, all of the flowers at the same height so that the inflorescence looks like an umbrella. In a compound umbel several inflorescent branches arise together at the apex of the stem, all subtended by an involucre of bracts; these branches each produce a simple umbel at their apex subtended by a ring of bracts which in this position is called an involucel. Though most flowers are regular, the peripheral flowers in compound umbels may be zygomorphic with enlarged outer petals. Each flower has five small white or pink, free petals, but the calyx of five small tooth-like lobes is not always present. There are five stamens on thin filaments, and an inferior ovary of two fused carpels with two locules each containing one ovule. Protandry is common in the family, so the flowers are usually cross-pollinated. The fruit is typically a schizocarp which splits when mature into two indehiscent, single seeded mericarps. The outer wall of a mericarp is variously ridged and furrowed, and it may be more or less bristly. Within the mericarp wall are canals and ducts filled with aromatic resins, gums and essential oils which give the fruits their characteristic flavours and smells.

Among the economically important members of the family are the carrot (*Daucus carota*) and parsnip (*Pastinaca sativa*) which are grown for their swollen tap roots and hypocotyls, parsley (*Petroselinum crispum*), grown for its leaves, and celery (*Apium graveolens* var. *dulce*), grown for its etiolated petioles and leaves. The important umbelliferous spices grown in the tropics are all ancient crops native to the eastern Mediterranean area, now grown in Europe, North Africa, the Middle East, China and India, and on a small scale elsewhere in the sub-tropics and warm temperate regions. They are cultivated most extensively in northern India where they are used chiefly as ingredients in curry powders. They are also used to flavour food, pickles, sauces and alcoholic spirits and liqueurs; the fruits of some species yield aromatic essential oils used in perfumes, soaps and in medicine.

Carum carvi, caraway, is a biennial herb with slender, furrowed stems which grow erect to about 0.5 m tall. The leaves are four to five times pinnately divided, with narrow, cylindrical feathery ultimate pinnae. The inflorescence is a terminal compound umbel of small yellow flowers which produce small brown mericarps. The crop is cultivated as an annual.

Coriandrum sativum, coriander, is an annual herb with slender, solid, smooth stems around 0.5 m tall. Its lower leaves tend to be broad and undivided, while the upper ones are finely dissected with narrow lobes. The compound umbels of small white or pink flowers have zygomorphic peripheral flowers. The whole plant and the unripe fruits have an unpleasant smell, but as they mature the indehiscent fruits develop a pleasant, delicate smell.

Cuminum cyminum, cumin, is a small annual about 30 cm tall with much branched, angular stems carrying dissected, almost sessile upper leaves and less divided lower leaves on long petioles. The inflorescence is a compound umbel of purple or white flowers.

Pimpinella anisum, anise, is a quick growing annual herb about 0.5 m tall with simple hairy leaves more or less entire at the base of the plant, but two to three times pinnate above. The compound umbels of small yellow flowers produce schizocarps which split into ovoid, ribbed mericarps.

Foeniculum vulgare, fennel, is a tall, variable perennial with much dissected, four to five times pinnate leaves on long petioles. It is grown in both temperate and tropical areas. The whole plant is aromatic and the leaves are used as a pot-herb. The inflorescence is a compound umbel of yellow flowers. As well as their use as a spice and source of essential oil the mericarps contain a semi-drying oil which is processed to yield lauric acid for use in making detergents, and adipic acid, which is used in nylon, in lubricants and in plasticizers.

Further reading

General

Anon. (1973). *Spices*, London: Tropical Products Institute. (Conference Proceedings, 1972).
Maistre, J. (1964). *Les Plantes à Épices*, Paris: Maisonneuve and Larose.
Parry, J. W. (1953). *The Story of Spices*, New York: Chemical Publishing Co.
Parry, J. W. (1969). *Spices*, 2nd edition, 2 vols, New York: Chemical Publishing Co.
Purseglove, J. W. (1968 and 1972). *Tropical Crops*, 2 vols, London: Longman.
Rosengarten, F. (1969). *The Book of Spices*, Wynnwood: Livingston Publishing Co.

Ginger

Kannan, K. and Nair, K. P. V. (1965). *Zingiber officinale* (ginger) in Kerala. *Madras agric. J.*, 52, 168—76.
Prentice, A. (1959). Ginger in Jamaica, *World Crops*, 11(1), 25—6.
Tewson, L. (1966). Australian ginger, *World Crops*, 18(3), 62—5.

Turmeric

Sopher, D. E. (1964). Indigenous use of turmeric (*Curcuma domestica*) in Asia and Oceania, *Anthropos*, 59, 93—127.

Cardamom

Mayne, W. W. (1954). Cardamoms in South Western India, *World Crops*, 6(10), 397—400.
Mukherji, D. K. (1973). Large cardamom, *World Crops*, 25(1), 31—3.

Cinnamon

Fock-Leng, P. A. (1965). Cinnamon in the Seychelles, *Econ. Bot.*, 19, 257—61.
Samarawira, I. St. E. (1964). Cinnamon, *World Crops*, 16(1), 45—9.

Pepper

Abraham, P. (1959). Pepper cultivation in India. New Delhi: Ministry of Food and Agriculture, *Farm Bull.*, 55.
De Ward, P. W. F. (1964). Pepper cultivation in Sarawak, *World Crops*, 16(3), 24—30.

Vanilla

Bouriquet, G. (1954). *Le Vanillier et la Vanille dans le Monde*, Paris: Lechevalier.
Chadwick, M. G. A., Orr, E. and Pope, R. (1961). The market for vanilla beans, *Trop. Sci.*, 3, 174—83.
Correll, D. S. (1953). Vanilla — its botany, history, cultivation, and economic importance, *Econ. Bot.*, 7, 291—358.
Dodson, C. H. and Hills, H. G. (1966). *The Biology of Orchids*, Nashville: Benson.

Clove
Tidbury, G. E. (1949). *The Clove Tree*, London: Crosby Lockwood.

Allspice
Ward, J. F. (1961). *Pimento*, Kingston: Govt. Printer.

Chilli peppers
Heiser, C. B. and Smith, P. G. (1953). The cultivated *Capsicum* peppers, *Econ. Bot.*, 7, 214–27.

Pickersgill, B. (1969). The domestication of chilli peppers, in: Ucko, P. J. and Dimbleby, G. W. (eds), *The Domestication and Exploitation of Plants and Animals*, London: Duckworth.

Chapter 10

Vegetable Fibres

The vegetable fibres are among the most important of the world's crops, and a valuable commodity in world trade because they are essential to man for the manufacture of much of his clothing, his cordage and coarse fabrics. The number of fibre crops important in commerce is small, most of them of ancient origin in the tropics where the bulk of the world's vegetable fibre requirements is now produced. Though commercial production from plantations, with their attendant processing factories, accounts for a large proportion of total production, especially of cordage fibre (see below), there are many tropical plants locally important to peasant farmers who grow them to meet domestic needs. Since the Second World War synthetic fibres have supplied an ever increasing proportion of the world's fibre requirements, but with notable exceptions (e.g. sisal before 1974) increasing world demand for fibres of all kinds in recent years has ensured that vegetable fibre production has also increased.

Vegetable fibres are classified into three groups according to their anatomical origin in the plant:

1. Surface hairs associated with the fruits and seeds of plants are single-celled outgrowths from the testa or from the ovary wall, which protect developing seeds. Cotton, which belongs to this group, consists of the surface hairs produced by the testa of the seeds of cultivated species of the genus *Gossypium*; it accounts for the greater part of world vegetable fibre production. Kapok, which is the floss filling the capsules of the tree *Ceiba pentandra*, consists of single-celled, lustrous hairs with a waxy coating which grow from the ovary wall; because it is very light and buoyant, kapok is used to stuff lifejackets, in heat and sound insulation, and to fill cushions and mattresses; after chemical treatment to roughen the smooth hairs, they can be spun into yarn. Kapok probably originated in Central or South America, then spread to Africa and Asia where different forms grow wild, or are cultivated in the rain forests or savanna woodlands. The tree is commonly 10–30 m tall, but var. *caribaea*, of the West African rain forest, may reach the astonishing height of 70 m. World production of kapok is about 25,000 tonnes each year, half of it from Thailand.

2. Phloem, bast or 'soft' fibres are schlerenchyma fibres associated with the phloem of the stems of plants. They are consequently rather easy to separate from underlying woody tissues as a constituent of the 'bark' which can be peeled from the stems. They arise with primary tissues from the apical meristem, or with secondary tissues produced by the lateral meristem, the cambium, associated with the vascular tissues of the stem. Each fibre cell (the 'ultimate' fibre) is 0.8–250 mm long, depending on the species in which it occurs, with tapering ends and cellulose primary walls on which varying amounts of lignin are laid down as a secondary wall thickening. The ultimate fibres occur in bundles of up to 700 individual cells which provide mechanical support for the stem and help to keep it rigid. It is the bundles of fibre cells which are extracted from the stems and which are commercially important; in jute each may be as long as 3 m. Though many plants contain phloem fibres, only species from which they can be extracted easily have become important in commerce. Flax is the ancient bast fibre of the Mediterranean area and of Asia and Europe, though the species *Linum usitatissimum* from which it is obtained is now more important as the source of linseed oil. Flax is of the highest quality among bast fibres and is used to weave linen. Other bast fibres, which are commercially important for the manufacture of cordage and sacking, are jute, hemp and kenaf.

3. The leaf, structural or 'hard' fibres are bundles of sclerenchyma fibres which occur in the leaves of some monocotyledons, notably sisal, abacá and New Zealand hemp. The ultimate fibres are generally coarser than those of bast fibre, so leaf fibres are used mainly to make cordage and coarse sacking.

The 'textile fibres' of trade include cotton, flax, jute, kenaf, aramina, hemp, roselle hemp, Deccan hemp and sunn hemp, while 'cordage fibres' are those coarser, tougher fibres such as sisal, New Zealand hemp, bowstring hemp and abacá, which are used mainly in the manufacture of ropes, cables and twine. The textile fibres are mostly 'soft', while the cordage group comprise the 'hard' fibres. Coir fibre, obtained from the fruits of the coconut, is used to make brushes, mats, coarse yarn and to stuff upholstery. The waste from fibre processing may be used to make paper.

The surface hairs

Cotton: *Gossypium* spp.

Cotton is the most important of the vegetable fibres. It is used for a variety of purposes, but especially to make textiles used in the manufacture of a large proportion of man's clothing, particularly in the tropics where the environment demands the lightest possible absorbent fabric for clothing purposes. Cotton is predominant as a textile fibre because, as they dry, the mature testa hairs twist in such a way that fine, strong threads can be spun from them. It belongs to the large family *Malvaceae* which includes some 50 genera and about 1,000 species, world-wide in distribution, but with a tendency to be concentrated in the tropics. The species vary from annual herbs to perennial trees and shrubs, and as a rule they bear alternate leaves which are commonly palmately lobed or divided. Many species have stellate hairs on their stems and leaves, and mucilage and resin sacs are common features of the family. Most species have large, showy flowers born singly in the axils of leaves. It is characteristic of the *Malvaceae* that the filaments of the anthers are fused to form a prominent staminal column bearing numerous, often coloured anthers, and that in many, but not all, species each flower is subtended by an epicalyx of modified leafy bracts which may serve the protective function of the calyx. The epicalyx often persists with the fruit, which is a dehiscent capsule, or a group of indehiscent schizocarps. As well as cotton, other economically important members of the *Malvaceae* are three species of *Hibiscus* (kenaf, okra and roselle), and *Urena lobata*, aramina fibre.

Gossypium is a large and very variable genus including numerous wild and several cultivated species which fall into two natural groups, depending upon chromosome number. All the wild species in both the Old and New World, and cultivated species which originated in the Old World, are diploids with 26 somatic chromosomes, whereas the cultivated species from the New World are tetraploids with 52 chromosomes. The number of species in the genus is considered by some authorities to be 20, by others 32. They are divided into eight sections, and it is convenient to discuss their origins and relationships under three headings:

1. Wild, lintless, diploid species ($2n = 26$)

The wild, ancestral *Gossypium* species probably evolved in southern Africa, and spread from there to the arid regions of Arabia, South-east Asia, Australia and America. With geographical isolation, perhaps during the Mesozoic era (around 10^8 years ago) different groups of species which are now genotypically distinct subsequently evolved in each of these areas. These wild cottons are lintless, perennial, xerophytic shrubs or small trees, often found growing on desert fringes, in dry river-beds or on rocky hillsides where vegetation is sparse. They all have strong, deep tap roots, and though their leaf area is not decreased as an adaptation to the arid climate, their leaves are commonly hairy or leathery, enabling the plants to survive in such rigorous localities. Though these wild cottons are phenotypically distinct, they all produce small dehiscent capsules which contain small seeds covered to varying degrees in most species with short, dark coloured surface hairs ('fuzz') which cannot be spun to make yarn.

The wild diploids constitute the first six sections of the genus, each of which includes species with more or less the same geographical origin and the same set of haploid chromosomes, or genome. For convenience, the different genomes in *Gossypium* are labelled with the letters A—E.

Section I	*Sturtiana*. Nine species in Australia	Genome C
Section II	*Erioxyla*. Three species in Mexico and southern California	Genome D
Section III	*Klotzschiana*. Two species in the Galapogos Islands, Peru and western Mexico	Genome D
Section IV	*Thurberana*. Four species in Arizona and Mexico	Genome D
Section V	*Anomala*. Three species in the South African desert fringes and along the borders of the Sahara	Genome B
Section VI	*Stocksiana*. Five species in the Sind, south-eastern Arabia, Somalia and parts of the Sudan (The Sind is a desert in Pakistan.)	Genome E

Fertile hybrids can be obtained from crosses between species within each of these sections, but they are difficult to obtain from crosses between species belonging to different sections.

2. The Old World linted diploid species ($2n = 26$)

Two diploid species, *G. herbaceum* and *G. arboreum*, each with perennial and annual races, constitute Section VII, *Herbacea* (genome A), of the genus. As well as fuzz, they all have long seed coat hairs (lint), and several races, especially of *G. arboreum*, were the ancient cultivated cottons of Asia and Africa. The mutation of the gene controlling the development and structure of the seed coat hair, which resulted in the production of spinnable lint in addition to fuzz, occurred in *G. herbaceum* race *africanum*, which is now to be found growing wild in southern Africa. Indeed, it may be the only truly wild diploid, linted cotton, for other apparently wild forms of both species in Section *Herbacea* have been closely associated with agriculture and are perhaps escapes from cultivation. We believe that all the diploid, linted cottons evolved from race *africanum*, and that the ancestral form was probably domesticated in Ethiopia or Arabia. The diploid species became of greatest importance as sources of fibre in Asia, especially in India, where races of the 'tree cotton', *G. arboreum*, were cultivated as long as 5,000 years ago. Indeed, races of *G. arboreum* and *G. herbaceum* still make significant contributions to Asian cotton production.

The tree cottons are much branched shrubs growing to a height of 2 m or more, with more or less hairy, five to six-lobed leaves and linear stipules. The bracts of the epicalyx are more or less triangular and enfold the flower bud closely; they have entire margins except at their apices where there are several teeth. The flowers have a long staminal column bearing anthers on short filaments throughout its length. The capsules, or bolls, of *G. arboreum* are usually trilocular and tapering, profusely pitted, with oil glands in the pits. They open widely when ripe, and contain up to seventeen seeds in each locule. The seed coat has fuzz as well as short, strong lint hairs.

Annual *G. herbaceum* cottons are found throughout Asia and in many parts of Africa. They are shrubby plants with thick, rigid stems up to 1.5 m tall, few vegetative branches and alternate, flat, lobed leaves which are more or less hairy. The bracts of the epicalyx differ from those of *G. arboreum*, for they do not enclose the flower bud closely, but spread widely from it, and are round and broadly triangular with six to eight definite teeth. The bolls are more rounded than those of *G. arboreum*, but have a 'beak' at their apex. They are smooth, with few pits or oil glands, and are divided into three or four locules, each containing up to eleven seeds which have both fuzz and lint. The bolls do not open widely when they are ripe.

The lint of the Old World cottons is strong, but shorter and of poorer quality than that of New World, tetraploid species.

3. The New World linted tetraploid species ($2n = 52$)

Three New World tetraploid species, *G. hirsutum, G. barbadense* and *G. tomentosum*, constitute Section VIII, *Hirsuta* (Genome AD), of the genus.

They arose in South America following the natural hybridization of an Old World, A genome, linted diploid of Section *Herbacea* (probably *G. herbaceum*), and a New World, D genome, lintless diploid (probably *G. raimondii* or *G. thurberi*, or both). The diploid hybrid (Genome AD) from this cross (or crosses) was infertile until its chromosome number was doubled; the New World linted cottons are thus allotetraploids with the genomic formula AADD. At meiosis normal pairing occurs between thirteen pairs of homologous A genome chromosomes and between thirteen pairs of homologous D genome chromosomes. Whether the cross between Old and New World diploids occurred only once, or whether *G. hirsutum* and *G. barbadense* arose independently from two or more natural crosses is not certain, nor is there general agreement about the way Old World linted diploid cotton reached South America. Several theories propose overland migration from Africa by way of Asia or Antarctica, or across ancient land bridges between Asia and South America. Alternatively, it has been suggested that seeds of the Old World species were carried by man across the Pacific Ocean, or that they floated across the south Atlantic. Perennial, linted, tetraploid cottons were cultivated in Mexico, South America and the West Indies long before Europeans reached the New World; indeed, the earliest recorded archaeological remains of cultivated cotton bolls are from Mexico, dated 3500 B.C. *Gossypium tomentosum* occurs wild in Hawaii but is not economically important.

The world predominance of New World cotton as a textile fibre is relatively recent, and the history of the dispersal of tetraploid species and of the development of the modern cotton industry is reasonably well known and very interesting. *Gossypium hirsutum* var. *latifolium* now accounts for most of the world production of about 13 m. tonnes of lint each year. It probably originated in Mexico, and perennial forms were introduced to the United States in 1700, where they were cultivated as annuals and consequently acquired the annual habit. They became widespread in what is now the 'cotton belt' of the USA, and were called 'Upland' cotton to distinguish them from annual cultivars of *G. barbadense* grown near the coast. The cotton industry of Lancashire in England, and cotton growing in the United States, expanded together during the nineteenth century, but supplies of high quality 'Upland' cotton lint to Lancashire were temporarily curtailed during the American Civil War. As a matter of deliberate British policy Upland cotton was distributed throughout the tropics, especially in the British colonies, during the late nineteenth and early twentieth centuries, with the intention of ensuring future supplies of high-quality lint for the Lancashire mills. The growing, spinning and weaving of Upland cotton has since become an important industry in several developing countries, where the crop has often been the first, and most important, cash crop introduced into peasant agriculture.

Perennial forms of *G. barbadense* were taken from the West Indies to the coastal areas of South Carolina in 1786, where they too became annuals under cultivation, and gave rise to cultivars which produce lustrous, strong lint of the highest quality called Sea Island cotton. This crop ceased to be much grown in the United States after infestation by an insect pest called the boll weevil became very severe, but it is now grown in the West Indies, and was involved in the breeding of cultivars which now constitute the Egyptian and Sudanese crops.

The earliest movement of cotton from the Old World was to West Africa in the sixteenth and seventeenth centuries. Perennial forms of *G. barbadense* gave rise to the 'Ishan' cotton of Nigeria, and they spread from there to the Sudan, and eventually to Egypt. In the mid-nineteenth century these perennial *barbadense* cottons in Egypt were crossed with annual Sea Island cotton from the United States, and from the progeny annual cultivars with the very high quality lint for which the Egyptian crop is renowned were selected; they now constitute the irrigated crops of Egypt and the Sudan Gezira. A second early movement from the New World to the Old was of *G. hirsutum* var. *punctatum* which reached West Africa towards the end of the seventeenth century and spread east across Africa south of the Sahara. It developed the annual habit, and eventually replaced Old World diploid cottons in much traditional African peasant farming.

Modern cultivars of *G. hirsutum* are shrubby annuals with few vegetative branches and green or brown woody stems about 1.5 m tall when they are grown with good husbandry. Their leaves are large, cordate, three- to five-lobed and often hairy. The flowers have large, pale, yellow, showy petals, usually without a spot at their bases. The bolls are 4–6 cm long, and rounded, with few oil glands; there are three to five locules, each with up to eleven seeds covered with fuzz and lint hairs. The average length of the longer lint hairs on a seed is called the 'staple length', and it has been established practice to describe it in inches rather than in metric units. An important classification of cultivars is based on staple length and is discussed later. In Upland Cotton it varies from about $1-1\frac{1}{8}$ in (25—28 mm); it is longer than lint from *G. herbaceum* or *G. arboreum*, but mostly shorter, and not so fine and lustrous, nor so strong as the lint from *G. barbadense*.

Cultivars of *G. barbadense* are annual shrubs up to 3 m tall with several strongly ascending branches. The leaves are fairly deeply divided into three to five lobes, and are usually glabrous. The flowers are large and showy, but the petals do not spread so widely as those of Upland cotton, they are a deeper yellow colour, and have a red or purple spot at their bases. The bracts of the epicalyx are large, and partially enclose the flower; each has ten to fifteen long, pointed teeth at its apex. The bolls are deeply pitted with oil glands and are darker green than those of Upland cotton. They have three to five locules, each containing five to eight or more seeds which have green or brown fuzz confined to one end, though the whole testa is covered with long, lustrous lint hairs. Sea Island cotton is creamy white with a staple length often longer than 2 in (50 mm); because it is stronger and finer than any other cotton it is used to spin the finest yarns which are used to weave the finest textiles. The Egyptian cottons, which are referred to this species, produce lint similar to Sea Island cotton, but shorter and not so fine, having staples varying from $1\frac{1}{4}-1\frac{3}{4}$ in (32—44 mm); nor is their colour so good as that of Sea Island cotton.

The cultivated cottons

Cotton is grown in every tropical country where conditions are suitable, as well as in many parts of the sub-tropics from latitudes 37° N. to 32° S., and its production exceeds that of all other vegetable fibres. With increasing world population the demand for cotton continues to increase, especially for lightweight clothing fabrics in the tropics. The United States is the leading producer with a crop of some 3 m. tonnes of lint each year, almost all from Upland cultivars, followed closely by Russia. These two countries, with mainland China and India, produce 75 per cent of the world crop of about 13 m. tonnes. Russian, Chinese and Indian cotton is partly from Upland and partly from 'Asian' (*G. arboreum* and *G. herbaceum*) cultivars. Other large producers with annual crops of more than 0.5 m. tonnes are Brazil, Pakistan, Egypt and Turkey. Including Egypt, the African continent produces less than 10 per cent of the world's cotton. Upland cultivars are grown in African countries south of the Sahara and in parts of the Sudan, but the main Sudanese and Egyptian crops, which account for more than half of Africa's production, are of high quality 'Egyptian' cotton from *G. barbadense* cultivars grown under irrigation. A very small proportion of the world's crop is of Sea Island cotton grown in the United States and the West Indies.

Cotton requires a frost-free growing period of about 200 days with

Fig. 10.1 *Gossypium hirsutum*: Upland cotton. A mature crop of upland cotton, Nigerian Allen, cultivar Samaru 26J in Nigeria. (By courtesy of the Cotton Research Corporation.)

mean temperatures greater than 22°C, and not less than 500 mm of rain, though in the hot tropics 1,000–1,500 mm of well-distributed rainfall are necessary. The vegetative and reproductive growth of cotton, and consequently lint yields, are greatly influenced by rainfall and temperature, by agronomic variables like sowing date, sowing density, rates and kinds of fertilizer application, and especially by the interaction of all these factors with the timing and intensity of insect pest attacks. For example, vegetative growth is promoted and reproductive growth is retarded by cool temperatures; though adequate water is necessary for vegetative growth, heavy rainfall when the crop flowers leads to boll shedding, and a deficiency of water when bolls are maturing decreases lint yields. Average lint yields in African and Asian peasant farming are less than 200 kg per hectare though research indicates a great potential for increased yields through the adoption of improved husbandry, especially 'early' sowing combined with the proper use of fertilizers and insecticides.

The cultivation of cotton on a commercial scale is relatively recent in most countries, but breeding and the selection of improved cultivars has given rise to a multiplicity of diverse types. The following description of the cotton plant is generalized, for it is well nigh impossible to describe the distinguishing features of the many cultivars. The main features outlined below refer to any of the annual cottons cultivated at the present time.

The plant has a long tap root, varying greatly in length depending on the type of soil and whether the crop is grown under conditions of rain cultivation or irrigation. The tap root is commonly 1.5–2 m long, though on the irrigated inland deltas of the Sudan and in many parts of India it may penetrate as deep as 6 m. It tapers fairly rapidly and gives rise to numerous laterals, usually arising in four rows opposite the protoxylem elements of the root. Lateral root development is also a function of soil and moisture conditions. The main root grows rapidly provided penetration is easy, and the root system is established early in the life of the plant.

Growth of the main stem is monopodial from the apical bud; it carries leaves and branches, but no flowers. The leaves are spirally arranged on the main stem and upon its vegetative branches, but appear to be alternate on reproductive branches. Each has a long petiole subtended by two small stipules, and a broad palmately-lobed lamina, dissected to varying degrees depending upon cultivar and species. The lamina commonly has three to five lobes, though in some cottons the leaves are almost undivided. There are three to nine palmately arranged veins, and sometimes a red spot at the junction of the petiole and lamina. The leaves bear multicellular stellate hairs derived from single cells of the epidermis, but the degree of hairiness is genetically controlled and vary variable. Dense, long hairs on the leaves of some cultivars render them almost immune to the attack of leaf-sucking insects such as jassids, which are important pests of commercial crops in some areas. Resistant Upland and Egyptian cultivars have been bred in Africa using hairy-leaved Indian cultivars as a source of jassid resistance. In some cultivars leaf hairs are confined to the undersides of veins. A small, extra-floral nectary occurs on the underside of the main vein; it is a gland-like cavity filled with multicellular hairs, but its function is unknown. Most aerial parts of the cotton plant are covered with oil glands which consist of a sac surrounded by flattened secretory cells. They are filled with a dark-brown fluid composed of oils and resins. The leaves of most cultivars have the typical anatomy of the dicotyledons, but some Asiatic cottons have palisade cells beneath both the upper and lower epidermis.

In each leaf axil are two buds, but only one usually develops. The lower branches on the main stem grow from the true axillary buds and repeat the general structure of the main axis, with no flowers; they are vegetative monopodia with spirally arranged leaves. Higher up the main stem a different kind of branch grows from the extra-axillary bud of the pair in each leaf axil. It is a fruiting branch with sympodial growth in which the terminal bud produces a flower, while vegetative growth is continued from the bud in the axil of the last leaf. Sometimes the extra-axillary bud in this last leaf axil may grow into a short, secondary sympodial branch terminated by a flower, but as a rule it remains dormant. The flowers conse-

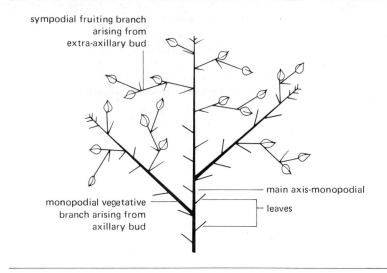

Fig. 10.2 *Gossypium* spp.: Cotton.   Branching habit.

quently seem to be borne opposite the leaves on a fruiting branch, but in reality they are terminal, and the continuation of the branch is by lateral bud development. Vegetative monopodial branches are usually carried at an acute angle to the main stem, whereas fruiting sympodial branches tend to lie more horizontally.

In all cottons there is a tendency for the lower branches on the main stem to be vegetative and for the upper ones to be reproductive. The position of the lowest reproductive branch varies between cultivars; as a general rule the lower the first reproductive branch is borne the earlier the maturity of the cultivar. The relative number of vegetative and reproductive branches also varies between cultivars. In Egyptian, *G. barbadense* cottons, as many as ten nodes on the main stem may produce vegetative branches, with fruiting branches above them. On the other hand, Upland cultivars may have no vegetative monopodial branches, or as many as four from the first four nodes of the stem, the upper nodes all bearing reproductive sympodia.

Growth of the main stem is usually fairly rapid, but growth rates vary widely and are much influenced by the environment, especially by temperature. In the hotter, drier parts of the tropics, where cotton thrives, growth of the stem usually occurs mostly at night, because water stress often inhibits growth during the day. While the length of the internodes is usually a function of the water supply to the plant, the number of nodes seems to be controlled, at least in part, by the availability of nitrogen. As the plant matures the growth rate declines, and finally stops as carbohydrates are increasingly diverted from the apical growing point to the developing flowers and fruits. The diversion of carbohydrates from all parts of the plant to the flowering axis is accompanied by decreased nitrogen uptake by the roots, which gradually ceases.

The flowers arise singly from the terminal buds of sympodia; they are pedunculate, the length of the peduncles varying with cultivar. Flowers or flower buds may be shed when there are insufficient nutrients to support all the flower initials produced, or when there is excessive rainfall at the time of flowering, but the amount of shedding varies considerably between cultivars and is due to genetic as well as environmental factors. Shedding is brought about by the development of an abscission layer, one cell thick, which is visible as an external ring near the base of the peduncle. The flower is subtended by an epicalyx of three to four large, leafy bracts which are commonly deeply divided or laciniate, though they are sometimes entire. They give full protection to young flower buds, and though they are deciduous in some cultivars, in others they persist with the fruit. The withering and shedding of deciduous epicalyx bracts are determined genetically, and are desirable traits because persistent, large bracts tend to be harvested with the lint (especially by cotton-picking machines) and decrease its quality, which is partly assessed on cleanliness. Within the epicalyx there is a small, cup-shaped calyx at the base of the corolla. It is light-green, dotted with oil glands, and has five indistinct lobes. A ring-shaped floral nectary occurs within the base of the calyx-tube, and three extra-floral nectaries occur outside it. Five large petals are closely rolled in the bud, fused together and to the staminal column at their base, but free above. They have a narrow base which widens rapidly to a broad, flat, upper lobe. Upland cultivars have creamy white or pale yellow petals when the flowers open. They turn pink and then red as they become older. Egyptian and Sea Island cultivars have deeper yellow petals with a red or purple spot at their bases.

There are 100–150 stamens in the cotton flower, united by their filaments into a thick tube around the style. This staminal column is massive and rigid, with the anthers on fine, short branches which arise all along its length, or only from its apex. The anthers are one-celled and kidney-shaped, and they and the pollen grains vary in colour from pale yellow to deep gold. Pollen is shed soon after the flower opens and is able to germinate immediately. The grains are large and spherical, with a markedly sculptured, spiny exine having several germ pores. The superior ovary is composed of three to five united carpels, each containing several ovules in axile placentation. A five-lobed stigma protrudes from the tip of the staminal column. About 1 month elapses between the appearance of the flower bud and the opening of the flower, which occurs at dawn. Wild, perennial cotton species are photosensitive, short-day plants, but the cultivars lost their photosensitivity when they acquired the annual habit; they are day-neutral with respect to flowering. The corolla, the androecium and the style wither before evening on the day the flower opens, and are shed 3 days later. The degree of opening of the flowers varies with

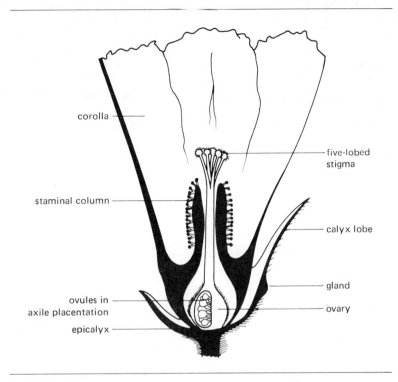

Fig. 10.3 *Gossypium* spp.: Cotton.   A longitudinal section through a flower.

species and cultivar; some have flowers with widely flaring petals, but in others the corolla remains closely folded. In general the flowers of Upland cottons open more widely than those of Egyptian or Sea Island cultivars. The stigma is receptive soon after the flower opens and when the pollen is shed, so that self-pollination is the rule; fertilization is usually accomplished about 30 hours after pollination. The pollen grains retain their viability for about 12 hours, and some 6–25 per cent cross-pollination does occur owing to the visits of insects to the flowers. To prevent cross-fertilization in breeding plots it is necessary to separate them from any other cotton by distances of the order of 100 m, or to place wire rings over the unopened buds, or to tie them closed with cotton thread. However, even when two types of cotton are grown side by side, the percentage of cross-fertilization may be small even though cross-pollination may have been effective, because the pollen tubes of different cultivars grow at different rates.

The cotton fruit, or boll, is a spherical or ovoid, leathery capsule; it matures about 50 days after the flowers open and splits along the carpel edges which are visible as furrows in its surface. The dry carpels open more

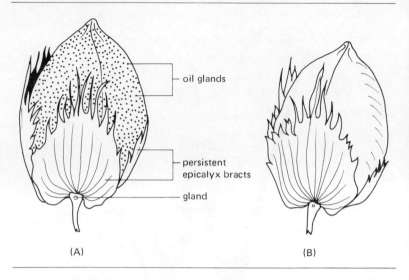

Fig. 10.4 *Gossypium* spp.: Cotton. (A) The immature fruit, or 'boll' of *Gossypium barbadense* (x1). (B) The 'boll' of *Gossypium hirsutum* (x1).

or less widely to expose the 'seed cotton', which consists of seeds with the lint and fuzz attached. Both the lint and fuzz begin to develop while the seed is very young; outgrowths from individual epidermal cells are visible on young ovules at the time of fertilization. The growth of lint hairs takes place mostly at night, and for the first 25 nights they increase only in length, with a maximum growth rate about the fifteenth night. After 25 nights they have reached their maximum length. Little thickening of the wall of the hair is laid down before the twentieth night after fertilization, but thereafter cellulose is deposited on the inner wall in increasing amounts until the boll opens. The important feature of the deposition of this secondary wall thickening is that it is laid down in spiral bands, and that the direction of the spiral may be reversed at any point. At these reversal points the wall of the lint hair is thin, and it is here that the mature lint twists as it dries and collapses into a ribbon when it is exposed to the air. The occurrence of twists in the lint hairs enables them to be spun into yarn. The mature lint hair thus consists of a collapsed tubular cell with a very small lumen, and thick walls composed of superimposed spiral bands of cellulose, but having thinner places where the direction of the cellulose spirals was reversed. Around 94 per cent of the dry weight of lint is cellulose.

During boll development several changes occur in the integuments of the ovule such that identifiable layers are produced in the testa. They are characteristic of the cotton seed and make it possible to identify cotton in seed mixtures. From the outside these layers are: (*a*) thick walled, irregular

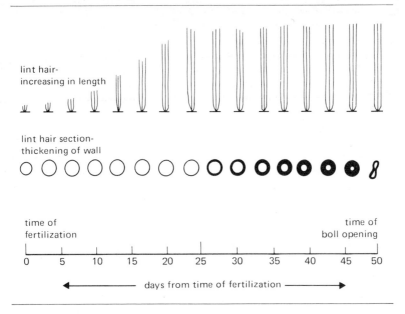

Fig. 10.5 *Gossypium* spp.: Cotton. A diagram showing the development of lint hairs in *Gossypium* during the period of maturation of the boll.

shaped, dark-brown cells with yellowish contents; (*b*) opaque cells containing crystals of calcium oxalate; (*c*) palisade cells; (*d*) dark-brown, dead cells which are very much flattened; (*e*) an inner layer of thick-walled cells with fungus-like growths from their walls into the cell cavities. The embryo develops slowly and is not visible to the naked eye until the fifteenth day after fertilization. The cotyledons are convoluted and eventually fill the seed almost entirely. Many of their cells contain resin which produces a blood-red coloration when crushed seeds are treated with concentrated sulphuric acid. This colour change is used as a test to identify the 'cake' obtained after cotton seeds have been pressed to express their oil.

The seed cotton is picked when the bolls are fully open and the lint has had sufficient time to dry out thoroughly. In most parts of the world it is picked by hand, which requires much labour and may take several weeks because all the bolls in a crop do not ripen together. Most of the crop is harvested by machine in the United States, where the very high cost of labour has provided a strong incentive to develop mechanical harvesters. The quality of machine-picked cotton is in one respect usually poorer than that picked by hand because it contains more 'trash'; that is, bits of leaves, bracts and carpels. It is one object of breeders to produce cultivars for mechanical harvest which can be effectively defoliated by chemical sprays, or which have deciduous leaves and epicalyx bracts. For mechanical harvest it is also important to grow cultivars which mature all their bolls together over the shortest time, and in which the lint is not shed from the open bolls before it can be picked. The lint is separated from the seed in a machine called a 'gin' at a factory called a 'ginnery'. Two kinds of gin are in common use. The saw gin was invented by Whitney in the United States in 1793, and consists essentially of several metal discs, with notched or 'saw-toothed' margins, mounted together on a common axle which rotates fast inside a closed box. One side of the box has slits through which the teeth of the discs protrude as they revolve. Seed cotton is fed against the teeth and lint is torn from the seeds, which are too large to pass through the slits with it into the box. Brushes or compressed air are used to remove the lint from the 'saw'. The roller gin damages the lint hairs less, though its output of lint in a given time is only one-tenth that of the saw gin; it is used mostly to gin high-quality Egyptian cottons and long staple Upland types such as those grown in Uganda. In the roller gin two cylinders covered with hard leather or rough hide rotate in opposite directions close to each other. Seed cotton is fed between them and the lint adheres to the rough leather and is pulled from the seeds more gently than it is torn from them in the saw gin. Lint from the gin, especially from the saw gin, is fluffy and of low density. For transport to the spinning factory or for export it is pressed into dense bales of varying weight depending upon the country, 181.4 kg net weight in Africa and India (the British bale), 227 kg gross weight in the United States and 327 kg gross weight in Egypt.

The quality of cotton lint is very important to the spinning and weaving industry and determines the use to which it is put, as well as influencing the price paid for the crop. Several characters are used to assess quality, some of them requiring sophisticated measuring and testing devices. They are determined genetically and by the conditions under which the crop is grown and harvested. Staple length is the average length of the longer lint hairs on a seed. It is an inherited characteristic of cultivars, which are classified into five groups from 'short' staple Asian cottons (less than $\frac{13}{16}$ in (21 mm)) to 'extra-long' staple Egyptian and Sea Island cottons ($1\frac{3}{8}$ in (35 mm) and longer). Some 80 per cent of world production is of 'medium' ($\frac{7}{8}-\frac{31}{32}$ in (22—25 mm)) and 'medium long' ($1-1\frac{3}{32}$ in (25—28 mm)) staple lint from Upland cultivars. The maturity of the lint is determined by the degree of secondary wall thickening laid down before it is picked, and therefore depends largely upon when the crop is harvested. Fully mature lint hairs have thick walls and a narrow lumen; they are strong and spin well. On the other hand immature hairs are not twisted and do not cling together when the lint is spun. Consequently they produce tangles and knots of lint hairs called 'neps' in the yarn, and 'neppiness' in the cloth woven from it. Lint with high tensile strength is desirable because it is less liable to break during ginning and spinning, and because it produces strong yarn. Fine lint hairs with a small diameter and fully developed walls are desirable because they produce the strongest yarns for a given staple length. Good quality cotton consists of long, fine,

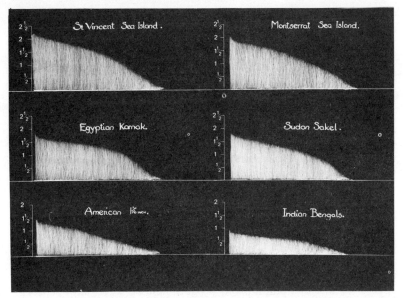

Fig. 10.6 *Gossypium* spp.: Cotton. Lint hairs combed out to illustrate variation in 'staple length'. The upper four cultivars are *G. barbadense*. 'American' is *G. hirsutum* (Upland cotton). 'Indian Bengals' is *G. arboreum* (Asian cotton). (By courtesy of the Cotton Research Corporation.)

strong lint hairs, and is free from trash and aborted ovules or immature seeds (called 'motes').

After the lint has been removed from the seed in the gin, the fuzz may be removed by further processing. The fuzz hairs are called 'linters'; they are used to make felt or cotton wool, or as a source of cellulose which is made into rayon (a synthetic fibre), explosives, paper and many other industrial products.

A small proportion of the cotton seed from the ginnery, with or without fuzz, is used to sow the next year's crop. It is an important feature of cotton improvement that growers sell their seed cotton. They are either given, or they buy seed to sow the next crop, so it is easy to introduce new cultivars, or to maintain the purity of existing ones. Cotton seed contains up to 30 per cent starch, 25 per cent of a semi-drying oil and 16–20 per cent of protein. The oil and protein content of the seed vary between cultivars and the environments in which the crop is grown. As a general rule seed from long staple cottons has a greater oil content than seed from short staple cultivars. For example, the average oil content of the seeds of 'Asian' cottons is about 18.5 per cent, that of Upland cultivars about 19.5 per cent and that of Egyptian and Sea Island types 22–24 per cent. Cotton seed oil is one of the most important of the world's semi-drying oils; each year almost 2 m. tonnes are extracted from cotton seed, mostly in the United States. After it has been refined the oil is used for cooking, as a salad oil, in the manufacture of lard substitutes and oleomargarines and in soap making. In the extraction process the seeds are cleaned and the testas may be removed in a process called decortication. They are then heated to about 105°C to assist expression of the oil in presses. The oil is filtered, refined by a caustic soda process, deodorized and bleached.

The seed cake remaining after extraction of the oil is a nutritious stock feed. The cake from decorticated seed contains up to 41 per cent (by weight) of protein, undecorticated seed cake about 21 per cent. It may also be used as a nitrogenous fertilizer. Raw cotton seeds contain a poisonous phenolic pigment called gossypol, which occurs in small amounts throughout the plant, but which constitutes 0.4–2 per cent of the kernel. The amount present in the seed seems to be related to the oil content, and is greatest in cultivars of Egyptian and Sea Island cotton. Although toxic in uncrushed seeds, gossypol is rendered harmless in the crushed seed cake by union with the protein in the seed.

The bast fibres

A large number of tropical plants produce bast, or 'soft', fibres of varying quality in their stems, but most of them have properties which hinder their commercial exploitation. In many cases it is difficult to extract the fibres, and the few species which have become important crops provide fibre which is not only of good quality, but which is also comparatively easy to extract from the stems. They are species which experience has shown can be grown and processed reliably in large quantities at comparatively low cost. The quality of the bundles of sclerenchyma fibres which constitute the fibre of commerce depends, like cotton, upon characters which determine the ease with which it can be spun into yarn, and the strength, durability and appearance of the yarn and the fabrics woven from it. Among these length, strength, fineness, suppleness, lustre and colour are important. The fineness of bast fibres depends upon the diameter of the sclerenchyma cells, and is measured as weight of yarn per unit of length; the standard measure, or 'Tex', is the weight (g) of 1 km of yarn. The best quality bast fibre is flax which contains about 80 per cent of cellulose and hemicellulose and only 2 per cent of lignin. The greater the proportion of lignin in the fibre the coarser it becomes (its 'Tex' value becomes greater). Sclerenchyma fibres become increasingly lignified with age so that it is very important to harvest crops when the best balance between yield and quality is attained.

Bast fibres, and some of the 'hard' leaf fibres, are extracted from stems (or leaves) after a process called retting. The stems are cut close to the ground at harvest, which is usually when they have grown to their full height, and when they have flowered or bear very young fruits, at which time the largest yields of good quality fibres are obtained. The stems are left to dry out slightly until their leaves fall or can be removed easily, then they are bundled together and held by weights under water for varying lengths of time, depending upon the crop species and upon the speed at which bacterial decomposition of the stems takes place. The activity of bacteria is influenced chiefly by the temperature of the water. Slowly flowing water is preferable to still water because it removes the products of bacterial action, which tend to slow down the retting process if they are permitted to accumulate. The stems gradually become saturated with water and their soluble contents are dissolved out as air is expelled from the tissues. Various aerobic bacteria develop on and in the stems at the beginning of the 'ret', and persist until all the air is driven out and all the oxygen is consumed. Eventually, anaerobic bacteria, especially *Clostridium*, bring about the gradual decomposition of the pectin material forming the middle lamellae of the cell walls so that the fibre bundles are released from the tissues around them. The bark is removed from the retted stems and beaten until the fibre bundles are separated from phloem, cortex and epidermal tissues. The retting process is such a costly and labour-intensive part of the bast fibre production that many attempts have been made to make it less so. The bark may be stripped from the stems in a 'ribboning' machine before retting, or retting may be omitted altogether by the mechanical extraction and cleaning of the fibres. All such attempts to decrease costs of production result in fibre of poorer quality than the retted product.

Jute: *Corchorus capsularis* and *Corchorus olitorius*

Jute is the most important of the bast fibres, and the second most important vegetable fibre after cotton. Though its fibre is somewhat coarse and lignified it can be spun into yarn by a technique developed in Dundee, Scotland, in about 1883. The yarn is used to weave hessian or burlap chiefly for the manufacture of bags and sacks. Until recently jute was cheap because labour in the main growing area of the Ganges and Brahmaputra river basins of India and Bangladesh was plentiful and cheap. Now these countries not only consume much of their home production of jute, but their increasing populations demand more land and labour to grow food which conflicts with increasing jute output. World production is about 2.7 m. tonnes each year, most of it from Bangladesh (1.0 m. tonnes) and India (1.1 m. tonnes). Although attempts have been made to grow the crop commercially in many other countries, it has become established only in Brazil, mainland China and Taiwan, on the alluvial plains of great rivers. Such locations fulfil the requirements of the crop for hot and humid lowland tropical climates with not less than 1,000 mm annual rainfall, though more than 1,500 mm is desirable. Furthermore, the extension of the crop into new areas is limited by the requirements of plentiful clean water after harvest for retting, and of plentiful cheap labour for weeding, thinning and retting. A world shortage of jute for sack and bag manufacture has been partly overcome by greatly increased production of other bast fibre crops in many tropical countries. Some of these are dealt with briefly later in this chapter.

Corchorus belongs to the family *Tiliaceae*, which is fairly closely related to the *Malvaceae*, but has free filaments and two celled anthers. There are about 40 species of *Corchorus*, fairly widely distributed in the tropics, but only *C. capsularis* and *C. olitorius* are economically important. Some 75 per cent of the world's jute is from *C. capsularis* ('white' jute), though spinners pay a slightly higher price for the fibre of *C. olitorius* ('tossa' jute). *Corchorus capsularis* is wild in China, and *C. olitorius* in Africa and Asia where short, branching forms are commonly grown as home garden, leaf vegetable crops. The fibre cultivars of both species are herbs up to 5 m tall with straight, cylindrical stems branching only at their tops when sown densely and thinned to a spacing of 10—15 cm between plants. The leaves are alternate along the stem on short petioles, subtended by two erect, narrow, pointed stipules. The lamina is 5—12 cm long, broad at the base but tapering gradually to a pointed tip, with serrated edges; the lowermost teeth of the leaf margin drawn out into long, fine points. The flowers are solitary, or in groups of two or three opposite the leaves; they have five free narrow sepals and five small, yellow petals about 5 mm long. The petals are narrow at their bases and alternate with the sepals. A very short corona separates the petals from the insertion of numerous stamens which have fairly short, stout filaments, and short, bilobed anthers. The superior ovary has five locules, a short-style with a flattened stigma, and numerous ovules carried in axile placentation. The fruits of the two cultivated species differ considerably, and provide the easiest means of distinguishing between them. Those of *C. capsularis* are small, almost globular, ridged and wrinkled capsules, 1—2 cm in diameter with flat tops; they are divided

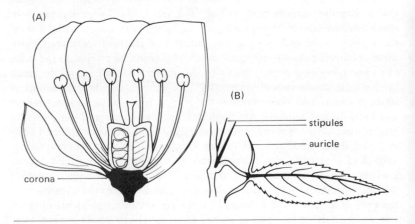

Fig. 10.7 *Corchorus capsularis*: Jute.   (A) A longitudinal section through a flower (x12).   (B) A leaf.

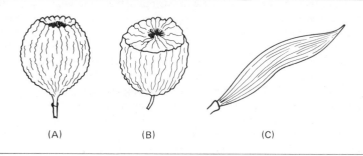

Fig. 10.8 *Corchorus* spp.: Jute. The fruits of: (A) *Corchorus capsularis*: Lateral view (x2). (B) *Corchorus capsularis*: From above (x2). (C) *Corchorus olitorius* (x1).

internally by five septa and dehisce into five segments. The seeds are 2–3 mm long, oval and pointed, more convex on one face than the other. They vary in colour from light brown to russet brown or even darker, depending upon cultivar. On the other hand, the fruits of *C. olitorius* are cylindrical capsules up to 10 cm long, extended into long beaks. They have four or five locules and contain numerous, small, pyramidal, dark-green seeds.

Jute is sown by broadcasting the very small seeds onto a well-prepared, moist seed bed, after which a weighted pole is dragged over the land to ensure that the seeds are in close contact with the soil. Jute seedlings cannot survive waterlogging, but once they are about 1 m tall plants of *C. capsularis* withstand deep flooding (*C. olitorius* cannot, and is sown on higher land). Sowing dates are chosen which ensure adequate but not excessive rainfall for seedling growth, followed by more intense rainfall as the crop develops. The seedlings are thinned by hand to leave very dense populations of up to 500,000 plants per hectare. The largest yields of good quality jute fibre are obtained when the crop is harvested at the 'early pod' stage of development 100–130 days after sowing. Harvesting at this time is a compromise between yield, quality and ease of fibre extraction, which are determined by the duration of vegetative growth and by anatomical changes which accompany flowering and fruiting. Throughout vegetative growth the vascular cambium proliferates secondary xylem and phloem tissues, including phloem fibres. At any time before flowering some 40 per cent of the sclerenchyma fibres are immature, and do not have fully thickened cellulose walls; they do not contribute to the yield of good quality fibre if the crop is harvested before it flowers. After flowering cambial activity decreases, and eventually stops when its cells become differentiated into xylem and phloem tissues. This not only brings to an end increase in stem diameter and in the quantity of fibre the stems contain. It also makes it difficult to separate the bast from the wood after harvest because, when present, the actively dividing cambium is a 'weak' zone along which the 'bark' is easily stripped from the wood. Once the crop has flowered and fruits develop the fibres already present continue to mature, and many of the older ones become increasingly lignified and coarse. Consequently, the greatest proportion of mature, but least lignified, fibres is obtained at an early stage of pod development when the cambium has ceased to divide, but is still present so that the bark can be stripped easily from the wood after the stems are retted.

At harvest the stems, which may be standing in deep water, are cut close to the ground with sickles and left to dry for a few days until the leaves have withered and can be shaken off. The leafless stems are then tied in bundles 15—25 cm in diameter and submerged horizontally in ditch, pond or river water 0.5—1 m deep. They are left to ret for a varying period of time depending upon the temperature of the water, about 10 days at 30°C, but as long as 5 weeks when the water is cooler. The lower and thicker parts of the stems, or 'butt' ends, ret more slowly than the thinner parts. To try to achieve uniform retting of the whole stem, and consequently more uniform fibre quality from it, the bundles of stems may be stacked vertically in water so that only the butt ends are submerged to ret for a few days before the whole bundles are retted. Alternatively, since they commonly produce inferior, coarse fibre, the butt ends are cut off before the stems are retted. After retting the bark is stripped by hand from the stems, the fibres are washed free of tissues still adhering to them, and hung to dry for several days. The individual fibre cells, or 'ultimate fibres', are 2—5 mm long, but the fibre bundles harvested from jute are up to 3 m long. They are weaker than flax or hemp (the fibre from the stems of *Cannabis sativa*), and not so durable. In particular jute is less resistant to deterioration and rotting in water than hemp. It takes dye well, but is difficult to bleach.

There are numerous cultivars of both species, but especially of *C. capsularis*, varying in degree of hairiness, stem colour, plant height and leaf shape.

Other bast fibres

Several other crops are cultivated for their bast fibre, and in general, though their soil and climatic requirements vary, the methods used to grow them and to extract their fibres are much the same as those employed to produce jute. In recent years successful efforts have been made to grow and process several species on a commercial scale in many tropical countries to meet large local demands for produce bags. There has been no corresponding increase in international trade with new or little known bast fibres, however good their quality, because there has been no market for them in industrialized nations with well and long-established fibre manufacturing industries based on better-known bast fibres.

As well as the individual soil and climatic requirements which may limit the range over which species producing bast fibre can be grown, important pests and diseases play a major role in determining the success of a species in any place. Of course, this is true of all crops, but especially so when they are introduced or expanded into large-scale commercial production. For example, stem borers restrict the areas over which kenaf can be grown in Nigeria, and nematodes have restricted the expansion of this crop in East Africa.

Kenaf: *Hibiscus cannabinus*

Kenaf is obtained from the stems of *Hibiscus cannabinus* of the family *Malvaceae*. It is a bast fibre much like jute, and though coarser and less pliable, it is also used principally to make bags, sacks and coarse fabrics, sometimes without being retted. Kenaf probably originated in Africa, where it occurs wild, and has been cultivated there and in Asia (especially in India) for centuries as a fibre crop for domestic use. Since the Second World War kenaf has spread to most tropical countries, but has not become important in world trade. It is less demanding in its soil and climatic requirements than jute, and is grown in a greater range of environments in the tropics and sub-tropics, from latitudes 45° N. to 30° S. It grows best on well-drained soils where there are 500–650 mm of rain in a 4–5 month period, and where temperatures are in the range of about 16°–28°C.

When sown densely for fibre production *H. cannabinus* is an erect annual up to 4 m tall, with a strong tap root and several branches from the

Fig. 10.9 *Hibiscus cannabinus*: Kenaf.   Flowers.

274 *Vegetable Fibres*

main stem. The stems are straight, more or less prickly, and red, purple or green in colour. The alternate leaves are borne on long, slightly prickly petioles. They vary greatly in form and degree of dissection, from cordate to deeply, palmately divided into five to seven lobes. The sessile flowers are single in the axils of the upper leaves on the stem; they have an epicalyx of five to ten linear bracts less than half the length of the calyx. The five sepals are broad, with a large gland about one-quarter of the way along a raised mid-rib. At the base they are fused together, and at the tip extended into long, fine points. The showy, pale yellow corolla is up to 10 cm across when open, and consists of five large petals, often with a crimson spot at their base, where they are fused together and to the staminal column. As well as the epicalyx, a second feature of the flower of kenaf which is characteristic of the family *Malvaceae* is the fusion of the filaments to form the staminal column around the style. Small anthers arise all along its length, or in three groups, one at the base, one in the middle and one at the tip. The superior ovary is composed of five united carpels terminated by a thin style which emerges from the top of the staminal column into a broad, five-lobed stigma. The ovary is covered with

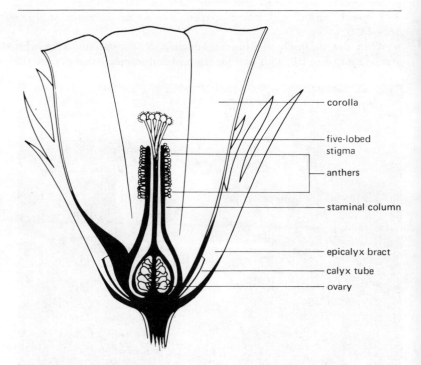

Fig. 10.10 *Hibiscus cannabinus*: Kenaf. A longitudinal section through a flower (x2).

downward pointing hairs which persist in the fruit, but are easily brushed off (though not without irritating the human skin). Placentation is axile. Kenaf is a reproductively photosensitive, short-day plant with varying day-length requirements of 12½ hours or less for flower initiation, depending upon cultivar. Consequently, at any latitude and within the limits set by rainfall distribution, the duration of the vegetative phase of growth, which is so important in bast fibre production, is determined to a large extent by sowing date. Provided other climatic conditions are favourable the crop should be sown at a time which will give the longest period of vegetative growth, and hence the greatest yield of long fibres, before flowers are initiated. Kenaf is self-pollinated and inbreeding, though up to 4 per cent cross-pollination is reported. The fruit is a more or less spherical, dark-brown capsule with an apical point. It is invested by the persistent, hard, spiny calyx and epicalyx. The capsule dehisces into five segments, to disperse the brown or grey, wedge shaped seeds. They contain up to 20 per cent of a semi-drying oil, rather similar to cotton seed oil, which when refined can be used for cooking, and in the manufacture of soap, paints and varnish. It has a milder smell than cotton seed oil and the cake remaining after oil extraction from the seeds is good stock feed.

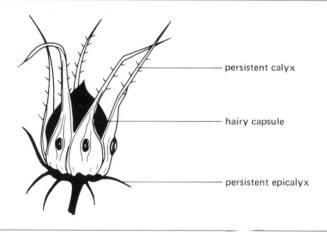

Fig. 10.11 *Hibiscus cannabinus*: Kenaf.   A fruit (x2).

Though hand labour is commonly used to harvest and ret kenaf, and to extract the fibres, much has been done to mechanize production, and so to decrease costs. It can be harvested by machine and the stems can subsequently be passed through a 'ribboner' to strip the bark from them before or after retting. Very coarse fibre is obtained by cleaning unretted ribbons of bark; it is used only to make very coarse bags, which none the less serve a very useful role in the local movement and storage of some produce (e.g. groundnuts), especially when the bag is used only once for this purpose.

A fibre called roselle, which is much like kenaf fibre, is obtained from *Hibiscus sabdariffa* var. *altissima*. It probably originated in West Africa and has been grown on a small scale in many tropical countries, notably in India. The crop is botanically very closely related and similar to kenaf, but it is glabrous and the calyx consists of five large, rounded sepals which tend to become fleshy and which do not have the prominent glands found in kenaf. The species is perhaps better known, and is more often grown, for the fleshy calyces obtained from var. *sabdariffa* which are used to make drinks and preserves.

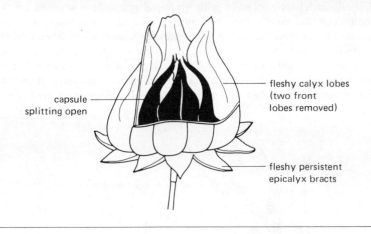

Fig. 10.12 *Hibiscus sabdariffa*: Roselle.   A fruit (x1).

Sunn or Sann hemp: *Crotalaria juncea*

Sunn hemp is grown on a fairly large scale for its bast fibre in India, which is the main source of the fibre in international trade, and on a small scale to supply local fibre needs in several other tropical countries. The fibre is coarser, stronger and more durable than jute, but not so strong as hemp. It is used principally to make twine and rope because it is much less suitable for the manufacture of sacks and bags than either jute or kenaf. It is favoured for making fishing nets.

Crotalaria is a member of the sub-family *Papilionaceae* of the *Leguminosae*, with around 550 species widely distributed in the tropics and sub-tropics. *Crotalaria juncea* probably originated in India, and though it is not known to occur wild, it is cultivated throughout the tropics, more often as a green manure or cover crop than for fibre. As a legume it nodulates freely, producing much-branched, lobed root nodules which may be 25 mm or more across when mature. Sunn hemp sown densely for fibre is a tall annual, reaching heights of 3 m or more with a strong, deep tap root and a well developed lateral root system. When they are sown at wider spacing the plants are shrubby and much branched.

The simple, spirally arranged leaves are narrow and acuminate, softly hairy and often deciduous; they are up to 12 cm long, with short petioles subtended by minute, pointed stipules. The flowers are in open racemes on short peduncles, each flower subtended by a short, pointed bract. The peduncle is covered with short, silky hairs and is at first flexible and pendulous, but becomes stiff when the fruit develops. Just below the flower on the peduncle are two small lateral appendages which subtend the five calyx lobes. The sepals are long and pointed, light green and covered with silky hairs. The three lower sepals are united at their tips and more or less support the keel of the corolla. The two upper sepals lie behind the standard petal and are also united at their tips at first, later becoming free. The corolla is bright golden yellow, with a large, almost round posterior standard petal, two short lateral wing petals, and a twisted keel made up of the two fused anterior petals. The ten stamens are free from each other except at the very base; five stamens with long filaments and small, round anthers alternate with five having short filaments and large, long anthers. The ovary is covered with forward-pointing hairs; the style is long and

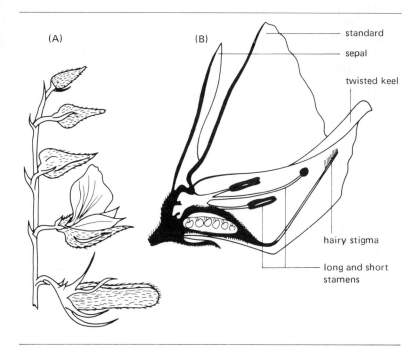

Fig. 10.13 *Crotalaria juncea*: Sunn Hemp. (A) Flowers and fruit (×1). (B) A longitudinal section through a flower.

sharply bent or kneed, with a densely hairy stigmatic surface only on the upper side. Sunn hemp has a typically papilionaceous, bee-pollinated flower, and although the large anthers shed their pollen before the flower bud opens, self-pollination is rare. The fruit is a short, inflated, light yellow or brown pod about 3 cm long and 1 cm in diameter. It is covered with soft hairs, grooved along the upper surface, and terminated by a short beak formed from the base of the style. The mature seeds are loose in the pod; they are dark grey to black, broad, flattened and slightly hooked.

In India sunn hemp is harvested at various stages of growth, from flowering to late pod development, depending upon which of three groups of cultivars is grown, and upon the location. The best fibre is obtained from plants harvested at the early pod stage. The husbandry of the crop, and harvest, retting and fibre extraction methods are much the same as in kenaf. The bundles of fibres harvested consist of interwoven, slightly lignified sclerenchyma cells with rounded ends; each bundle is 1–1.5 m long.

Three other bast fibres are worthy of brief mention, though they have not become important tropical crops. The true hemp of commerce, sometimes called Italian hemp, because Italy was once the chief exporting country, is obtained from *Cannabis sativa* of the family *Cannabidaceae*. Hemp probably originated in Asia and is one of the oldest of cultivated plants. It occurs throughout the tropics and sub-tropics, but grows best in humid, temperate climates, so that most hemp fibre production is now from Russia, southern Europe and South America. There are three main groups of cultivars, one grown for bast fibre, one for the seed which contains up to 35 per cent of a drying oil and the third, which grows best in the tropics, for its medicinal and narcotic properties. Indeed, *C. sativa* has recently become best known in the world not for its fibre, but as the source of 'bhang', 'hashish', 'marijuana', 'pot' or simply 'cannabis', which are some of the names of various narcotic preparations made from the plant to drink or smoke. The narcotic properties of hemp are due mainly to the presence of cannabidol in a resin which occurs in glandular hairs on the leaves, stems and inflorescences. Hemp fibre is valued for its length and strength; the ultimate fibres are up to 55 mm long and taper towards the ends, which are forked. The fibre bundles are up to 4 m long, strong and durable, less flexible than flax, but unaffected by water. They are used for ropes, twines and canvas, but chiefly as a substitute for flax to make textiles. The fibre cultivars are annual, herbaceous, dioecious plants with slender stems 1–5 m tall, more or less branching, depending upon the density of sowing. There appear to be no differences in the quantity or quality of the fibres obtained from male and female plants.

Ramie: *Boehmeria nivea*

The bast fibre called ramie is obtained from *Boehmeria nivea* of the nettle family, *Urticaceae*. It is a native of Eastern Asia where it has been cultivated for thousands of years, and where it is still chiefly grown, though it has been tried experimentally elsewhere. Ramie is a rhizomatous, perennial herb with bulbous storage roots and a number of tall, slender stems arising from the rootstock. It produces one of the best vegetable fibres, prized for its strength, lustre and durability. The ultimate fibres are up to 400 mm long, with an average length of about 140 mm; the fibre has up to three times the tensile strength of cotton and is even stronger when it is wet, but it lacks elasticity and is too smooth to be spun into fine yarn. It is used for a great variety of purposes, especially for twines and yarns or fabrics in which strength, especially strength in water, is important. Ramie has not achieved the world importance its quality suggests it might, partly because it is costly to extract and clean the fibres, and because it cannot be spun into fine yarns. The fibres are coated with a pectic, gummy substance which cannot be removed by retting, but must be scraped from unretted bark and fibres by hand or machine. Up to four crops may be harvested from each plant in one year as a succession of stems is produced from the rootstock, and harvesting may continue for several years until soil nutrients are so depleted that both yield and quality of fibre deteriorate. The crop is usually propagated vegetatively with pieces of rhizome or, less commonly, stem cuttings, and rarely from seed. Two botanical varieties of ramie are known; *B. nivea* var. *nivea* is the true ramie or 'China grass' of Eastern Asia; the under surface of its leaves are matted with short, white hairs. *Boehmeria nivea* var. *tenacissima* is known commonly as rhea, and originated in Malaysia; it does not have a mat of hairs on the under surface of its leaves.

Aramina: *Urena lobata*

Aramina fibre or 'Congo jute' is obtained from the stems of the malvaceous plant *Urena lobata* which may have originated in China, though it is now found throughout the tropics and sub-tropics. In Africa and Asia fibre is gathered from wild plants for domestic use, and the Malagasy Republic has an aramina fibre industry based on wild plants. The crop has been tried in several countries, but is grown on a large scale only in Brazil and Zaire. It is a short lived perennial very variable in form from low-growing, shrubby weeds to little branched cultivars which may grow to heights of 4 m. Aramina is a hairy plant with all the characteristic features of the *Malvaceae*; it has alternate, simple leaves; hypogynous, regular, pink or violet flowers on short peduncles, solitary, or in small groups in the leaf axils; an epicalyx and a staminal column; and five indehiscent, one-seeded schizocarps. For fibre production it is grown as an annual, for though it will produce a second 'ratoon' crop after the first harvest, and is perennial when wild, fibre yields are poor if the same crop is cut repeatedly. It requires a hot, humid but sunny climate, and deep, well-drained, fertile soil. Aramina is grown and processed like kenaf and produces a fibre of good quality which can be used as a substitute for jute to spin fine yarns on jute spinning machinery (unlike most other jute substitutes). The fibre is fine, soft and lustrous, but less durable than kenaf or roselle.

The structural fibres

The structural, or leaf, fibres are the long bundles of sclerenchyma which occur in the leaves of some monocotyledons. They are extracted without retting, and are used principally to make rope and twine.

Sisal: *Agave sisalana*

Agave is a large genus in the monocotyledonous family *Agavaceae* with about 300 species, most of them monocarpic, perennial, xeromorphic plants native to tropical Central and South America. Three species are cultivated, and some are harvested growing wild, for the hard structural fibres in their large, often spiny, leaves, but only *A. sisalana*, sisal, and *A. fourcroydes*, henequen, have become important fibre crops. Sisal is the world's foremost cordage fibre, and accounts for some 65—70 per cent of total cordage fibre production. Cantala or maguey fibre is obtained from *A. cantala* which is cultivated in India, Java and the Philippine Islands. A stiff fibre called ixtle (or istle) is obtained from wild plants of *A. lecheguilla* in Mexico and used to make brushes.

Agave sisalana originated in Central America and Mexico where it now occurs as a wild plant, as well as in cultivation. The crop, and the species, get their name from the small Mexican port of Sisal, from which the fibre was first exported. At the end of the nineteenth century it was introduced into East Africa and to Brazil. Tanzania became the leading producer of sisal, followed by Brazil, both with outputs of more than 200,000 tonnes each year in the 1950s. The fibre was used mostly for binder twine to tie harvested temperate cereals into sheaves before they were stacked; then, when combine harvesters replaced binders, as baler twine to tie bales of straw. During the 1960s increasing competition from synthetic fibres caused a large fall in the world market price for sisal, and consequently decreased production. A dramatic recovery took place in 1974 after the price of mineral oil, which is the basic raw material from which many 'man-made' fibres are synthesized, increased four-fold. As well as cordage, sisal is used to pad upholstered furniture, for coarse-produce bags and in the manufacture of fibre boards.

Though it is a xerophyte, and can withstand long periods of drought, sisal grows best as a crop in savanna zones which have an annual rainfall of 1,000—1,300 mm. With greater rainfall, or on soils subect to waterlogging, the swollen stem, or bole, tends to rot. The best yields of fibre are obtained when the crop is grown on fertile, freely draining sandy loams, though it will grow on poor, shallow soils. In East Africa sisal is grown on plantations in the hot, humid coastal belt and in the relatively dry, cool climate of the central highlands of Kenya at up to 1,800 m above sea level.

The condensed, swollen stem may be 1 m tall and 20 cm in diameter by the time a plant flowers. It is a food-storage organ, and provides nutrients for the very rapid growth of the inflorescence. Rhizomes grow from leaf axils below the soil surface and spread widely from the bole to give rise to new plants, called suckers, at their tips. The roots are adventitious from the base of the bole and from the rhizomes; they too extend widely in the top 40 cm of the soil. Sessile leaves arise in a dense spiral from the apical meristem of the bole, at first vertically, but gradually assuming a greater angle from the stem as younger leaves replace them. The leaves are lanceolate and commonly about 120 cm long, though rarely as long as 180 cm, 10−15 cm wide in the middle, triangular in section, and tapering to a very sharp, lignified spine at the apex. Cultivars have a smooth leaf margin, but occasional plants with spiny margins occur; they are undesirable because the crop is harvested by hand and spiny leaves slow down the work and may injure the workers. The leaves are dark green and covered with a waxy bloom. The vegetative phase of growth, during which 200−250 leaves are produced, commonly lasts 7−10 years, and is terminated when the apical meristem becomes reproductive and produces an inflorescence. Though the length of the vegetative phase varies, and is said to last up to 20 years in Mexico, the total leaf number remains more or less constant, which suggests that it in some way determines the time of flowering. Though sisal can withstand drought, leaf production stops if drought is prolonged. Consequently the number of leaves produced each year is fewer in areas with monomodal rainfall distribution and a long dry season than where rainfall is bimodally distributed, as it is in East Africa.

Sisal plants flower only once, then they die. The inflorescence is a very large panicle carried on a flowering shoot called the 'pole' which reaches a height of 5−6 m with scale-like bracts along its length. The panicle has 25 or more primary branches, which branch repeatedly at their ends; the ultimate branches carry flowers in clusters. The regular perianth consists of six segments fused along most of their length to form a tube 5−6 cm long with six short lobes at the top. Six free stamens are inserted at the base of the perianth tube; they have long filaments which expose versatile anthers beyond the mouth of the tube. The ovary is inferior with an awl shaped style and a three-lobed stigma. The flowers are protandrous and insect pollinated; the anthers dehisce 2−3 days before the stigma of the same flower is receptive. Sisal rarely sets seed because most flowers wither, and are shed before the fruits, which are capsules, develop. Reproduction is predominantly vegetative, either from the rhizomes, or from bulbils which grow in the axils of bracts just below the insertion of each flower. Each bulbil is a small plant with a stem apex, several small leaves and a few adventitious root buds. They are shed from the panicle and germinate in moist soil to produce new plants. Bulbils are used as the planting material for the vegetative propagation of the crop. They are gathered, sorted into

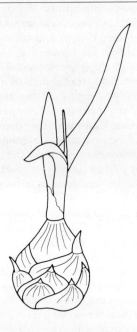

Fig. 10.14   *Agave sisalana*: Sisal.   A bulbil (x½).

uniform size groups, and planted in a nursery. Young plants of the same size, about 30–40 cm tall, are selected from the nursery for field planting. Uniformity is important to ensure that the crop develops evenly, and that all, or most, plants are ready for the first harvest of leaves at the same time, which is 2–4 years after planting. Bulbils are preferred to suckers for propagation because more uniform crops are grown from them. The crop is usually sown at populations of 5,000 plants per hectare in pairs of rows with 0.8 m between plants in a row, 1 m between the rows of a pair and, to permit access, 4 m between pairs of rows. It is also common practice to grow other crops, especially legumes, cotton or maize, between the pairs of rows until the sisal is ready for the first harvest. During early growth the crop is kept free of weeds, and suckers are removed.

The first harvest is taken when the oldest leaves begin to wither, or when they bend to touch the ground. The number of leaves taken at each harvest is less important than the requirement that enough are left for the plant to continue normal growth. In East Africa the first harvest is taken when plants each have about 120 leaves; 25 are left on each plant. After subsequent harvests, which are taken once each year, only 20 leaves are left, and when the plants have flowered all the leaves are removed. Each leaf is cut off close to the bole by hand. The leaf spine is removed and bundles of leaves are taken to the decorticating factory where the fibre is extracted by machine. As the leaves are passed through the decorticator large quantities of water, as much as 45,000 litres per hour, are sprayed on to the fibres to wash them and to carry off the waste. A single modern decorticator can process the leaves from 1,500—3,000 hectares of sisal each year, but to do so (and to establish a sisal estate at all) an ensured supply of large quantities of water must be available. After the fibre bundles have been extracted and washed they are dried in the sun or in drying machines, brushed to achieve uniformity of length, and to remove short, broken fibres, or 'tow', which constitutes about 5 per cent of the fibre extracted from the leaf, and graded according to length, colour and cleanliness.

The fibres in the sisal leaf are of two kinds:

1. Mechanical, supporting fibres found mostly in the outer tissues of the leaf close to the epidermis, constitute about 75 per cent of the total in each leaf. In cross-section these bundles are horseshoe shaped; they vary in length from a few centimetres to 1.5 m and are made up of ultimate fibres 2—5 mm long. Characteristically, mechanical fibres do not split in the decorticating and cleaning processes and they largely determine the fineness of the sisal.

2. Ribbon fibres are those associated with the vascular tissues of the leaf, and are most frequent in its centre. The bundles are crescent shaped in cross-section and may be as long as the leaf. Ribbon fibres associated with the xylem are thin walled and weak; they are removed during the decorticating and washing process. Those associated with the phloem are of good quality, but tend to split during processing. They extend beyond the tip of the leaf to form the sharp leaf spine.

Though sisal rarely produces seed and it is impossible to synchronize flowering deliberately in plants of different species or cultivars, plant breeders have obtained hybrids and have succeeded in producing some improved cultivars. The objectives of breeding are, of course, to increase the yield and quality of fibre, but also to achieve early maturity to first harvest and a high rate of leaf production throughout the life of the crop, resistance to diseases and to the sisal weevil, and good adaptation to local environments.

Henequen: *Agave fourcroydes*

Agave fourcroydes is a Mexican species which has been cultivated there as a source of henequen fibre for centuries. Annual production of henequen in Mexico is about 140,000 tonnes each year; it is also cultivated on a much smaller scale in Cuba and El Salvador. Henequen plants closely resemble sisal in appearance, but have grey leaves with spiny margins. The fibre is hard and wirey, very durable and elastic, but less attractive than sisal. It is used almost entirely to make twine.

Abacá, Manila hemp: *Musa textilis*

Abacá fibre is obtained from the sheathing leaf bases of *Musa textilis*, a member of the banana family, the *Musaceae*. It is the strongest of the structural fibres and, because it does not deteriorate or rot in fresh- or saltwater, and is elastic, light and durable, it is used mostly for the manufacture of ships' cables and ropes, for strong sacking, coarse fabrics and strong paper. Abacá was probably first cultivated in the Philippine Islands, which have always been the world's leading producer with an output which has risen to about 60,000 tonnes each year at the present time. It is from the city of Manila in the Philippines that the name 'Manila hemp', or simply 'Manila', comes, though abacá fibre is not a hemp, but gradually replaced hemp (from *Cannabis sativa*) in the cordage industry of Europe during the early nineteenth century. The facts concerning the origin of the crop are obscure, and there is doubt whether it now occurs wild at all; like banana, abacá is thought to be hybrid in origin. Abacá and banana plants look sufficiently alike to be confused, though abacá has a more tufted growth habit with more slender, pale green false stems (not grey as in banana), and smaller, narrower leaves than banana. Unlike banana, abacá regularly produces viable seeds.

Musa textilis is a tufted, herbaceous perennial with a shallow adventitious root system in the top 30 cm of the soil, though the depth of rooting varies between cultivars. The crop grows best in the lowland, wet tropics where annual rainfall is more than 2,000 mm, evenly distributed throughout the year, and on friable, free-draining soils rich in organic matter. It is susceptible to short periods of drought, and to excessively wet, waterlogged soils. Cultivars with shallow roots are liable to lodge and are not grown in places subject to seasonal typhoons. The adventitious roots arise from a cylindrical corm, and from the many short rhizomes which grow from it. Aerial 'suckers' produced by the rhizomes form a clump of as many as 30 'pseudostems', each consisting of the sheathing bases of 16—25 leaves. The leaf bases are tightly rolled around each other to form a false stem which eventually grows to 30—40 cm in diameter and 5—8 m tall, with the large laminas clustered at the top. Successive leaves grow inside, and longer than those preceding them. The lamina is 1—3 m long and 25—30 cm wide, with a broad thick mid-rib from which parallel veins arise at right angles; it splits easily between the veins when buffeted by the wind. Eventually, after 2 or 3 years' vegetative growth, the apical meristem of the condensed stem becomes reproductive. Instead of leaves, the inflorescence, a large spike, is pushed up through the cylinder of leaf bases on an unbranched, white peduncle 5—8 cm in diameter. The axis of the spike carries closely overlapping, reddish-brown to green bracts about 10 cm long, each with two rows of flowers concealed in its axil. Flowers at the base of the spike are female, those at the top male as a consequence of the early abortion of either the androecium or the gynaecium. The proportion of female and male flowers in each inflorescence may be determined by the nutrition of the plant, but there are usually more male flowers than female. The bracts at the apex of the spike remain tightly folded, never exposing the flowers beneath them; the rest become reflexed as the flowers and fruits develop. The perianth consists of six segments, five of them fused and bearing short, horn-like hooks, the sixth free. The perianth segments are not differentiated into sepals and petals. Female flowers have an inferior ovary of three united carpels with numerous ovules in axile placentation. The male flowers have five free stamens and a sterile organ interpreted as a staminode or as a rudimentary pistil. The fruits are small, hard, green berries, 5—8 cm long, which curve upwards when they are mature. They are inedible, and contain numerous large, black seeds. Abacá flowers are cross-pollinated by bats.

A large number of abacá cultivars is grown in the Philippines and a few in Central America. They differ in agronomically important characters such as tolerance of heavy clay soils and resistance to disease; in rooting depth and consequently resistance to lodging and drought; in earliness of maturity, frequency of production of suckers and tolerance of repeated harvesting which together determine the productive life of the crop; and in ease of fibre extraction, fibre yield and quality. The crop is propagated vegetatively with suckers, or with whole corms which become established more quickly than suckers. The first harvest of pseudostems for fibre extraction can be taken when the crop is 2 or 3 years old, and harvesting may continue for 5—15 years or even longer, depending upon cultivar and growing conditions. Eventually, replanting becomes necessary when the yield of fibre decreases to uneconomic levels.

Individual pseudostems are harvested just before, or soon after, the inflorescence is exserted from the pseudostem, when the fibres have the greatest tensile strength. On one plant there may be three or four pseudostems ready for harvest at one time, and many other younger ones in various stages of vegetative development. Each plant is harvested at intervals of 4—6 months. The pseudostems are cut off close to the ground, the laminas are removed and discarded with the stalk of the inflorescence because they contain no commercially useful fibre, and the leaf-sheaths are separated into three or four groups which produce different qualities, or grades, of fibre. The best quality is from the three outer, oldest sheaths, and the poorest, of soft, weak fibre, from the innermost seven to eight sheaths. The edges of each sheath, which constitute about 15 per cent of its weight, are the only parts containing useful fibre; the centre is soft, parenchymatous tissue without fibre. The fibrous edges are removed by hand in strips 5—8 cm wide called 'tuxies', and from these the abacá fibre bundles are removed by drawing them over a knife blade, mechanically or by hand. A large amount of water is needed for washing the fibre. The fibres are dried quickly in the sun and pressed into bales.

The ultimate fibres of abacá are up to 12 mm long, with uniformly thin walls, a large lumen and tapering ends. They contain about 60 per cent cellulose, 20 per cent hemicelluloses and 5 per cent lignin. The fibre bundles are 1–3 m long, white or reddish yellow, light, stiff and lustrous. They are markedly hygroscopic, and absorb up to half their weight of water in a saturated atmosphere.

Other structural fibres

Among a wide variety of tropical plants from which hard fibres for local use are obtained, the following are some which have achieved minor commercial importance. The first three to be discussed here belong to the same family as sisal, the *Agavaceae*, and in many respects resemble sisal in form and habit, being xerophytes with short stems and a rosette of large, fleshy, simple leaves. They are also monocarpic perennials, and flower only once after periods of vegetative growth which may last for many years.

Furcraea gigantea is native to tropical South America, and is cultivated on the island of Mauritius for its leaf fibre, Mauritius hemp, which is used there mainly to make sugar bags. It is one of the largest members of the *Agavaceae*, and from its short, stumpy stems produces leaves up to 2 m long with prickly margins and an apical spine.

Phormium tenax is the source of New Zealand hemp. Unlike most other members of the *Agavaceae*, it is not from the New World, but is native to New Zealand. It is cultivated there and in South America, South Africa, Japan and in St Helena where it is the chief export. Its fibres are softer and weaker than sisal or abacá, and though it is now used mainly for cordage, it was used by the Maoris of New Zealand to weave fine fabrics. The plant has short, fleshy, creeping, branched rhizomes bearing tough, sword-shaped leaves up to 4 m long and 8–10 cm wide, without spines.

Bow-string hemp consists of strong, white elastic fibres obtained from the leaves of several species of *Sansevieria*, another member of the *Agavaceae*. The genus includes about 60 species widely spread throughout tropical Asia and Africa. Fibres for local use are extracted from some species with large leaves. In Africa the fibre is used to make bow strings, from which the name is derived. *Sansevieria* fibre attained some prominence during the early years of this century when considerable work was done to determine the fibre-producing potentialities of various species, but it has not become an important crop anywhere. Cultivated species are herbaceous perennials with creeping horizontal rhizomes. Their thick leathery leaves vary in length up to 1.8 m, and each one is subtended by a large bract. The leaves are flat, concave or cylindrical in cross-section and their surface is often variegated with blotches or stripes of grey. Some species with variegated leaves are popular ornamental plants.

Palm fibres: Many of the palms of the tropics are used locally as sources of coarse, tough fibres which come from the leaf bases or the leaves, and are extracted by a process of beating and retting. They are used to make ropes, rough bags and matting which have many local uses in building, furniture making, etc. The majority of these fibres never enter world trade, but piassava and coir are produced on a large enough scale to meet a small, but definite, world demand. Piassava consists of long, wiry, dark brown, somewhat flexible fibres used in the manufacture of brooms and brushes. They are extracted in various parts of the world from a number of different palms including *Raphia* spp. on the coast of West Africa and *Attalea funifera* in the Amazon and Orinoco regions of South America. The leaf bases are retted in running water and then beaten to extract the fibres. The fibre called raphia is obtained by removing thin strips from the upper surface of very young leaflets as they are unfolding.

Coir fibre is obtained from the mesocarp of the coconut fruit, and is produced as a subsidiary industry wherever coconuts are grown for their copra, which is the source of coconut oil. After the husks have been split and removed from the fruits they are retted in salt or brackish running water for a period of 6–9 months or more, then the fibres are cleaned, dried and sorted into different lengths which are used for a variety of purposes, principally to make coarse ropes, matting and brushes. Large quantities are also used as stuffing for furniture and mattresses. South India and Sri Lanka are the largest producers.

Further reading

General

Dempsey, J. M. (1975). *Fibre Crops*, Gainesville: University Presses of Florida.
Kirby, R. H. (1963). *Vegetable Fibres*, London: Leonard Hill.
Purseglove, J. W. (1968 and 1972), *Tropical Crops*, 2 vols, London: Longman.

Cotton

Balls, W. L. (1953). *The Yields of a Crop Based on an Analysis of Cotton Grown by Irrigation in Egypt*, London: Spon.
Brown, C. H. (1953). *Egyptian Cotton*, London: Leonard Hill.
Brown, H. B. and Ware, J. O. (1958). *Cotton*, 3rd edition, New York: McGraw-Hill.
Cardozier, V. R. (1957). *Growing Cotton*, New York: McGraw-Hill.
Empire Cotton Growing Corporation (subsequently Cotton Research Corporation). *Empire Cotton Growing Review and Progress Reports from Experimental Stations.*
Hutchinson, J. B., Silow, R. A. and Stephens, S. G. (1947). *The Evolution of Gossypium*, London: Oxford University Press.
Hutchinson, Sir Joseph (1962). The history and relationships of the world's cottons, *Endeavour*, 21, 5–15.
Lagiere, R. (1966). *Le Cotonnier*, Paris: Maisonneuve and Larose.
Meyer, V. G. (1974). Interspecific cotton breeding, *Econ. Bot.*, 28, 56–60.
Munro, J. M. (1966). Cotton and cotton research in Africa, *Field Crop Abstracts*, 19, 173–82.

Prentice, A. N. (1972). *Cotton with Special Reference to Africa*, London: Longman.
Saunders, J. H. (1961). *The Wild Species of Gossypium and their Evolutionary History*, London: Oxford University Press.
Sethi, B. L. et al. (1960). *Cotton in India*, 2 vols, Bombay: Indian Central Cotton Committee.
Smith, C. E. and Stephens, S. G. (1971). Critical identification of Mexican archaeological remains, *Econ. Bot.*, 25, 160–8.

Jute

Banjeree, B. (1955). Jute – especially as produced in West Bengal, *Econ. Bot.*, 9, 151–74.
Kundu, B. C. (1956). Jute – world's foremost bast fibre. I. Botany, agronomy, diseases and pests, *Econ. Bot.*, 10, 103–33. II. Technology, marketing, production and utilization, *Econ. Bot.*, 10, 203–40.
Sarma, M. S. (1969). Jute, *Field Crop Abstracts*, 22, 323–36.

Kenaf, aramina and ramie

Haarer, A. E. (1952). *Jute Substitute Fibres*, London: Wheatland Journals.
Haarer, A. E. (1953). Congo jute and its possibilities, *World Crops*, 5(2), 54–5.
Willimot, S. G. (1954). Ramie fibre. Its cultivation and development, *World Crops*, 6(10), 405–8.
Wilson, F. D. and Menzel, M. Y. (1964). Kenaf (*Hibiscus cannabinus*) Roselle (*Hibiscus sabdariffa*), *Econ. Bot.*, 18, 80–91.
Wilson, F. D. (1967). An evaluation of kenaf, roselle and related Hibiscus for fibre production, *Econ. Bot.*, 21, 132–9.

Sisal

Lock, G. W. (1969). *Sisal*, 2nd edition, London: Longman.
Weink, J. F. (1969). Long fibre Agaves, in: Ferwerda, F. P. and Wit, F., *Outlines of Perennial Crop Breeding in the Tropics*. Wageningen: Veenman and Zonen.

Abacá

Spencer, J. E. (1953). The abacá plant and its fibre, Manila hemp, *Econ. Bot.*, 7, 195–213.

New Zealand hemp and sansevieria

Critchfield, H. J. (1951). *Phormium tenax* – New Zealand's native hard fibre, *Econ. Bot.*, 5, 172–84.
Gangstad, E. O., Joyner, F. J. and Searle, C. C. (1951). Agronomic characters of *Sansevieria* spp., *Trop. Agriculture, Trin.*, 28, 204–14.

Chapter 11

The Vegetable Oils and Fats

Most plants store some of their food reserves as oils or fats, especially in their seeds where they are a concentrated source of energy for germination containing about twice as many calories per gram as carbohydrates or proteins. The oils and fats are chemically similar, but they differ physically; at normal temperatures oils are fluid, whereas fats are solid, and become fluid only when they are heated. A third group of compounds in plants, chemically related to the oils and fats, are the waxes whose chief characteristic is that they are impervious to water. Waxes are a constituent of the cuticle where they prevent water loss from the plant surface. Oils and fats are produced in plants through the synthesis of fatty acids from carbohydrates, and the combination of three molecules of fatty acid with glycerol to form triglycerides. Their composition varies greatly throughout the plant kingdom; even among species of the same genus different kinds of oil may be stored. This variation is due largely to the fact that each vegetable oil is a mixture of different fatty acids which vary in the degree to which they are saturated; that is, they differ in the number of double-bonds in their molecules. The less saturated the fatty acids in the oil, the less stable it is in air, and the more readily it is oxidized to form a thin, elastic, waterproof surface film.

Two kinds of oil are produced by plants. The fixed, non-volatile oils which are a food reserve are by far the most important in commerce and are the subject of this chapter. The 'essential', volatile oils are aromatic, and are commercially important in scents and perfumes. They do not serve as reserves of stored energy in plants and their true function is not known; they are discussed in a later chapter. The fixed vegetable oils are in ever increasing demand in world markets. Millions of tonnes of them are consumed as food or used in industry each year. How they are used depends upon their various chemical and physical properties, but the largest proportion of world production is used in the manufacture of edible fats (e.g. margarine). Edible, fluid oils are converted into edible fats by a process of catalysed hydrogenation in which relatively unsaturated oils become more saturated by combining with hydrogen. Large quantities of vegetable oils are also saponified to make soap. When the oils are mixed with sodium or potassium hydroxide (other metallic salts are sometimes used) the fatty acids combine with the metal to produce soap, leaving glycerin which is a valuable by-product of soap manufacture. The fatty acids from vegetable oils are also used in the manufacture of detergents. Vegetable oils are also used as lubricants, and the least saturated of them (the drying oils) are used in paints and varnishes and to make linoleum.

Though the production of fixed oils by plants and their storage as a food reserve is widespread in the plant kingdom, only about 30 species have been exploited commercially for the extraction of oil, and only twelve of these provide 90 per cent of the total world supply. Most vegetable oil comes from tropical plants, but commercially important 'oil seeds' are not confined to the tropics. The oils and fats are classified in several ways, but for the agricultural botanist it is convenient to divide them into four main groups, as follows:

1. Drying oils are those which absorb oxygen rapidly on exposure to air, and dry to form a thin, elastic film. They react in this way because they contain a large proportion of unsaturated fatty acids, especially linolenic acid with three double bonds in its molecule, linoleic with two, as well as others with three, or even four, double bonds which are conjugated (adjacent to each other). The rapidity with which an oil dries in air is correlated with its ability to absorb iodine, which is referred to as its 'iodine number'. Drying oils have a high iodine number, greater than 130, and include linseed, tung, soyabean, safflower, hemp seed and Niger seed oils. They are used in the manufacture of paints and varnishes, though soyabean oil is equally important as a food, and is sometimes classified as a semi-drying oil. Soyabeans are one of the world's chief grain legumes and have been discussed under that heading in Chapter 4.

2. Semi-drying oils form a soft surface film only after long exposure to air. They contain a large proportion of linoleic and oleic acids, but little linolenic acid, and have iodine numbers of 100–130. The most important oils in this class are from the seeds of cotton, maize, sunflower, sesame and melons. They are used principally as foods, either cooking and salad oils, or, after their unsaturated fatty acids have been hydrogenated, as cooking fats.

3. Non-drying oils remain liquid, and do not form a film on exposure to air; they react slowly or not at all with oxygen, and so have low iodine numbers less than 100. They are characterized by a large content of saturated oleic acid, and include olive oil and groundnut oil which are edible, and castor oil which is used chiefly as an industrial raw material. The edible non-drying oils are hydrogenated to make cooking fats, and all non-drying oils are saponified to make soap. The olive, *Olea europaea*, is not discussed in detail here because it is a crop of warm temperate climates. Groundnuts, though an extremely important oil seed, are also an important grain legume, and were described in Chapter 4.

4. Fats are those vegetable oils which are solid at ordinary temperatures, e.g. coconut oil, palm oil and palm kernel oil, cocoa butter and shea nut butter.

Oil is extracted from seeds commercially by various processes, the commonest being by means of mechanical pressure applied to heated or unheated, crushed seed. The pressure is applied gradually and the expressed oil is refined to remove impurities such as water, dirt and vegetable matter. Pre-heating the seeds with steam before they are pressed yields greater amounts of oil, but its quality may be poor compared with the oil from cold-pressed seeds. Seed oil is also extracted in chemical fat solvents such as benzene, carbon disulphide, trichlorethylene and petroleum fractions. It is separated from them by distillation at temperatures of $60°-110°C$. In local use throughout the tropics oil seeds are crushed in a mortar with some form of pestle, then the crushed seeds are boiled in water until the oil rises to the top where it can be skimmed off. Hand- or animal-powered grindstones are also used to express the oil. Oil seeds are commonly eaten without any prior oil extraction, and some of them are used to make sweetmeats.

The residue of the seeds after the oil has been expressed is the seed cake. It is a nutritious stock feed, commonly rich in protein; but some oil seeds produce cake which contains toxic substances, and though such cake is useless as stock feed it can be used as a nitrogenous fertilizer.

Drying oils

Safflower: *Carthamus tinctorius*

Carthamus tinctorius is a member of the *Compositae*, a large and distinctive family of some 20,000 species, most of them herbs or shrubs, scattered throughout the world. The family includes other important crops such as sunflower (*Helianthus annuus*), pyrethrum (*Chrysanthemum cinerariifolium*), lettuce (*Lactuca sativa*) and Niger seed (*Guizotia abyssinica*), many well known ornamental plants, and a great many common weeds. The members of the *Compositae* are easy to recognize by their characteristic inflorescence, the capitulum or head, which consists of many small flowers, each one subtended by a bract, all clustered together on a flattened receptacle which is surrounded by an involucre of bracts. To the casual observer the inflorescence may appear to be a single flower, especially if it is very small. The flowers are called florets, and are of two kinds. Actinomorphic, tubular or disc florets with an inconspicuous tubular corolla are usually hermaphrodite; they are packed closely together over the centre of the head and are commonly surrounded by a margin of zygomorphic ligulate or ray florets which have a conspicuous, strap-like corolla, and which are usually male or sterile. In a few species the inflorescence has only tubular, or only ligulate florets.

The regular disc florets have five petals fused to form a narrow tube with five small lobes at the top. Five epipetalous stamens have free filaments, but their anthers may be fused into a disc at the mouth of the corolla tube. The inferior ovary is unilocular with a single ovule; the fruit is an achene. The calyx in hermaphrodite flowers is often represented by scales or hairs which arise around the top of the ovary. It may persist with the fruit to form the pappus of hairs which is effective in assisting the wind dispersal of the small fruits of many species. Hermaphrodite flowers are protandrous, and their pollen is shed into the corolla tube before the stigma is receptive. None the less, cross-pollination by insects is common in the *Compositae*.

The irregular ray florets have five petals which are united into a short tube at their base, but which expand into a long, strap-like lobe above. They serve to attract insects to the inflorescence.

Safflower belongs to a group of species in the *Compositae* which are commonly called 'thistles'. It is a much branched herbaceous annual, 0.5–1.5 m tall, with a strong, somewhat thickened tap root, and many thin lateral roots which grow more or less horizontally in the upper layers of the soil. During slow early growth a rosette of leaves is produced close to the ground. Later several erect branches grow from leaf axils in the rosette, and these in turn give rise to more branches higher up the plant. The stems are stiff and cylindrical, fairly thick at the base, but thinner above; they are smooth and glabrous, light grey to white, and marked with fine longitudinal grooves. The leaves are inserted spirally, but at uneven intervals on the stems and branches, and may appear to be opposite. The internodes become progressively shorter at the tops of the stems and branches until the leaves merge into the involucre of bracts around the terminal inflorescence. The leaves are simple, sessile, dark green and glabrous, about 12 cm long and 3 cm wide, with a few short spines along their margins. The broad mid-rib is extended as a short spine from the accuminate tip of the leaf.

The inflorescence of safflower is a dense capitulum about 4 cm across, closely invested by numerous bracts, the lower ones free from each other and more or less leaf-like, the upper ones fused, triangular, spine tipped and covered with soft white hairs. The involucre is conicle, broad at the base, but tapering to a very small opening at the top through which the long corolla tubes of as many as 90 flowers emerge. All the flowers are tubular, about 4 cm long, with yellow to deep orange petals where they are exposed beyond the involucre of bracts. The structure of these flowers is typical of the family, with the exceptions that there is no pappus in the domesticated species, and the fused anthers are exposed beyond the mouth of the corolla tube.

The achenes are smaller than the better-known fruits of the sunflower, but the same shape; they are about 8 mm long, and light grey. The bracts of the involucre and the upper leaves become very spiny by the time the fruits are mature so that the crop is difficult to handle at harvest. Though

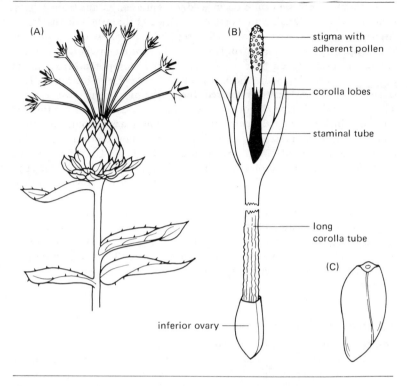

Fig. 11.1 *Carthamus tinctorius*: Safflower. (A) An inflorescence (with most flowers removed) (x¾). (B) A single flower (x1½). (C) An achene.

spineless cultivars have been bred to overcome this problem, they tend to yield less oil than spiny types. The achenes consist of 40–60 per cent by weight of pericarp, 26–37 per cent of oil and 20–55 per cent of protein. The percentages of oil and protein are very variable because they depend upon the thickness and weight of the pericarp. Cultivars with thin pericarps, or with achenes which are easily decorticated in processing, yield a larger percentage of oil and protein. The oil is a good drying oil with iodine numbers up to 149. It contains little linolenic acid, but the proportions of linoleic and oleic acids vary widely between cultivars, linoleic from 9–85 per cent and oleic from 9–87 per cent. The oil is valued in the manufacture of white paint because it retains its colour well, and does not yellow with age. In the United States safflower oil is used mostly for salad oil and to make margarine, in India as a food oil or illuminant and to make soap. Safflower seed cake is useless as stock feed unless the fruits have been decorticated, because the fibrous pericarps make it unpalatable. Decorticated seed cake contains up to 50 per cent protein.

Safflower is not known to occur wild, except as an escape from cultivation. It is most variable in north-east Africa (Ethiopia) and the Iran–Afghanistan area where several wild species of *Carthamus* are indigenous. The crop may have originated from *C. oxyacantha* in the Afghanistan centre of diversity, or from *C. lanatus* in the Mesopotamian centre of origin of crop plants. It has been cultivated in Egypt and the Middle East since ancient times, first for the red dye carthamin, or safflower carmin, which is obtained from the dried flowers and used as a substitute for true saffron (obtained from *Crocus sativus*). Safflower has become an important oil seed only in relatively recent times, and is now cultivated for oil in the semi-arid parts of North Africa, India, China, the United States and Australia. As a crop it is tolerant of heat and drought, and is not subject to lodging in strong winds, but it is not a success in the lowland, humid tropics. Young plants require a period of cool temperature to become established, and hot, dry weather when the fruit is maturing. Safflower can be grown in areas with 500 mm annual rainfall, or even less, provided there are no long, rain-free periods during the growing season. It is propagated by seed and harvested about 4 months after sowing. The fruits are closely invested by the involucre of spiny bracts which protect them from bird damage.

Niger seed: *Guizotia abyssinica*

Guizotia abyssinica is also a member of the family *Compositae* cultivated for the drying oil in its seeds. It is not important in world trade, and is cultivated on a large scale only in India and in Ethiopia, where it probably originated. The oil is edible, with a 'nutty' flavour, and much of it is consumed in the countries where it is produced, either as food, or for soap, paint and illumination.

Niger seed is an annual herbaceous plant up to 1.5 m tall, with softly hairy stems and branches, and sessile, lanceolate leaves up to 20 cm long and 3 cm wide, covered on both surfaces with soft hairs. The leaf margins are dentate, not spiny, and the tip is acuminate. The inflorescence is 2–3 cm in diameter with about eight female ray florets surrounding numerous hermaphrodite disc florets, both with yellow petals which are very hairy at the base. The achenes are black, up to 5 mm long with no pappus. The seeds contain 38–50 per cent by weight of a light-yellow drying oil which is about 70 per cent linoleic acid. The cake obtained after oil extraction is good stock feed with about 33 per cent protein.

Niger seed is propagated by seed. It is often grown on poor soils where annual rainfall is 1,000 mm or less. The crop is harvested 3–4 months after sowing.

Tung: *Aleurites montana* and *A. fordii*

Tung oil is a quick-drying oil, in many respects superior to linseed oil. It

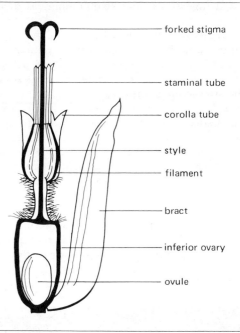

Fig. 11.2 *Guizotia abyssinica*: Niger seed. A diagrammatic longitudinal section through a disc floret (×6).

dries more quickly and evenly without the skin which develops on films of linseed oil. It is resistant to weathering and water and is consequently in demand for the manufacture of outside, protective, oil-based paints and for waterproofing; it is used in enamels and lacquers, linoleum and waterproofed fabrics, in the manufacture of brake linings, pressed boarding and printer's inks. Though the oil has no toxic properties it has a laxative action and is therefore not used as an edible oil.

The tung oil of commerce is obtained from the fruits of two species of the genus *Aleurites* of the family *Euphorbiaceae*. The genus is small with six species which are all native to the tropical and sub-tropical parts of Eastern Asia and Malaysia. They are all trees of varying size, and though all are cultivated for their fruits which contain large amounts of oil, only *A. fordii* and *A. montana* are important sources of tung oil. Both species are natives of China, *A. fordii* from the cooler parts of western China, *A. montana* from the warmer, sub-tropical parts of China and Burma. *Aleurites fordii* has been introduced into the United States, and is the chief source of tung oil in mainland China which supplies more than half of total world production of 120,000 tonnes each year. *Aleurites montana* has been introduced to several tropical countries, but grows well only at high elevations. After mainland China, the leading producers of tung oil are Paraguay and Argentina (about 20,000 tonnes each year) and Malawi (2,000 tonnes).

Aleurites montana is a much-branched tree of the hillsides of southern China where it thrives on poor soils and with little attention. The tree may be 20 m tall, and is deciduous. It grows quickly to produce a small first crop in as few as 4 years, but never attains a great size or age, up to 30 years being the average life span. It is a more or less umbrella-shaped tree, with large ovate leaves, usually partly divided into five acuminate lobes. The leaves are alternate on long petioles, with five to seven veins arising from the base. The petioles have two stalked nectaries at their junction with the blade. The flowers are unisexual, but the two sexes are carried together in the same inflorescence. The inflorescence is a raceme or corymb, with varying numbers of flowers, and is produced terminally on a new season's growth. The proportions of male and female flowers in an inflorescence and on different trees are variable; some trees are predominantly male and bear little fruit, others predominantly female. Female flowers are carried at the apex of the inflorescence, male flowers at the base. Both are bell-shaped with a calyx of two to three lobes, and five white petals exceeding the length of the calyx lobes. The male flowers have a variable number of stamens up to twenty, arranged on a conical

Fig. 11.3 *Aleurites montana*: Tung.   Flowers.

receptacle in three whorls, the five outer stamens alternating with five glands on the disc of the flower. The female flowers have a three- to four-celled ovary, each locule with a single ovule, and a thick, fleshy style with two branches. The fruit is a drupe with a hard, woody pericarp; it is spherical or egg-shaped, up to 6 cm in diameter, marked externally by three longitudinal ridges and numerous transverse ones. The drupe contains three, rarely four, broad, oval seeds about 3 cm long.

Aleurites montana will grow well on soils too poor for other agricultural use, but thrives best on well-drained, deep, slightly acid soils. It will not tolerate alkaline soils, nor waterlogging, and requires at least 1,000 mm of rainfall during the growing season, followed by a cool dry season such as is found in the tropics only at altitudes greater than about 600 m. The fruits are harvested from the trees, or they may be allowed to fall to the ground where they are left to dry before they are gathered. The woody pericarp is removed by hand or mechanically, the testas are removed and the kernels are crushed, then pressed to express the oil. Air-dried, mature kernels yield about 60 per cent of tung oil. The seed cake is used as a nitrogenous fertilizer. The oil consists of up to 75 per cent of eleostearic acid and has an iodine number of 163.

Other drying oils

Though linseed is principally a crop of temperate agriculture, and the botany of soyabean is discussed in detail elsewhere in this book, both deserve brief mention here because they are so important as oil seeds. Linseed, *Linum usitatissimum*, of the family *Linaceae*, was one of the first crops cultivated for its oil, as well as for its bast fibre, flax. Linseed oil was for many centuries the most important drying oil in industry, though in recent years other vegetable oils and synthetic protective coating materials have been used in greater amounts. Present world production of linseed, which contains up to 43 per cent drying oil, exceeds 2.5 m. tonnes each year, most of it from North America, Russia, Argentina and India. The oil consists of about 75 per cent linoleic and 17 per cent linolenic acids; it is a yellow-brown liquid which sets quickly to a hard, elastic film when exposed to the air. As well as its use in paints and varnishes, it is used in printer's ink, soft soap, oil cloth and linoleum, and as a sealer sprayed on to the surface of newly laid concrete. The oil seed cultivars are bushy, much branched plants unlike the taller, little branched flax cultivars.

World production of the seed of soyabean, *Glycine max*, of the family *Leguminosae*, is now more than 60 m. tonnes each year. In terms of its contribution to human oil and protein requirements it is the most important legume of all. Some two-thirds of world production is from the United States (43 m. tonnes) where soyabeans are grown for the extraction of their oil and for their protein. The development of the soyabean industry in the United States began in the 1930s, and expanded rapidly after the Second World War. Previously the crop was important only as a grain legume in the Far East, especially in mainland China, which remains the second largest producer (11.8 m. tonnes each year) and where the seeds are still a major source of protein in human diets.

Soyabean seeds contain 13—25 per cent of an edible oil, classified by some as a drying oil, by others as a semi-drying oil, and up to 50 per cent protein. The oil contains about 50 per cent linoleic, 30 per cent oleic and 7 per cent linolenic acids. It is used extensively in the manufacture of drying oil products, and for margarine or as salad oil.

Semi-drying oils

The most important tropical crops from which semi-drying oils are obtained are cotton, maize and sesame (*Sesamum indicum*). Sesame is described below, and though the botany of cotton and maize has been discussed elsewhere, their importance as sources of vegetable oil in the United States deserves emphasis here. Only a small proportion of the United States maize crop goes for the extraction of oil from the embryos, but since production is around 140 m. tonnes of grain each year, maize makes a substantial contribution to American cooking and salad oil requirements. In contrast, some 90 per cent of the 4.5 m. tonnes of cotton seed from United States ginneries goes for oil extraction, yielding around 2 m. tonnes of oil each year, most of which is used to produce margarine and cooking fats. Two other crops, sunflower (*Helianthus annuus* of the *Compositae*) and rape (*Brassica* spp. of the cabbage family, *Cruciferae*), are major sources of semi-drying oil not dealt with in detail here because they are usually classed as temperate crops. Annual world production of sunflower seed is around 9 m. tonnes, 5 m. from Russia; Africa produces less than 200,000 tonnes, and Asia less than 700,000. Sunflower probably originated in Peru or Mexico and reached Europe in the sixteenth century. It is not known as a wild plant. The cultivars are very large herbaceous annuals, rarely as tall as 5 m, with a capitulum as much as 40 cm in diameter. Many hermaphrodite disc florets are spirally arranged on the receptacle and surrounded by a row of bright yellow, sterile ray florets. The achenes are about 1 cm long, variously coloured from white to black. They normally contain around 30 per cent oil, though cultivars bred in Russia may contain up to 50 per cent. The composition of sunflower seed oil is more dependent than that of many other seed oils on the environment in which the crop is grown, especially variation in temperature. The oil from crops grown in very hot countries may contain as little as 20 per cent of linoleic acid and as much as 65 per cent of oleic acid; that from sunflowers grown in the coolest climates in which seed can be ripened may contain 70 per cent of linoleic and only 15 per cent of oleic acid, and is consequently a better drying oil (with a larger iodine number).

The Brassicas, species belonging to the genus *Brassica*, are erect, branched, glabrous plants with strong, deep tap roots. Their pale yellow, regular flowers are carried in long, simple racemes. They have four erect sepals, four long-clawed, free petals, six stamens and a superior ovary of two fused carpels with ovules in parietal placentation. The fruit is a type of pod, a siliqua or silicula, in which dehiscence occurs from the base upwards when the two valves (the dry carpels) separate and curl apart, leaving a false septum in position between them. The small, spherical, black seeds contain 30—40 per cent oil and around 20 per cent protein. The oil is used extensively in India for food and illumination, and elsewhere as a lubricant, edible oil, in synthetic rubber and for fuel. The chief oil seed Brassicas are *Brassica napus*, rape, *B. campestris*, field mustard, and *B. juncea*, Indian mustard. Rape is a crop of temperate climates and is grown chiefly in Europe and Canada as a fodder crop and for its seed oil. Field mustard is a source of cooking oil in India, but is less important there than Indian mustard (*rai*) which provides the chief cooking oil in northern India, Pakistan and Bangladesh, and is grown elsewhere in Asia and in parts of Europe and Africa. Its seeds contain around 35 per cent of a semi-drying oil.

Many other species are of minor importance as semi-drying oil seeds, some of them only after the fruits have been processed for other purposes. For example, the seeds of *Citrus* species and tomatoes (*Lycopersicon esculentum*) are processed for oil extraction after the juice or pulp has been taken from the fruits; in Japan even the bran of rice is utilized as a source of semi-drying oil obtained by solvent extraction. Several members of the family *Cucurbitaceae* produce seeds containing sufficient oil to make extraction a worthwhile undertaking. Chief among these are the water melon, *Citrullus lanatus*, with bitter fruited wild forms in Africa, and the melon, *Cucumis melo*.

Sesame, simsim, benniseed: *Sesamum indicum*

Sesamum indicum is perhaps the most ancient crop cultivated for its oil and was known in Iran at least 4,000 years ago. All but one of the twenty species in the genus are indigenous to Africa; this and other evidence suggests that sesame was domesticated in Africa, probably in Ethiopia. The crop was spread very early to India where it came to be cultivated to a much greater extent than in its place of origin. India, mainland China and Burma now produce almost half the world crop of 2 m. tonnes of seed each year, whereas Africa (principally the Sudan) produces only 0.5 m. tonnes. Sesame is a member of the family *Pedaliaceae*, which is closely related to the *Scrophulariaceae*, but its members have a unilocular ovary with two parietal placentas which grow inwards, producing two false locules with axile placentation. Further intrusive placentas may develop forming an apparently four- or even eight-loculed ovary.

Sesame is an erect, more or less branched annual, covered with glandular hairs, and has square, grooved stems commonly around 1 m tall, but sometimes as tall as 3 m. A strong tap root penetrates as deep as

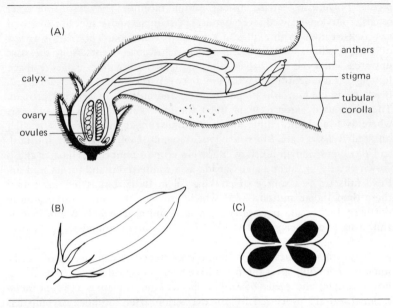

Fig. 11.4 *Sesamum indicum*: Sesame. (A) A longitudinal section of a flower. (B) A capsule. (C) A transverse section through a capsule.

90 cm, and a dense, much branched lateral root system spreads in the surface soil. The leaves are opposite near the base of the stem but alternate above; they are carried on light-green petioles around 5 cm long without stipules. They vary in shape from lanceolate to palmate, and have coarse teeth along their margins. A pallisade layer occurs beneath both the upper and lower epidermis. The flowers are normally single, but rarely in groups of two or three in the axils of leaves on the upper parts of stems and branches. At the base of the very short peduncles is a pair of prominent, cup-shaped extrafloral nectaries, subtended by deciduous bracts which also appear to subtend the flower. The flower is irregular, with a two-lipped tubular, bell-shaped corolla 3—4 cm long, which widens gradually from the base into five lobes at the mouth of the tube; it is white or pink, with red or purple spots inside the tube, and hairy outside. The short, hairy calyx lobes are united at the base, narrow and acuminate above. Four stamens are arranged in pairs, one pair shorter than the other, lying against the upper lip of the corolla tube. The connective of the yellow anthers is prolonged into a short, swollen beak. The superior ovary of two united carpels eventually become four-, or even eight-loculed with the intrusive growth of the parietal placentas. The anthers dehisce before the flower opens, so that self-pollination is the rule; and though a very small amount of insect cross-pollination occurs (less than 1 per cent of seeds come from cross-fertilization), cultivars are more or less true breeding, pure lines. The

fruit is a rectangular, deeply grooved capsule up to 3 cm long with a short beak. It dehisces by two apical pores through which seed may be lost before or during harvest, though indehiscent forms have been bred. The seeds are about 3 mm long, flattened and ovate, with a smooth or reticulate white, yellow, red, brown or black testa. Seeds with white testas yield the best quality oil.

The seeds contain between 45 and 58 per cent of a very good semi-drying oil, mostly oleic and linoleic triglycerides. It is used mainly for food, but also in the manufacture of soap, and locally in India and China as an illuminant. In Africa and Asia the seeds and young leaves are eaten in stews and soups. The seeds are often made into confectionery, and when decorticated are used as an ornamental and flavouring dressing on cakes and pastry. In Europe and Asia the oil is usually expressed in three stages. The first pressing is made cold, and the oil, after filtration, is suitable for use without further treatment. The two subsequent pressings are made using the heated residues from the first. They are subjected to greater pressure, and yield inferior, darker-coloured oils which need to be refined and deodorized before use. The fine oil from the first pressing is a clear yellow colour, and may be used as a substitute for olive oil. An important characteristic of sesame oil is that it is stable and does not become rancid. The cake remaining after extraction is a very good cattle food. It contains about 35 per cent protein and appreciable quantities of calcium and phosphorous. The cake is also used as fertilizer, and the dry stems are used for fuel.

Sesame is mostly grown as a rain-irrigated crop of the semi-arid tropics. It is drought resistant and will succeed where there are as few as 400 mm of rain during the growing season, though it is more often grown in areas with monomodal distribution of 500–1,100 mm summer rainfall. The crop is sensitive to excessive rainfall and waterlogged soils, and does best under rain irrigation on deep, free-draining, sandy soils. In India it is grown with irrigation in some areas, in rotation with rice and vegetables. A very large number of cultivars is known, some very localized, a common situation where ancient, inbred crops are part of traditional peasant agriculture. They vary in height, degree of branching, number of locules per capsule and time to maturity. Some cultivars are insensitive to day-length for flowering, while in others flowering is hastened by short days.

Non-drying oils and fats

The non-drying oils do not form films on exposure to air, but retain their physical properties for considerable periods, and are fairly stable chemically. They are used mostly as salad and cooking oils, to make margarine and soap, and as lubricants and illuminants. The most important non-drying oils are olive, castor, groundnut and palm oils, the last usually classified as fats because they are solid at temperatures below around

24°C, though they are chemically similar to the fluid oils. Olive oil is the most valued of all vegetable oils for use as a salad or cooking oil in Europe and North America. It is obtained from the drupes of the warm temperate plant *Olea europaea* of the family *Oleaceae*, a small evergreen tree which has been cultivated in the Mediterranean basin for its oil since ancient times. World production of olive oil is around 1.5 m. tonnes each year, most of it from Spain, Italy, Greece and Turkey. Olive trees may grow well in some tropical climates, but produce little fruit. Groundnuts, or peanuts, are described fully in the chapter on legumes, but deserve mention here because they are a major source of non-drying oil, with world production of seeds 'in shell' of about 17 m. tonnes each year, equivalent to 12 m. tonnes of seed or 'kernels', half from Asia and one-third from Africa. The kernels contain about 50 per cent by weight of oil with an iodine number around 90, and up to 30 per cent protein, and the crop is perhaps more important as a source of vegetable protein in human nutrition than it is as an oil seed. Most oil expression is from kernels imported into Europe, where the seed cake is an important stock feed; 90 per cent of United States production of 1.6 m. tonnes ('in shell') is used to make peanut butter or as confectionery peanuts for human consumption, and accounts for some 5 per cent of North American protein intake.

Non-drying oils are extracted from the seeds or fruits of a great many other plants, usually for local domestic use and not in economically significant quantities. A few among these are the oil from the fruits of avocado, *Persea americana*; cashew nut oil from the kernels of *Anacardium occidentale*; ouricuri kernel oil from the seeds of *Cocos coronata* (syn. *Syagrus coronata*); sawarri or souari oil from the fruits and seeds of *Caryocar* spp.

Castor oil: *Ricinus communis*

Castor oil is obtained from the seeds of *Ricinus communis* of the family *Euphorbiaceae*, which grows wild in Africa, Asia and America, but which probably originated in Ethiopia, or at least in north-east Africa. With sesame it was one of the ancient oil crops of early Mediterranean civilizations, and spread to India and China before recorded history. The earliest use of the oil was probably as a medicinal purgative, as a base in cosmetics and as an illuminant; in recent years it has been used industrially as a high-grade lubricant in precision piston engines, in hydraulic fluids, in soap manufacture, and, after dehydration, in the drying oil industry as a good substitute for tung oil. In paints dehydrated castor oil has the advantage of drying to a flexible film which does not yellow with age. World production of castor seed is estimated to be more than 0.8 m. tonnes each year with an average oil content of 44 per cent. Brazil produces most, around 0.4 m. tonnes, then India (0.14 m. tonnes), but both these countries use all they grow. The major exporters are mainland China and Thailand; other countries where castor is a major crop are Russia, Tanzania, Sudan, Ecuador and Paraguay.

The genus *Ricinus* is monotypic, *R. communis* being the only species, though it is very variable and was in the past divided by some authorities into three separate species, or five sub-species. As a wild plant in Africa it is a short-lived, woody perennial and may attain the size of a small tree, exceptionally up to 12 m tall. Cultivars, though naturally short-lived perennials, are cultivated as annuals, some of them dwarf, and grow to heights of 1–7 m. They have a well-developed tap root with thick, horizontal laterals which branch profusely to produce a characteristically quick-growing mat of roots in the surface soil. The erect, glabrous stems are often much branched, though in some cultivars branching is confined to the upper parts of the stem. Branching is sympodial. The apical meristem produces an inflorescence, and axillary buds below it grow into branches which in turn produce apical inflorescences and sub-terminal branches. The stems are variously coloured depending upon differences between cultivars, and are green or reddish, with well defined nodes and prominent leaf scars. The internodes are shortest at the base of the stem, progressively longer above. The stems become hollow with age. The leaves appear to be alternate but are arranged spirally in two-fifths phyllotaxy; they are borne on stout petioles up to 50 cm long, and are subtended by a pair of deciduous stipules which envelop and protect the leaf bud. The petioles are rounded in section, broadly clasping at the base, green or reddish like the stems, and have conspicuous oil glands. The dark, glossy-green laminas are deeply palmately lobed, and up to 75 cm across with prominent veins on the under surface; they have five to eleven lobes which are more or less acuminate, and serrated margins. The terminal dense panicles are up to 40 cm long with unisexual flowers, about half of them male in cymes at the base of the inflorescence, the rest female at the top. The flowers have no petals, but are protected in the bud by three to five sepals which remain united at the base when the flower opens. The male flowers have numerous branched stamens, each branch bearing a small, almost spherical pale yellow anther which dehisces explosively when it dries, scattering the pollen grains. Pollination is by wind, or by insects attracted to plants by the nectar from glands on the leaves and petioles. Castor is protogynous and cross-pollination is usual. Female flowers have a superior, three-celled ovary terminated by three styles, each divided into two fleshy, pink lobes with papillate surfaces. The ovary is covered with fleshy, green spines, tipped by transparent bristles which break off as the fruit develops. The fruit grows quickly after fertilization to form a globular capsule covered with the spines of the ovary wall, and with the persistent remains of the styles. Each locule contains a single seed which may be pale buff or almost black, or commonly mottled black on a buff background. The seeds are 0.5–1.5 cm long, more or less oval and flattened, with a thin, brittle testa and a large, yellowish-white caruncle. Most of the seed is occupied by the large, oily endosperm, and the embryo is small with two thin cotyledons. The seeds look like well-fed ticks, and the generic name of castor comes from the Latin name for that insect.

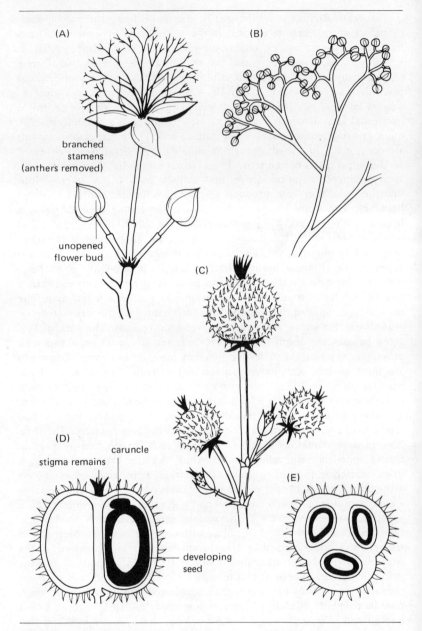

Fig. 11.5 *Ricinus communis*: Castor.   (A) A male flower with the anthers removed (×3).   (B) The branched stamen of a male flower.   (C) A branch of the inflorescence bearing female flowers and young fruits.   (D) A longitudinal section through a young fruit.   (E) A transverse section through a young fruit.

Fig. 11.6 *Ricinus communis*: Castor.　　Inflorescences and immature fruits on dwarf castor in Brazil. (By courtesy of the FAO.)

Castor requires 350–500 mm of rain during the growing season, but provided that it is grown on deep, well-drained soils where there is extensive root growth it is drought resistant. It is most common where temperature during the growing season is 20°–30°C; seed set is poor in temperatures hotter than 40°C. Wild plants and most cultivars set fruit first at the base of the inflorescence, and progressively later higher up the axis. The fruits shatter when they are mature, so repeated hand picking is necessary to avoid seed losses, and such cultivars are unsuitable for mechanical harvest. Breeders, especially in the United States, have produced dwarf, non-shattering hybrid cultivars and inbred lines which mature in as few as 150 days and are suitable for mechanical harvest. Some of these new cultivars consist of predominantly female plants and it is necessary to sow rows of pollinators among them to obtain good yields.

The seeds contain 40—55 per cent of oil with an iodine number of about 85. It consists of up to 90 per cent ricinoleic acid and has chemical properties which make dehydration possible. Industrially the oil is extracted by cold pressing, but the ancient method, and that still used where castor is grown for local domestic use, was to boil crushed seeds in water and skim the oil from the surface. The seeds contain enough of the albuminoid toxin ricin to cause serious illness, or death, if only a few are eaten. The toxin remains in the seed cake when the oil is expressed, and the cake can be fed to livestock only after treatment to render the ricin harmless. The cake, or 'pomace', is used as fertilizer in India.

Coconut oil: *Cocos nucifera*

The coconut is the most important palm of the wet tropics, and until recently was the major source of vegetable oil imported into the industrialized nations of Europe and North America. It supplies a great variety of the needs of the inhabitants of many tropical coastal areas, especially those bordering or in the Pacific and Indian oceans. The fruits are very large drupes. They provide both food and drink, and are economically important for the oil extracted from their dried endosperm, which is called copra. Coconut oil is especially valued for making soap, but also for margarine, in confectionery, cooking and cosmetics. The unopened flowering spathe, when wounded, yields a delicious sap rich in sugars and vitamin B, called toddy, from which sugar syrup and sugar can be obtained. An intoxicating drink called arrack is prepared by distilling fermented toddy. The hard endocarps of the fruits supply fuel and charcoal; coir fibre from the mesocarp is used to make mats, brushes and ropes, and to stuff mattresses, while the leaves and trunks of the palm provide a continuous supply of materials for every purpose in building houses suited to the equable climate where coconuts grow.

Cocos is a monotypic genus in the sub-family *Cocoideae* of the *Palmae* (Monocotyledons), with a pantropical distribution, predominantly in coastal areas latitude 20° N. and S. of the equator. Though *C. nucifera* grows best at low altitudes, and especially near the coast, it is not halophytic (salt-loving), but tolerates saline soils. It is most common, and grows well, on sandy coastlines, because there its requirements are met for a humid atmosphere, moderate temperatures in the range 27°—32°C and a

Fig. 11.7 *Cocos nucifera*: Coconut.   Coconut palms growing in Jamaica.

loose, free-draining and well-aerated soil. On coastlines there may also be an ensured supply of fresh ground water provided by drainage to the sea from higher land nearby, which is necessary for the growth of coconuts if annual rainfall is less than 1,000 mm. They will tolerate as much as 2,500 mm of rain each year, but only where the soil is free draining. The crop has a long history in both the eastern and western hemispheres, and the fact of its widespread and apparently ancient occurrence has been the chief cause of uncertainty about its centre of origin. Palms of the *Cocoideae*, closely related to *C. nucifera*, have a centre of diversity in north-west South America, which has been suggested as the area in which coconuts were domesticated. The current view differs from this and suggests that, though the wild ancestor may have been South American, it was dispersed widely before coconuts were eventually domesticated in the Papua New Guinea–Fiji region of the Pacific, between longitudes 145° and 180° E. The means and the routes by which the crop has spread so widely from its centre of origin have been the subject of controversy for many years. It seems most likely that it was carried by man as a source of food and drink on sea voyages, and that the fruits also floated in ocean currents and germinated after they were washed ashore in new locations.

There are two kinds of modern coconut. Short-lived, dwarf forms are referred to var. *nana* and do not contribute significantly to world production, which comes mostly from tall forms referred to var. *typica*, which may live as long as 100 years. The mature coconut palm of var. *typica* is a very handsome unbranched tree, often 20—30 m tall with a cylindrical, light-grey trunk indistinctly marked with ring-like leaf scars. The trunk may be up to 50 cm in diameter, but since the stem has no cambium, its diameter does not increase with age. Consequently variations in diameter record past favourable or unfavourable growing conditions. The palm has no tap root, but as many as 7,000 adventitious roots are produced throughout its life from the swollen base of the stem, or bole, which is around 60 cm long. The roots are brittle, woody but elastic, and do not produce root hairs. Water is absorbed in the region just behind the swollen root apex and the rest of each root has a hypodermis of thick cells impermeable to water. The roots have no cambium and rarely exceed 13 mm in diameter; as they age they become hollow and the red hypodermis becomes exposed. Soft, white protruberances from the hypodermis serve the function of lenticels. Most roots grow in the top metre of the soil, and may extend for 10 m around the plant.

The leaves of seedlings and 1- or 2-year-old plants are entire; in older plants, though each leaf is entire in the bud, it splits between the veins to form up to 250 linear, lanceolate pinnae about 5 cm wide, 60 cm long at the base of the leaf, 1 m or more long in the middle and progressively shorter from the middle to the tip. The number of vascular bundles in the stem is fixed by the number arising from the apical meristem, and remains constant because there is no lateral meristem (or cambium) to produce secondary vascular tissues. This restriction limits the supply of nutrients which can be transported to the stem apex and consequently the number of leaves, which remains more or less constant. On average, twelve leaves fall and twelve new ones expand each year, and about 30 mature leaves are carried in a graceful and beautiful crown at the top of the trunk. The apical meristem is within this crown of leaves, protected by the sheathing bases of young leaves which do not expand for more than 2 years after they are initiated; so the apical bud is large, and though its removal kills the palm, it is sometimes eaten as a delicacy.

Tall coconut palms (var. *typica*) begin to flower and to bear fruit after 6 or 7 years' growth, and reach full bearing after 10 or 12 years. Dwarf palms of var. *nana* flower earlier. The inflorescences are produced singly in the axils of leaves, and each year the number of mature inflorescences and of new mature leaves is the same. The whole inflorescence is at first completely enclosed in two sheathing spathes. The inner spathe and the inflorescence inside it burst through the outer, which remains small. The inner spathe is a thick, leathery, cylindrical envelope, about 1.5 m long and tapering at both ends. It splits open along one side to expose the inflorescence which carries flowers on up to 40 lateral branches from the main axis. The coconut palm is monoecious. Female flowers occur singly, or rarely in groups of three or four, at the base of each branch, the total number in each inflorescence varying from 20—40, whereas there may be 200—300 male flowers on the upper parts of each branch. The male

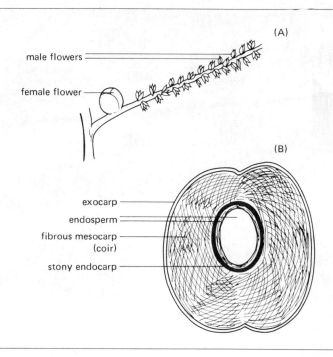

Fig. 11.8 *Cocos nucifera*: Coconut. (A) Part of a branch of the inflorescence. (B) A diagrammatic section through a fruit.

flowers have six small, pointed, yellow perianth segments which spread widely when the flower opens to expose six pointed anthers on short filaments and a rudimentary style and stigma. The female flowers are globose and about 2 cm in diameter with two small bracts inside their base. Flowering begins with the male flowers at the tip of each branch, and pollen production may continue for 2 weeks from one inflorescence, after which most male flowers are shed before any female flowers open. This usually ensures cross-pollination because it is rare for two spathes to be open at the same time on any one palm. Under some conditions and in some cultivars the male and female phases of flowering within one inflorescence sometimes overlap. Self-pollination and successful self-fertilization may then occur. The pollen is dispersed by wind and by insects which visit the nectaries in both male and female flowers.

After fertilization two of the carpels in the ovary abort. The fibrous drupe which develops from the remaining carpel is one of the largest fruits in the plant kingdom. It is up to 30 cm long, slightly less broad than long, and may weigh as much as 2 kg. Usually fewer than six fruits develop to maturity in each inflorescence, so that each palm may yield 60–70 fruits each year under the best growing conditions (but exceptionally as many as

Fig. 11.9 *Cocos nucifera*: Coconut.   Coconuts.

400). The exocarp of the drupe is a smooth, shining green skin on the surface of young fruits, but becomes yellow as the fruit ripens. In the mature fruit the mesocarp is a thick fibrous layer from which the coir of commerce is obtained. Embedded in it lies the 'nut', which is the very hard endocarp enclosing the single seed. When mature the nut is 10—15 cm in diameter and weighs up to 1.5 kg. The testa of the seed is a brown, paper-thin tissue closely attached to the endocarp and the endosperm which begins to develop when the fruit is about 6 months' old. At first the endosperm is a translucent, whitish-grey jelly surrounding a very large cavity which develops from the embryo sac of the ovule, and which contains up to 0.5 litre of sap or 'coconut water'. The endosperm grows, but never fills the cavity, and eventually becomes a thick, hard white layer rich in oil which accumulates in it only during the later stages of fruit development. The fruits are mature about 12—14 months after fertilization. The small embryo is located at the basal end of the fruit beneath the soft 'eye' of three which are visible as circular depressions on the outside of the endocarp.

The seed germinates slowly, and it may be 4 months after planting before the shoots appear. The single cotyledon surrounds and protects the plumule and radicle. Part of it grows with them through the soft 'eye' into the mesocarp while the rest swells and elongates into the cavity of the endosperm where it enlarges into a haustorial organ, known as the 'apple', which takes food from the endosperm and passes it to the embryo. The

Fig. 11.10 *Cocos nucifera*: Coconut.   Sun drying copra in Kenya. (From Child, R. (1974). *Coconuts*, 2nd edition, Longman: London.)

plumule and radicle burst from the protecting sheath of the cotyledon and grow through the mesocarp to emerge from the fruit. Food materials are extracted from the fruit by the seedling for several months after it has emerged, both from the endosperm and from the mesocarp where the primary root obtains mineral salts, especially potassium. The primary root is eventually replaced by adventitious roots.

The dried endosperm extracted from the fruits is called copra. It is economically the most important product of the coconut palm, and around 4 m. tonnes are produced each year, 1.8 m. of them from the Philippine Islands, large amounts from Indonesia, India, Sri Lanka, Mexico and Papua New Guinea, and smaller quantities (in total less than 0.5 m. tonnes) from most countries where the crop will grow. Harvesting and the extraction and drying of the copra are all done by hand. The mesocarp is first removed, then the nut is split into halves which are dried in the sun for 2 days before the endosperm is scraped out. The endosperm is then dried quickly in the sun, or in ovens, often fuelled by the shells of the nuts, until the moisture content has fallen from the original 50 per cent to less than 7 per cent. Well-dried copra does not become rancid in store or during transportation, it remains free from moulds and is less liable to insect pest damage than insufficiently dried copra. Good copra is obtained only from fully mature but ungerminated nuts, and contains 60—70 per cent by weight of oil which is extracted by hot pressing, refined by treatment with alkali and deodorized with superheated steam. The press cake is used as stock feed. The oil is solid at temperatures cooler than 24°C. It contains more than 80 per cent of saturated fatty acids, mainly lauric, myristic and palmitic, it has a very low iodine number around 10, and has always been important in soap manufacture because the soaps made from it lather freely. The fresh endosperm may be ground and pressed to yield 'coconut milk', or dried to produce dessiccated coconut for use in confectionery and cooking.

Oil-palm: *Elaeis guineensis*

The oil-palm is native to the humid tropics of West Africa from 16° N. in Senegal to 15° S. in Angola. It occurs wild along the banks of rivers and streams in the transition zone between rain forest and savanna where there is plentiful water, but where the palms are not shaded by a dense growth of forest trees. These requirements for light and water confine the commercial exploitation of extensive groves and plantations to areas of cleared rain forest. Most of the oil-palms in West Africa were planted by farmers in forest clearings, but they form naturalized communities in which the individual trees may live for 200 years. The density of palms in these communities was determined by the density of the human population and the extent to which the rain forest had been cleared when they were planted. Naturalized communities are most extensive and dense in eastern Nigeria where palms in the groves may be so close together that no other crops can be grown. Oil-palms extend east across the central rain forest belt of Africa between latitudes 3° N. and 7° S., as far as the borders of Zaire and Uganda. They were introduced to South America from West Africa at the time of slave trade, and became established in Brazil; they also occur in scattered locations in East Africa and Madagascar. The main event in the history of oil-palms in Asia was the establishment of four trees in a botanic garden in Java in 1848. The seeds from one of them were selected and gave rise to a productive and true breeding population called 'Deli' palms which were later used to establish large plantations in Indonesia and Malaysia. The crop grows best and is most productive where the annual rainfall is 2,000 mm or more, and where there is no 'dry season'.

Oil-palms produce more oil per hectare than any other oil seed crop, but they are locally valued as well for the wine obtained by natural fermentation of the sap tapped from the male inflorescences. Two kinds of oil are extracted from the fruits. Palm oil is obtained from the fleshy, orange mesocarp of the drupes, and much of it is used for food where the palms are grown. Palm kernel oil is expressed from the seeds in countries which import them after they have been removed from the stony endocarps. These two oils have distinct properties which are described later. World production of palm oil is greater than 2.5 m. tonnes each year, most of it from Nigeria and Malaysia (each producing about 0.7 m. tonnes),

Non-drying oils and fats   313

Fig. 11.11 *Elaeis guineensis*: Oil-palm.   A naturalized community of oil-palms near Ibadan, Nigeria.

Indonesia (0.3 m.) and Zaire (0.2 m.), though the crop is established in many countries which produce less than 0.1 m. tonnes each year. Annual world production of palm kernels is about 1.4 m. tonnes, of which about 50 per cent by weight is oil.

Like the coconut palm, *E. guineensis* is a member of the sub-family *Cocoideae* of the *Palmae*. It is a stately erect tree up to 30 m tall with a stout trunk covered with persistent leaf bases which often provide suitable conditions for the growth of epiphytes. Adventitious roots grow from the base of the stem, the bole, and from aerial parts of it up to 1 m above the soil, though roots initiated far above ground may not reach the soil. A few primary roots penetrate deeply to anchor the plant, but most grow horizontally, and as far as 20 m away from the bole. From these secondary branches are produced, some of which also grow deeply, but most remain in the surface layers of the soil and branch freely to produce a dense mat of roots. Rooting depth varies with the depth of the ground water table and is least when the water table is high. The roots have no root hairs, and absorption occurs where the hypodermis is least developed behind the tips of the ultimate branches of the root system. Like other palms, *E. guineensis* has pneumathodes on its roots which are assumed to assist gaseous exchange between the root tissues and the soil atmosphere. They consist of vascular and cortical tissues which burst through the hypodermis like very short root branches. They eventually become suberized and lignified and presumably cease to function.

The apical growing point of the stem is situated in a depression at the centre of a mass of up to 50 developing leaves which form the apical bud.

314   The Vegetable Oils and Fats

Fig. 11.12 *Elaeis guineensis*: Oil-palm.   A young specimen tree grows as an ornamental in Ibadan, Nigeria.

After they are initiated, the leaves remain in the bud for up to 2 years before they are mature. They expand from the bud first as a vertical, tightly rolled 'spear' which unfolds into a very large pinnate leaf about 7 m long. As each new leaf expands, so the oldest leaf in the crown at the top of the palm is shed, and the mature leaf number remains more or less constant from 40–50. Each leaf has a massive petiole 2 m long, with a broad base clasping the stem. Small, sharp spines derived from the lower pinnae of the leaf are closely crowded along both sides of the base of the petiole, and merge into the shorter pinnae higher up. There are 100–160 pairs of pinnae in the oil-palm frond, lying more or less in two rows one along each side of the flattened petiole. The largest are in the middle of the frond and may be 1 m long; those above are progressively shorter and are carried at a more acute angle pointed towards the apex of the leaf. The leaf is terminated by a blade made up by the fusion of the last pair of pinnae and the single terminal leaflet, though they tend to split apart as the leaf grows old. The pinnae of each pair do not always arise opposite each other, and alternate pairs tend to be carried in different planes. Each pinna has a prominent mid-rib, raised on its upper surface, and numerous parallel veins.

The oil-palm begins to flower when it is 4—6 years old, and in full bearing produces two to six bunches of fruits each year. The inflorescences are compound spikes, one borne in the axil of each leaf, but because some abort, especially during periods of water stress, there are fewer inflorescences each year than new leaves. Like the leaves, each inflorescence develops in the apical bud for 2 years, and it does not expand and flower until about 9 months after its subtending leaf has opened. Each flower is potentially hermaphrodite, but at a very early stage in development one sex in each inflorescence usually aborts, so that the oil-palm is monoecious. Normally a series of up to ten consecutive inflorescences on one palm are of the same sex, followed by a similar sequence of inflorescences of the opposite sex. Consequently, oil-palms are not only monoecious, but behave as though they were dioecious, and cross-pollination is ensured. On young palms there may be an occasional hermaphrodite inflorescence, especially at the end of a phase during which a series of monoecious inflorescences has been produced. The sex of flowers in a monoecious inflorescence is determined when they are initiated, that is some 2 years before they are mature. Bright, sunny weather apparently favours the development of female flowers and the abortion of males, whereas male inflorescences tend to be those initiated during the wettest, cloudiest months.

The inflorescence is a compound spike, or spadix, enclosed by two sheathing spathes which protect young flowers until they are mature, when the inflorescence bursts through them. The male inflorescence has a short, stout peduncle, rarely more than 10 cm long and 5 cm broad, which supports a cylindrical mass of brown spikes, each with as many as 1,200 male flowers sunk in the axis of the spike. Each male flower is subtended by a small bract, and has six small, free perianth segments, six stamens with arrow shaped (sagittate) anthers on short filaments which are fused at the base, and the rudiments of the aborted gynaecium. The peduncle of the female inflorescence is shorter and stouter than in the male, and carries up to 150 branches, the whole forming a more or less rounded, compact inflorescence, up to 30 cm long. The female flowers are in groups of 12—30 called spikelets, each group in the axil of a large bract which is drawn out into a long terminal spine. Each female flower in a spikelet is subtended by a pair of small bracts, and has six broad perianth segments in two whorls, each about 2 cm long. The superior ovary is derived from

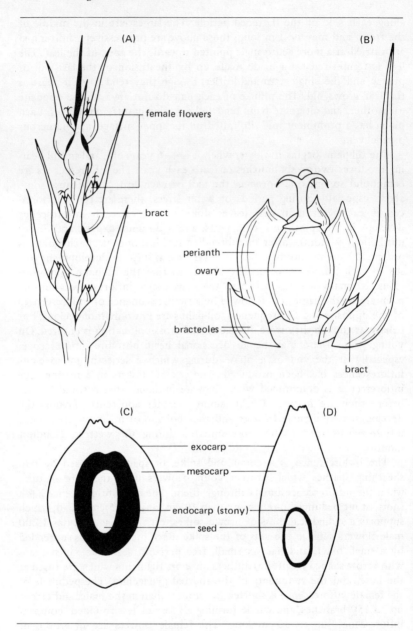

Fig. 11.13 *Elaeis guineensis*: Oil-palm. (A) A branch of the female inflorescence (×½). (B) A diagrammatic dissection of a single female flower (×1½). (C) A longitudinal section through the fruit of a thick-shelled type (×1). (D) A longitudinal section through the fruit of a thin-shelled type (×1).

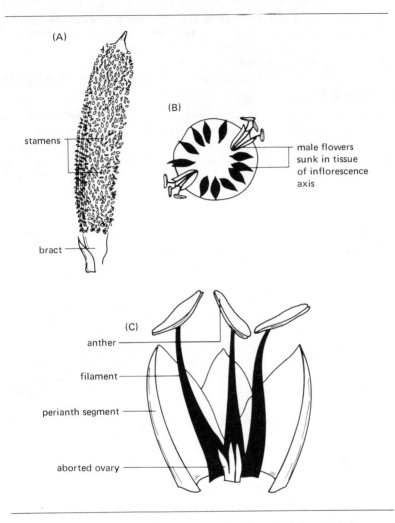

Fig. 11.14 *Elaeis guineensis*: Oil-palm. (A) A branch of the male inflorescence (x½). (B) A transverse section through a branch of the male inflorescence. (C) A diagrammatic dissection of a single male flower. Three stamens and three perianth segments removed (x8).

three carpels, but only one of them develops. It is about 6 mm long, and surmounted by a short, thick, white style divided at the tip into three stigmatic lobes. The ovary is unilocular and contains a single ovule.

The fruits mature 5–9 months after flowering. They are sessile, oval drupes about 4 cm long and 2 cm broad, with a pointed apex. Variation in fruit colour is used to classify them as:

Nigrescens: deep violet to black at the apex and colourless towards the base before ripening. On ripening two *Nigrescens* types are recognized, those with a black and those with a brown apex.
Virescens: green when unripe, orange when ripe.
Albescens: a rare form without carotenoid pigment in the mesocarp. On ripening the fruits are ivory or pale yellow with a dark cap.

The fruits are carried in large bunches of as many as 200 on each female inflorescence; well-spaced oil-palms may bear as many as five such bunches and 1,000 fruits each year. A single bunch from West African palms weighs around 18 kg, while those of the Deli palms grown in Indonesia and Malaysia may weigh as much as 25 kg, of which 60—65 per cent consists of fruit. The drupes have a leathery, fairly thin exocarp surrounding the fleshy mesocarp which constitutes 35—95 per cent of the weight of the fruit. Much of this variation is associated with variation in the thickness of the endocarp, and indeed its presence or absence, which is due to genetic differences between cultivars determined by a single pair of alleles, Dd. It is used to classify cultivars into three groups:

1. *Dura* cultivars are homozygous dominant at the D locus (DD). They have an endocarp 2—8 mm thick, and a mesocarp which constitutes 33—55 per cent of the fruit weight. Within this group a substantial proportion of West African palms have endocarps with large average thickness (6—8 mm); they are called *macrocarya* palms.
2. *Tenera* palms with a thinner endocarp (0.5—3 mm) are heterozygous at the D locus (Dd). Their mesocarp accounts for 60—95 per cent of fruit weight and is characterized by a faintly visible ring of fibres around the endocarp.
3. *Pisifera* cultivars have no endocarp; they are homozygous recessive at the D locus (dd). They tend to set little fruit and are of little importance, except in oil-palm breeding.

Oil-palms are propagated from seed. The percentage germination of untreated seed is usually very small, and an elaborate series of treatments is necessary to ensure a large percentage of vigorous and even germination. The mesocarp is cleaned from the endocarp, often by retting; various methods are used to heat the seeds before they are germinated, and only when they have done so are they sown in polythene bags or baskets in the nursery. There are many variations of these techniques, all very much more complex than this brief statement can convey. The early growth of seedlings is slow, and they are transferred to a second nursery before they are eventually planted in the field 1½—2 years after germination. After planting out the primary root of the seedling is soon replaced by adventitious roots, but it is 3 years or more before a trunk is formed.

Palm oil is pressed from the mesocarp, usually in the countries where the trees are grown. It is a rich source of carotene, a precursor of vitamin A in diets, and it is estimated that as much is eaten locally as is exported

Fig. 11.15 *Elaeis guineensis*: Oil-palm. A bunch of ripe fruits (a single female inflorescence) gathered together after harvest in Nigeria.

from producing countries. In industry palm oil is used principally in the manufacture of soap, but also in the production of margarine, lubricating oils and candles, and in the tinplate and sheet steel industries. It is one of the world's most important vegetable oils, solid at ambient temperatures, and ranging in consistency from that of soft butter to a hard, tallow-like substance, depending upon the method used to extract it from the fruit. The best quality palm oil has a free fatty acid (FFA) content around 3.5 per cent, and is obtained by expression in hydraulic, often power-operated, presses from fruits which have been processed quickly after they are mature. Poor quality oil obtained using traditional methods in Nigeria may contain as much as 85 per cent FFA. In mature whole fruits the FFA content increases very slowly unless the fruits are bruised; if they are the activity of the enzyme lipase causes the rapid breakdown of fat into FFA (mostly palmitic acid) and glycerol. The enzyme is inactivated by heating, so during the extraction process fruits are heat sterilized and the expressed oil is kept hot. Even in sterilized oil, especially if it is contaminated with 'dirt', the percentage of FFA may rise during storage or transit due to the action of micro-organisms unless the amount of water in it is less than 0.1 per cent. Palm oil varies in colour from pale yellow to deep orange; the darker coloured oils have the largest carotene content. The mesocarp contains 50—65 per cent of oil (even larger amounts in some cultivars) with iodine numbers in the range 44—58.

Palm kernel oil is obtained from the seeds. It is white or pale yellow with iodine numbers in the range 14—22, and is used principally to make soap and margarine. The press cake left after oil extraction is a good livestock feed.

Cocoa butter: *Theobroma cacao*

Cocoa butter is a pale yellow, rather brittle solid obtained from the seeds of the cocoa tree, *Theobroma cacao*. The seeds are commonly called 'beans'; they contain 50—57 per cent of oil which is solid at temperatures cooler than about 30°C. The fat keeps well, but tends to become rancid, and its colour becomes lighter when it is exposed to air and bright sunlight. Cocoa butter is composed mostly of the glycerides of palmitic, stearic and oleic acids, and is produced as a by-product of the manufacture of cocoa. It is used in the production of chocolate, confectionery, cosmetics and in some medicinal preparations. The botany of *T. cacao* is discussed fully in Chapter 8.

Shea butter: *Butryospermum paradoxum*

Butryospermum paradoxum is a medium-sized deciduous tree which is most common as a natural constituent of the more open savanna woodlands of West Africa where annual rainfall is around 1,000—1,300 mm. It also occurs eastwards across the African savanna zone to northern Uganda and the southern Sudan where a distinct type, var. *nilotica*, is recognized. The tree is protected and encouraged to grow where it occurs naturally. Its seeds contain 45–60 per cent of oil called shea butter, which is solid at temperatures cooler than 32°C. This fat is used locally in foods, cosmetics and for illumination, but seeds are exported from West Africa to Europe where shea butter is used as a substitute for cocoa butter in confectionery, and to make candles, soap and cosmetics. The press cake contains little protein, but is used in Europe as a constituent of compound feed cakes for cattle.

The tree grows up to 10 m tall with a thick trunk and a spreading crown of branches. The oblong leaves are clustered at the ends of stout, young branches; they are 10–25 cm long on petioles about 10 cm long. Sweet scented, white flowers occur in dense groups in the axils of the leaves. There are eight to ten sepals, petals and stamens. The sepals are in two whorls, the outer reddish-brown, the inner green. The petals are fused below and free above, with dentate margins. The stamens alternate with petaloid staminodes. The superior ovary has eight to ten locules, but only one develops to produce a dry, egg-shaped fruit about 5 cm long surmounted by the remains of the style. The single large seed is embedded in a sweet pulp within the fruit; it has a shiny brown testa and a long, white hilum. There is no endosperm.

Further reading

General

Corner, E. J. H. (1966). *The Natural History of Palms*, Los Angeles: University of California Press.
Godin, V. J. and Spensley, P. C. (1971). *Oils and Oil Seeds*, London: Tropical Products Institute.
Purseglove, J. W. (1968 and 1972). *Tropical Crops*, 2 vols, London: Longman.
Swerne, D. (ed.) (1964). *Bailey's Industrial Oil and Fat Products*, 3rd edition, New York: Wiley.
Vaughan, J. G. (1970). *The Structure and Utilization of Oil Seeds*, London: Chapman and Hall.
Weiss, E. A. (1971). *Castor, Sesame and Safflower*, London: Leonard Hill.

Safflower

Beech, D. F. (1969). Safflower, *Field Crop Abstracts*, 22, 107–19.
Knowles, P. F. (1955). Safflower – production, processing and utilization, *Econ. Bot.*, 9, 273–99.

Tung

Foster, L. J. (1962). Recent technical advances in the cultivation of the tung oil tree, *Aleurites montana*, in Nyasaland, *Trop. Agriculture, Trin.*, 39, 169–87.

Fry, V. K. (1974). Factors contributing to the demise of tung production in the United States, *Econ. Bot.*, 27, 131–6.
Goldblatt, L. A. (1959). The tung industry. ii. Processing and utilization, *Econ. Bot.*, 13, 343–64.
Hill, J. and Spurling, A. J. (1966). A note on the classification of montana tung (*Aleurites montana*), *Trop. Agriculture, Trin.*, 42, 311–21.
Potter, G. F. (1959). The domestic tung industry. I. Production and improvement of the tung tree, *Econ. Bot.*, 13, 328–42.

Sesame

Naylor, N. M. and Mehra, K. L. (1970). Sesame: its use, botany, cytogenetics and origin, *Econ. Bot.*, 24, 20–31.
Tribe, A. J. (1967). Sesame. *Field Crop Abstracts*, 20, 189–94.

Castor

Narain, A. (1974). Castor, in: Hutchinson, Sir Joseph, *Evolutionary Studies in World Crops. Diversity and Change on the Indian Subcontinent*, Cambridge University Press.

Rape and mustard

Narain, A. (1974). Rape and Mustard, in: Hutchinson, Sir Joseph, *Evolutionary Studies in World Crops. Diversity and Change on the Indian Subcontinent*, Cambridge University Press.

Coconut

Charles, A. E. (1961). Selection and breeding of the coconut palm, *Trop. Agriculture, Trin.*, 38, 283–96.
Child, R. (1974). *Coconuts*, 2nd edition, London: Longman.
Frémond, Y., Ziller, R. and De Nuce de Lamothe (1966). *Le Cocotier*, Paris: Maisonneuve and Larose.
Piggott, C. J. (1964). *Coconut Growing*, London: Oxford University Press.
Purseglove, J. W. (1968). The origin and distribution of the coconut, *Trop. Sci.*, 10, 191–9.

Oil-palm

Hartley, C. W. S. (1976). *The Oil-Palm*,. 2nd edition, London: Longman.
Rees, A. R. (1965). Evidence for the African origin of the oil-palm, *Principes*, 9, 30–6.
Sucre, C. and Ziller, R. (1963). *Le Palmier à Huile*, Paris: Maisonneuve and Larose.
Zeven, A. C. (1972). The partial and complete domestication of the oil-palm (*Elaeis guineensis*), *Econ. Bot.*, 26, 274–9.

Chapter 12

The Essential Oil Crops

In several thousand species of plants distributed throughout the world there occurs a group of highly aromatic volatile organic substances known collectively as the 'essential oils'. These differ both in physical properties and chemical structure from the fixed plant oils which have already been discussed in Chapter 11. Their existence in certain plants has been known for thousands of years, and the method whereby they could be obtained from plant tissue by a process of distillation in steam was known to the ancient Egyptians. It must also have been practised in Persia and India since early times. During the Middle Ages throughout Europe pharmacists explored the medicinal properties of a wide variety of plants, and began the search to discover the essential principle of plant tissue which they believed was responsible for the curative effects on the human body. Again the process of steam distillation was widely employed and a number of pleasant smelling, highly volatile liquids was obtained from different plants; these were considered to be the essential constituents of the plants, and hence the term 'essential oil' which has persisted to the present day. These essential oils were used extensively in medicine during the sixteenth century, and largely owing to the work of the pharmacists of that period various methods of extraction and purification of plant oils became widely known. It was not, however, until the nineteenth century (when the chemistry of the terpenes was first investigated) that the chemical nature of the volatile oils became known. Since that time most plant oils have been analysed, and the majority have been shown to contain a mixture of organic substances, some of them highly volatile and possessing distinctive odours and tastes. The most important chemical constituents are the terpenes, which, although having little if any smell themselves, act as carriers for the aromatic substances. Terpenes are hydrocarbons, being made up of isoprene units (C_5H_8), and existing as unsaturated straight chain molecules, or as ring structures, readily combining with other organic groupings. Most essential oils also contain camphors, and the more odoriferous compounds present consist of oxygen derivatives of terpenes, alcohols, esters, aldehydes and ketones. The essential oils are thus quite different in nature from the fixed plant oils which are true glycerides of saturated and unsaturated organic acids.

It is still largely a matter of conjecture as to how these substances are produced in the plant, nor is their function in plant physiology understood. They are secreted in the plant into external glands or epidermal hairs outside the main plant body, or into internal glands which may develop in any part of the plant tissue. When produced inside the plant the oil is secreted between the cells where increasing deposition brings about a separation of the walls of adjacent cells with the result that a cavity develops between them; several such cavities may eventually link up to form a short canal. The cells around the cavities or canals break down to form a definite boundary, protecting living tissues from the oil sac, which thus becomes completely isolated. Oil formation seems to be greatest in regions of photosynthetic activity, but the oil as such is not transported about the plant body. The composition of the oil changes during the development of the plant, and in some species oils of quite different chemical nature are produced in different parts of the same plant. Various suggestions have been made as to the function of essential oils in plants. It has been postulated that they may promote insect pollination because their scent attracts insects to the plant, that they may provide some form of protection against animals and parasites, that they may act as wound fluids or as reserve food substances, or that they may assist in water conservation by reducing the transpiration rate due to their heat screening effect. It seems most likely, however, that these substances, themselves toxic to living tissue, are by-products of the general metabolism of the plant, and that having fulfilled some function (e.g. acting as hydrogen donors in oxidation-reduction reactions) are then stored by the plant in such a manner and in such tissues as to isolate them from the main centres of metabolic activity.

During the latter part of the nineteenth century the use of essential oils for medicinal purposes declined considerably, but they became much more widely employed in the perfume industry and for scenting soaps, cosmetics, etc. Today there is a steady and quite considerable demand for essential oils from a variety of sources for these and other industries. The various plants from which they are obtained are not usually cultivated on any large scale, and many of the commercial essential oils are still produced in small areas by primitive means of distillation. On the other hand, sophisticated methods of steam distillation, rectification and solvent extraction are used in the industrial extraction of some essential oils. To collect and concentrate unstable essential oils which, when naturally vapourized, constitute the scent of some flowers, a complex, labour-intensive method called the 'cold-fat' or 'enfleurage' process is used in France. The 'scents' are concentrated by absorbing them on to layers of fat from which they are later extracted in alcohol. However, the majority of the essential oils of commerce are obtained from the tropics where cultivation of the particular plants is usually very localized, and where the oils are often produced on the spot using small, easily portable stills. In many cases they provide valuable exports, but the quantities involved are never great. Apart from their commercial value the plants are often grown for local medicinal purposes, and some of them are used in various religious ceremonies and rituals. The tropical spices, many of which owe their properties to the presence of essential oils, have already been discussed in Chapter 9.

Citrus oils: *Citrus* spp.

Most of the members of the family *Rutaceae* produce essential oils, but the best known and most important of them are obtained from species of *Citrus*. Citrus oils differ in composition depending upon the species from which they are obtained, and often oils with different characteristics are obtained from one species depending upon whether they are extracted from the fruits, flowers, leaves or stems. The largest quantities of citrus oils are by-products of the citrus fruit industry; they are expressed under pressure or by steam distillation from the waste left after fruits have been processed for their juice and flesh, and from the rind or flavedo. Essential oils in the fruit are most concentrated in the flavedo where they accumulate in ductless glands between the cells; the glands are surrounded by a protective layer of dead cells, but the oil in them is under pressure due to the turgor of nearby living cells. Consequently, when the peel of a citrus fruit is bent or broken oil droplets are released with some force and may squirt a considerable distance. Fruit waste and flavedo are bruised or macerated before essential oils are extracted from them. Citrus oils from fruit waste (especially waste from processing sweet oranges and grapefruits) are used industrially in the production of organic chemicals, while the oils from the flavedo are important as flavours and in perfumes.

Most of the lime oil of commerce is distilled from the expressed juice of fresh, undamaged green fruits of *Citrus aurantifolium*, and is used to flavour drinks and confectionery. Lime oil expressed from the peel of fruits has a different composition. Lemon oil is expressed from the peel of *Citrus limon* and is used as a perfume. Several essential oils which are important in the manufacture of perfumes and toilet waters are obtained from the sour or seville orange, *Citrus aurantium*, and the sweet orange, *Citrus sinensis*. Orange oil is expressed from the peel of the fruits of both species, while bergamot oil is obtained only from the rind of *C. aurantium* subsp. *bergamia*; petitgrain is obtained by distillation from the leaves of sour orange (or the sub-species *amara* of sour orange), mostly in the Mediterranean area, but also in Paraguay; oil of neroli or orange blossom oil is an important ingredient in some perfumes and toilet waters, and is distilled from the flowers of sour oranges which are cultivated for this purpose in the south of Europe, especially in France.

Geranium oil: *Pelargonium* spp.

Geranium oil is distilled from the leaves of some species of *Pelargonium*, and is much used in the perfume and soap industries as an 'extender' for more costly perfume oils, and as a substitute for rose oil. The chief constituent of geranium oil is an oxygenated derivative of a monoterpene, geraniol, but there are also varying proportions of esters in the oils from different *Pelargonium* species. The genus is a member of the *Geraniaceae* and includes more than 200 species, though many of them are probably of hybrid origin. The majority of species are native to South Africa, but many have been widely disseminated as garden and pot ornamentals, while three or four species whose leaves contain valuable essential oils have become of some agricultural importance, especially in North Africa and southern Europe. *Pelargonium* differs from the other large genus of the *Geraniaceae*, *Geranium*, in its irregular flowers where the two upper petals are usually larger than the rest, and also in the presence of a spur at the base of one of the sepals. This spur acts as a nectar tube, but is not easily visible unless the flower is sectioned, because it is adnate to the peduncle of the flower.

Pelargonium graveolens is the species most commonly cultivated in the tropics. It is a much branched shrub, up to 1 m tall and becoming more or less woody at the base with age. The stems are grey-green in colour and somewhat hairy and glandular, with opposite leaves on long petioles subtended by a pair of stipules. The leaf is deeply palmately lobed with toothed margins. Small umbels of numerous small, pink or pale purple flowers arise from the leaf axils. The flower has a hairy calyx tube divided about halfway along its length into five lobes, and five entire petals, the upper two of which are longer than the rest. There are ten stamens in the flower, partially united at the base, but only six to seven of them carry anthers, the rest being sterile. In the centre of the flower is the five-lobed superior ovary, with a single pendulous ovule in each lobe; the short style is five-lobed at the tip. The nectar-tube extends from one of the sepals and is adnate to the pedicel. The plant is propagated from stem cuttings, and persists for a number of years; the young shoots are harvested three times each year. Like most *Pelargonium* species, *P. graveolens* grows best in the cool tropics, or in warm temperate regions.

Pelargonium odoratissimum is also cultivated for an essential oil with similar properties to that obtained from *P. graveolens*. It has a short, thick, fleshy stem, slender branches, and rounded, cordate leaves with indented margins borne on very long petioles. Its white flowers are smaller than those of *P. graveolens*, and are carried in groups of five to ten opposite the leaves.

Pelargonium capitatum and *P. roseum* are also cultivated for essential oils which are characterized by a greater ester content than the oils from the two species already described.

Camphor and camphor oil: *Cinnamomum camphora*

Camphor oil is a general term used to describe a number of essential oils obtained from plants, all of which contain camphor which separates out as a transparent, colourless, crystalline solid when warm camphor oil is cooled. Camphor is chemically closely related to the terpenes, being a ketone derivative of a dicyclic terpene. Natural camphor was widely used in industry for the production of celluloid and various nitro-cellulose compounds, and in modern and very ancient medicines for the treatment of colds. Now most camphor is synthesized. The bulk of the world's supply of natural camphor is obtained from the wood of the camphor tree, *Cinnamomum camphora*, of the family *Lauraceae*, a native of Japan and Taiwan and a close relative of the cinnamon tree which has been described elsewhere in this book. The tree has been introduced to various parts of the tropics but has nowhere been cultivated on any large scale. It is not demanding in its environment or soil requirements and cannot be considered as a strictly tropical plant because it grows well at elevations of 1,000 m or more and at latitudes 45° N. and S. of the equator.

Cinnamomum camphora is a large tree when growing in its natural habitat, reaching a height of 20—30 m with a very thick main trunk. In cultivation, and when grown away from its natural home, it is a smaller, densely branched tree with small, dark-green, spirally arranged leaves which smell strongly of camphor if they are bruised. The small yellow flowers occur in dense axillary panicles which rarely exceed the length of the leaves. The small, spherical fruits are dark green, becoming dry and black as they ripen. Seeds may not be produced until a tree is 25 years old; their percentage germination is small, and they retain their viability for a very short time.

The yield of camphor and camphor oil is very uncertain. Some trees yield only crystal camphor with little oil, while others produce mainly camphor oil. The total camphor content of trees also varies considerably, as a consequence not only of genetic variation, but also because of differing environmental conditions. Camphor accumulates in oil cells found throughout the tissues of the plant; with increasing age camphor is deposited as a solid in these cells, and the colourless oil gradually disperses into other tissues. Extraction is carried out by a process of steam distillation of the leaves and young twigs, or of macerated older wood. As the camphor oil cools solid camphor crystalizes out, then the residual oil is distilled again, several fractions being collected. The most important of these is a heavy oil known as artificial sassafras oil, whose main constituent is the substance safrole. Safrole is a more valuable product than camphor, though it does not occur in such large quantities, and is produced only in the stems and roots. Safrole is used for the production of heliotropin for which there is a demand in the perfume industry; it is also used for flavouring.

An essential oil containing 60 per cent or more of camphor is also obtained from a small herbaceous plant, *Ocimum kilimandscharicum*, grown in parts of East Africa where it occurs naturally. The plant is a member of the family *Labiatae*, and is closely related to basil which is described later in this chapter. It is a perennial, low-growing shrub from which the leaves and stems may be harvested two to three times each year, then they are distilled together, though most oil is obtained from the leaves; the yield of oil is small, and it does not contain safrole. Essential oils which do contain safrole are obtained from the roots of *Sassafras albidum* cultivated in temperate regions, from the wood of *Ocotea pretiosa* and *Ocotea cymbarum* which are trees of the Brazilian forests, and from the bark and leaves of the Australian tree *Doryphora sassafras*.

Ylang-ylang and cananga oils: *Cananga odorata*

Ylang-ylang and cananga oils are valuable essential oils obtained from the flowers of the South-east Asian tree *Cananga odorata*, and used extensively in perfumery. The Philippine Islands were once the chief source of these oils, but most of the world's supply now comes from the island of Réunion in the Indian Ocean, where the tree was introduced in 1770. It is a member of the *Annonaceae*, the custard apple family, and is a large tree reaching a height of up to 30 m in the wild state, though in cultivation it is pruned down to about 3 m for ease of harvesting and to increase the amount of branching. The branches are dragged down and pegged to the ground so that in cultivation the tree has a peculiar, distorted appearance. Large, yellow-green strongly scented flowers hang in clusters from the axils of the alternate leaves, or from older leafless branches. Each flower has three hairy sepals, six almost lanceolate petals 5—7 cm long, and numerous stamens in two whorls with the connective of each anther prolonged as a beak. The flower has several free carpels, each of which develops into a separate green, fleshy fruit. The trees flower all the year round, with peak flower production between November and March.

The flowers are harvested for oil extraction as they reach maturity and change colour from green to yellow, a process which normally takes 2—3 weeks. The first crop of flowers is taken when trees are 1½—2 years old, and thereafter the flowers may be picked twice each year. They are usually gathered in the very early morning, even in the dark, so that their delicate perfume is not dissipated by the heat of the sun; and they are picked carefully because damaged blossoms yield an inferior oil. The flowers have also been harvested by treating trees with ethylene gas which causes the flowers to be shed on to matting spread beneath the tree. The oil is most commonly extracted by steam distillation, and several fractions are obtained, the first consisting mainly of the more volatile esters with a small admixture of terpenes, and the later ones consisting mostly of terpenes and sesquiterpenes. The first fractions are the most valuable and form the ylang-ylang oil of commerce. The last fraction is called cananga oil, which is also obtained from *C. latifolia*, a species closely akin to *C. odorata*, but bearing flowers with an inferior perfume. Solvent extraction yields a better product than steam distillation, and can be effected using alcohol, ether, chloroform or petroleum fractions. The main use of ylang-ylang oil is in the preparation of perfumes, while cananga oil is used to make scents of poorer quality.

Patchouli oil: *Pogostemon cablin*

Patchouli oil is distilled from the leaves and young shoots of *Pogostemon cablin*, a member of the family *Labiatae* native to the Philippine Islands. It is widely used in perfumes, as a scent in soaps and as a fixative for other essential oils, and for centuries its characteristic fragrance was associated in Europe with shawls and carpets from the 'Orient'. In many parts of Asia where the plant grows it is used as incense in various religious ceremonies. *Pogostemon cablin* has been introduced to several parts of the tropics and is cultivated in the Seychelle Islands, the Malagasy Republic and Brazil, but the centre of production is in Malaysia and Indonesia.

In cultivation the plant rarely flowers and is propagated vegetatively with stem cuttings which strike easily and grow rapidly to form a much branched bush about 1 m tall. It grows best at higher elevations in a warm, moist climate with evenly distributed rainfall, and on free-draining soils rich in organic matter. The stems are square in cross-section and carry opposite, ovate leaves with serrate margins on long petioles. Small flowers are usually in spikes in the leaf axils or at the ends of branches, but the inflorescence may be a more complex panicle of spikes. The sepals are united into a tube with four or five unequal teeth closely appressed to a two-lipped corolla tube which is white with purple streaks; each lip of the corolla has two lobes, those of the upper lip being longer than those of the lower. There are four stamens with straight bearded filaments, and a deeply four-lobed superior ovary of two united carpels with a long style arising from its centre. Each lobe of the ovary contains a single ovule, and the fruit consists of four smooth, ovoid nutlets.

Young shoots are harvested at intervals of about 6 months over a period of 2–3 years. They are cut from the plants, dried in the sun and then baled for export. The youngest leaves contain most oil, and oil from the stems is of inferior quality, so only young shoots are harvested. Care is necessary when the shoots are dried, for they disintegrate if they become too dry, and ferment if they are too wet. Although most oil is extracted in the producing countries, the best quality patchouli oils are obtained by distillation from leaves imported into Britain and the United States.

Bay oil: *Pimenta racemosa*

Bay oil is distilled from the leaves of the bay tree, *Pimenta racemosa*, of the family *Myrtaceae*, a plant closely related to *Pimenta dioica*, allspice, which has already been described in Chapter 9.

Though the bay tree has been introduced to other parts of the tropics, bay oil production is still centred in the West Indies where the tree is indigenous. The oil used to be extracted from the leaves there by distilling them in a mixture of rum and water, a practice which gave rise to the name 'bay rum' for a hair tonic containing the oil which is best known in the United States. Bay oil is now extracted by distillation in alcohol.

Wild bay trees grow as tall as 20 m, but in cultivation they are pruned to 5 m or less to facilitate harvesting the leaves. They are evergreens with dark-green, shiny, elliptical leaves about 12 cm long carried on short petioles. The leaves are gathered by hand, often from wild trees, then stored for a few days before they are bruised and placed in the still. The chief constituent of bay oil is eugenol; in toilet waters bay oil has a soothing effect on the skin as well as a pleasant odour.

Basil oil: *Ocimum basilicum*

Basil oil comes from *Ocimum basilicum*, basil, of the family *Labiatae*, and is one of 50—60 low-growing shrubby species in the genus, all containing more or less strongly scented essential oils. They are found distributed throughout the Old World tropics, from Central Africa through Asia to parts of Australia, and some have become important weeds of cultivation. For example, basil, though cultivated in some parts of the world for the extraction of basil oil, is one of the commonest weeds of the Sudan Gezira. Many of the species are used as pot herbs or as local medicines, and *O. sanctum* is the sacred bush of the Hindus, grown in front of most Hindu houses and temples. *Ocimum americanum* and *O. viride* are the fever plants of West Africa where infusions of their leaves are drunk to relieve high temperatures. Basil is cultivated in several parts of the tropics, but the production of basil oil is centred in the Mediterranean region. In India it is grown mainly as a pot herb, though some oil is distilled there. It is a slightly hairy annual with much branched, angular stems carrying opposite, ovate leaves which are usually less than 1 cm long, purplish in colour and borne on fairly long petioles. Small white flowers, often tinged with purple, are in more or less open, terminal and axillary racemes. The calyx tube is about 6 mm long and deeply divided at the top into two unequal lobes, the upper lobe with two indistinct teeth whereas the lower is more clearly divided into three unequal teeth. The corolla tube is longer than the calyx, and has two lips, the upper one with four lobes. There are four stamens and a long style cleft at the tip into two stigmas. The four-lobed ovary is similar to that described for *Pogostemon cablin*, and typical of the *Labiatae*.

Ocimum basilicum exists in a diversity of forms and cultivars, not all of which are good oil-yielding types. Cross-fertilization is common and hybridization between types or even with other species has complicated the picture. The oils from these different types vary considerably in composition, and basil oils from different regions of cultivation are not all of the same physical and chemical nature. Basil oils are used in perfumes and to give scent to soaps.

Cymbopogon spp.

The genus *Cymbopogon* includes a number of species which produce essential oils of some commercial importance. The genus is a member of the sub-family *Panicoideae* of the *Gramineae*, and belongs to the tribe *Andropogoneae*. It includes around 60 species of perennial, coarse-growing, tufted grasses which are widely distributed throughout the tropics, occurring especially as components of the great areas of mesophytic grassland savanna of the Old World tropics. The species intercross readily and may exist in a variety of forms; in addition some of the cultivated types seldom flower, so they are difficult to identify. The genus *Cymbopogon* resembles other members of the *Andropogoneae* in producing spikelets of two kinds in pairs, one member of a pair sessile and perfect, the other pedicelled and often male. It differs from other members of the tribe in the form of the inflorescence: the paired spikelets are in short racemes which also occur in pairs, each pair of racemes subtended by a large semi-sheathing spathe. The spathes, with their paired racemes, are borne in large, branched panicles which in some species are fairly open and spreading, but in others more compact. Many species of the genus are highly aromatic due to the presence of essential oils in their stems and leaves. Some of these are cultivated in various parts of the tropics for oil extraction, and some oils are extracted commercially from wild species.

Lemon grass oil: *Cymbopogon citratus*

Lemon grass is a tufted perennial with numerous stiff stems arising from a short rhizomatous root stock, though it is not so tall, nor so strongly growing as some of the other cultivated species. It is not known to grow wild, but is cultivated in many parts of the tropics for oil extraction, especially Indonesia, Malaysia, India and Burma and Guatemala. In parts of Asia the leaves are used to flavour food. Lemon grass rarely flowers, and is propagated by division of the root stock. Harvest of the leaves begins about 1 year after planting and may continue over a period of 3–4 years, with several harvests each year. The leaves contain around 0.5 per cent of oil, and though the amount decreases as they grow older, the quantity of citral in the oil increases with leaf age. Citral is an aldehyde derivative of a terpene and constitutes 70–80 per cent of lemon grass oil; it is extracted as a thin yellow oil, and is used mainly in the production of ionone, a synthetic perfume with the scent of violets, and in the synthesis of vitamin A. Thus lemon grass oil is important in the perfume and soap industries, and in medical and pharmaceutical chemistry.

Malabar lemon grass oil: *Cymbopogon flexuosus*

Cymbopogon flexuosus is native to southern India, and is cultivated there to some extent, especially on the poorer soils of forest clearings. Two

332 The Essential Oil Crops

Fig. 12.1 *Cymbopogon* sp.: Lemon grass.

cultivated types are recognized, one with red stems which is the source of essential oil, the other white-stemmed and not important in oil production. The red-stemmed type has been introduced to several tropical countries, but the bulk of the world's supply of Malabar lemon grass oil is produced in India and exported to Europe and North America for use in perfumery, soap making, the preparation of insect repellents and for the

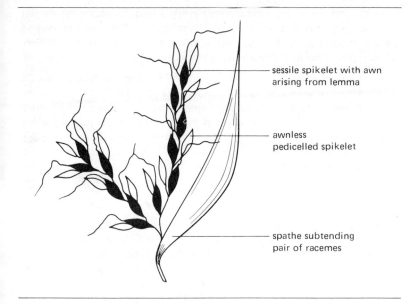

Fig. 12.2 *Cymbopogon* sp.: Lemon grass.

extraction of citral which constitutes around 80 per cent of the oil. *Cymbopogon flexuosus* is one of the few species of *Cymbopogon* normally propagated from seed; it grows as a tufted perennial up to 2 m tall, and once established its leaves are harvested every 6–8 weeks over a period of about 6 years. The essential oil is extracted locally in India by steam distillation.

Citronella oil: *Cymbopogon nardus*

Cymbopogon nardus is a strongly growing perennial grass reaching a height of 1 m or more, with broad leaves arising from numerous narrow stems which grow from short stolons. It exists in a variety of forms, some wild or semi-wild, others known only in cultivation. The cultivated types yield a greater proportion of oil from their leaves than the wild types, and the oil from them is of better quality. The plant is cultivated in Taiwan, Sri Lanka, Malaysia, Java, Fiji, East Africa, the Seychelle Islands, Zaire, Brazil, Uganda and Guatemala, and grows well in many other tropical countries. It is propagated by division of the root stock, and its leaves are harvested several times each year. The oil is extracted by steam distillation. Two distinct groups of cultivars occur in cultivation:

1. The Java types give large yields of an oil containing around 45 per cent of citronella and 40 per cent of geraniol. The grass grows in large

clumps and the leaves are harvested every 3 months beginning about 1 year after planting. The Java type is not hardy and is replanted every 4 years or so; it grows best in moist climates at elevations up to 1,000 m. Java type citronella is given varietal status as *C. nardus* var. *mahepengiri*.

2. The 'Ceylon' type, *C. nardus* var. *lenabatu*, is much hardier, with a deeper root system and narrower leaves than the Java types, and grows well in poorer soils over a range of environments. It is not so important in commerce as Java citronella and yields a poorer quality oil used in inexpensive preparations. The 'Ceylon' type is first harvested when plants are about 8 months old, and thereafter at intervals of 3–4 months over a period of 10–15 years.

Palmarosa oil, gingergrass oil, rosha oil: *Cymbopogon martinii*

This species is a perennial grass occurring in the drier parts of India, and cultivated to a small extent in Java, Malaysia and the Seychelle Islands. There are two morphologically indistinguishable types which grow best in different environments, have different growth forms and produce distinct types of essential oil. *Cymbopogon martinii* var. *motia* grows in separated clumps, and is wild in dry, hilly open localities. It yields palmarosa oil, or East Indian geranium oil, which contains 90–95 per cent of geraniol. *Cymbopogon martinii* var. *sofia* grows at lower altitudes, usually where rainfall is plentiful and the soil is poorly drained; it is not tufted. Gingergrass oil, which is inferior to palmarosa oil, is extracted from its leaves. Both oils are used as adulterants in other, more expensive oils, and are used especially in perfumes produced in the Middle East. In Europe and North America they are used as a source of geraniol.

Vetiver oil, khuskhus: *Vetiveria zizanioides*

The genus *Vetiveria* is a member of the tribe *Andropogoneae*, and includes about ten species. *Vetiveria zizanioides* is a densely tufted perennial native to India, Burma and Sri Lanka where its aromatic roots have been used for many centuries to make mats with a pleasant odour. Vetiver oil is distilled from the roots for use in perfumery, largely as a fixative for more volatile essential oils, and it is also used to scent soaps and cosmetics. The plant is cultivated in Java, Réunion, the West Indies and India and in parts of Africa and South America. It is commonly grown to control soil erosion. It grows wild in India, often on the rich soils of river banks in fairly hot, damp environments.

The plants grow in large clumps with stems 1–2 m tall rising from a much-branched, spongy rhizome. Narrow, rigid leaves are up to 1 m long. When they are young they provide good grazing, and old leaves are used for thatching and similar purposes. The leaves are not aromatic and do not contain oil. Some of the cultivated types never flower, but most produce a long terminal panicle of slender racemes. The spikelets are in pairs, one sessile, perfect and more or less spiny, the other pedicelled with two male flowers.

Further reading

Barber, L. A. and Hall, M. D. (1950). Citronella oil, *Econ. Bot.*, 4, 322–36.
Guenther, E. (1948–1952). *The Essential Oils*, 6 vols, New York: Van Nostrand.
Guenther, E. (1952). Recent developments in essential oil production, *Econ. Bot.*, 6, 355–78.
Haagen-Smit, A. J. (1949). Essential oils, a brief survey of their chemistry and production in the United States, *Econ. Bot.*, 3, 71–83.

Chapter 13

Rubber

Natural rubber is an amorphous hydrocarbon, polyisoprene $[(C_5H_8)_n]$, which has the property of being highly extensible, but which quickly and forcibly regains its original shape after distortion. Rubber accumulates, perhaps as an excretory product, in colloidal suspension in the latex produced by some plants, but its function in them is not known. Many compounds with the same or similar elastic properties, but with different chemical composition (they are co-polymers of butadiene and styrene), have been synthesized and are also called rubber; they are used instead of natural rubber in many of its industrial applications, especially in the United States. The production of latex is a characteristic feature of many plants, but latex containing rubber in large quantities occurs only in species of the families *Moraceae, Euphorbiaceae, Apocynaceae* and *Compositae*. The latex accumulates in special cells called laticifers, which are usually tubes ramifying through all the tissues of the plant, though they sometimes occur as individual cells. Tubular laticifers are of two kinds, simple or non-articulated, and compound or articulated; both types may be branched or unbranched, they never have secondary wall thickening and they remain alive when mature. Each simple laticifer (latex tube) is derived from a single cell which occurs in the embryo or in the apical meristem. This cell elongates and reaches enormous length as it branches and ramifies through the tissues of the plant. Compound laticifers (latex vessels) are derived from a series of cells between which the intervening end walls break down to form a continuous tube in much the same way as xylem vessels are derived. Compound laticifers arise from the cambium and are usually associated with the phloem, both primary and secondary; they are the source of rubber bearing latex in *Hevea braziliensis*, the para rubber tree. Latex accumulates in the laticifers under pressure and flows from them when they are cut. It is this property which makes possible the collection of latex by 'tapping' rubber-bearing plants. Latex varies in composition depending upon the species in which it occurs. As well as rubber particles protein, waxes, resins, essential oils, and sometimes starch, may also be suspended in it, and inorganic salts occur in solution. Rubber occurs as minute particles suspended in the watery latex, each with a covering of proteinaceous material which prevents them coagulating, though if untreated latex is allowed to stand after it has been gathered the products of natural fermentation eventually cause the rubber particles to coagulate. In commercial production they are separated from the latex by causing coagulation with the addition of acids or salts. The rubber particles vary in size, shape and concentration in different plants; in *Hevea braziliensis* they constitute about 30 per cent by weight of the dry latex.

When a mixture of raw rubber and sulphur is heated the rubber loses some of its elastic properties and retains a stable shape. Unlike raw rubber, it neither becomes 'tacky' when hot nor brittle when cold after this treatment. It is then said to be 'vulcanized', and can be manufactured into stable, useful shapes, or used industrially in many more ways than raw rubber. It was the discovery of the effects of vulcanization by Goodyear in 1839 that gave the first major impetus to the industrial use of rubber and its collection from wild plants. Much later the demand for pneumatic rubber tyres for motor vehicles was the main stimulus for increased natural rubber production from cultivated plantations of *Hevea braziliensis*. Thus until the beginning of the twentieth century natural rubber was obtained from wild plants of several different species and though *Hevea* is the main subject of this chapter other rubber-producing plants are of historical interest and deserve brief mention.

'Castilloa' or 'Panama' rubber was obtained from a tall forest tree native to Mexico and Central America, *Castilloa elastica* of the family *Moraceae*. When the trunk of the tree is tapped a thin watery latex flows very freely from the latex tubes (simple laticifers) which are slow to heal after they have been wounded. Consequently the wild trees were tapped infrequently, or they were felled before their latex was gathered. Rubber from this source was used in pre-Columbian Mexico, and may have been the first rubber seen by Columbus in the West Indies. Several other species of *Castilloa* in Central and South America yield rubber, and were once sources of minor importance. Another New World plant, *Manihot glaziovii*, a small tree related to cassava in the family *Euphorbiaceae*, is the source of 'ceara' rubber, and a few other species of the same genus have been tapped for rubber in Brazil. *Manihot glaziovii* is native to the drier parts of Brazil, and was cultivated in plantations in East Africa at the beginning of this century. It is difficult to tap the trees and their thick latex flows slowly. For these reasons, and because rubber yields were small, the production of ceara rubber was not a commercial success in competition with para rubber after about 1910. 'Guayule' rubber comes from *Parthenium argentatum*, a low-growing, much-branched shrub in the family *Compositae* which grows wild in the semi-arid parts of the south-western United States and in Mexico. The rubber is a constituent of latex which occurs in individual cells, especially within the medullary rays of the stems and roots. The latex is extracted mechanically by macerating whole plants, and most production has been from wild stands, though *Parthenium* was cultivated in the United States during the Second World War. Another member of the *Compositae*, *Taraxacum kok-saghyz*, though not a tropical crop, is an interesting minor source of rubber in Russia. Latex is extracted from its roots by maceration, though unlike *Parthenium* the latex occurs in articulated laticifers, not in individual cells. Until the development of *Hevea* plantations in Malaysia and other parts of South-east Asia, the chief source of rubber in Asia was *Ficus elastica*, a large tree in the family *Moraceae* native to Burma, India and Malaysia. 'India' rubber was obtained from its latex which was gathered by tapping the trunks and roots of wild trees. Several West African plants yield rubber-bearing latex. *Funtumia elastica*, a member of the *Apocynaceae*, was the source of 'Lagos' rubber. It is a large tree growing to heights of 15 m or more in the rain forests, and towards the end of the nineteenth century it was planted on a commercial scale in parts of West Africa. At one time considerable quantities of rubber were also obtained from wild plants of various *Landolphia* species in the forests of West Africa, especially from *L. heudelotii*, *L. kirkii* and *L. owariensis*. These are woody climbing plants with a growth habit which makes them unsuitable for large-scale cultivation. Two members of the family *Asclepiadaceae*, *Cryptostegia grandiflora* and *Cryptostegia madagascariensis*, were introduced from tropical Africa and the Malagasy Republic to the New World during the Second World War as sources of 'palay' rubber. Both species are quick-growing climbing plants which yield latex from the cut tips of their young stems and from macerated leaves and other plant parts.

Para rubber: *Hevea braziliensis*

Hevea braziliensis is the primary source of the world's natural rubber, and has become increasingly important since the beginning of this century as its superiority in cultivation over other potentially useful rubber-producing plants has been realized. Annual world production of para rubber now exceeds 3 m. tonnes, more than 90 per cent of it from South-east Asia, India, Sri Lanka, Thailand and Indochina. Malaysia produces about 1.5 m. tonnes annually, and Indonesia about 0.8 m. tonnes. The tree is cultivated in parts of Africa (Nigeria, Liberia and Zaire) with annual production from that continent around 0.2 m. tonnes, and insignificant amounts are still collected from wild and semi-cultivated trees of the Amazonian forests of South America, the original home of *Hevea*. It is a member of the family *Euphorbiaceae* which includes several economically important tropical species discussed elsewhere in this book, e.g. *Ricinus communis*, castor oil; *Manihot esculenta*, cassava; *Aleuritis* spp., tung. Nine species of *Hevea* are now recognized, though they hybridize readily where their ranges overlap in Brazil. All the species yield latex, but only *H. braziliensis* is important as a source of rubber, and it is the only species which has been exploited commercially; indeed it provides more than 90 per cent of the world's natural rubber. All of the *Hevea* species are native to the rain forests of the basin of the Amazon River, occurring as trees or shrubs over a range of elevations, and in regions with different amounts of annual rainfall. *Hevea*

Fig. 13.1 *Hevea braziliensis*: Rubber.   A rubber plantation in Malaysia.

braziliensis is a lowland type, requiring 1,900–2,500 mm annual rainfall fairly evenly distributed throughout the year, except during a short dry season, and temperatures in the range 24°–32°C or warmer.

Although para rubber originated in Brazil, the centre of production is in South-east Asia. The crop has never been cultivated on a large scale in its original home, nor has it become commercially important there except during two brief periods, one around 1910 and the other during the Second World War, when the world price for rubber was high enough to make the collection of latex from wild trees in remote areas worthwhile. The initiative to establish a rubber industry in Asia came from British administrators in India who financed attempts to collect several rubber-bearing plants from the New World during the 1870s. The most important of these efforts was the collection of 70,000 seeds of *H. braziliensis* in Brazil by H. A. Wickham in 1876. He sent these seeds to the Royal Botanic Gardens at Kew, England, where they were planted, and though only a small proportion of them germinated, some young plants were later sent from Kew to Sri Lanka, Java and Singapore. It was largely due to the unaided efforts of the Director of the Botanic Garden in Singapore, H. H. Ridley, that the potential of *Hevea* as a commercial source of rubber was realized. By careful husbandry and experiment he developed a system of management for the new crop and successfully encouraged the establishment of the first commercial plantations in Malaysia using material derived from the stock originally collected by Wickham in Brazil. One of Ridley's major contributions to the development of the Malaysian rubber industry was his demonstration of a method of tapping trees that enables a uniform regrowth of bark to occur for subsequent tapping, and which gives greater and more consistent latex yields than other haphazard tapping methods. The modern development of Ridley's method is described later; suffice it to say here that it gives large yields of latex because it involves repeated cutting across the ends of the same laticifers which appears to stimulate latex production in them.

Hevea braziliensis is a tree which commonly grows up to 25 m tall with a straight trunk covered with smooth, light-grey bark 6–15 mm thick. Comparatively slender branches form an open, leafy crown with spirally arranged trifoliate leaves. The thin, dark green, elliptical leaflets are around 15 cm long when mature, narrow at the base and pointed at the tip; extra-floral nectaries occur where the leaflets are attached to the petiole. The long petioles have a pair of deciduous stipules at their base. The tree grows in flushes in which the young leaves are small and bronze-red in colour, becoming larger and green as they expand and mature. The inflorescence of para rubber is a short axillary panicle with hairy branches bearing small, strongly scented, greenish-white, unisexual flowers. The larger female flowers are at the ends of the inflorescence branches, the males near their base, and for each female flower in an inflorescence there may be as many as 80 males. The flowers have no corolla, but a five-lobed calyx forms a conical envelope which protects the flower in bud, then splits into five recurved lobes as the flower opens. Male flowers have ten anthers which are sessile on a slender central column in two whorls of five, one whorl above the other. In female flowers the gynaecium is composed of three united carpels which form a three-lobed, three-celled, more or less conical ovary with a two-lobed sessile stigma. Each locule of the ovary contains a single ovule. The inflorescence matures over a period of about 2 weeks during which time some of the male flowers open first, then the females, and lastly the rest of the males. Pollination is by insects, mostly small

Fig. 13.2 *Hevea braziliensis*: Rubber. (A) Fruits. (B) Seeds.

midges and thrips, and though the trees are self-compatible, seed production appears to be greatest in groups of trees which are cross-pollinated. The fruit is a large, deeply three-lobed, dehiscent capsule containing three large (2—3 cm long) brown, mottled seeds. The seeds lose their viability soon after they are mature. They are poisonous because they contain a cyanogenetic glucoside, though this can be removed by boiling them in water. They also contain a dark-red drying oil which has been used as an illuminant.

In the early years of the Asian rubber industry the crop was propagated entirely from seed which led to the establishment of plantations of heterogeneous mixtures of trees of different quality, even though they were all descended from the original Wickham stock. Modern plantation rubber is propagated vegetatively by budding scions from selected, high-yielding trees on to vigorous root stocks to establish high-yielding clones. Yield has not been the only factor considered in rubber improvement, which has included breeding. Resistance to crippling leaf blights, the absence of undesirable low branching, the efficiency of replacing wounded bark and resistance to wind damage are all desirable characters in a good rubber tree. In the New World, but fortunately not in Asia, a major factor limiting rubber production is South American leaf blight caused by the fungus *Microcyclus ulei*. To achieve some degree of resistance to this and other leaf diseases a system of 'double' or 'crown' budding is practised. This involves budding a scion with good latex yield on to a vigorous root stock, and when this has attained sufficient size to form the trunk of the future tree a second scion from rubber with disease-resistant foliage is budded at its top to form the 'crown'. Plantation trees are grown at a density of around 270 plants per hectare, for although denser planting yields more latex per hectare it yields less latex per man employed in tapping. The planting density used is chosen to give the greatest money profit. It is normal practice to sow leguminous cover crops in rubber plantations, mainly to control soil erosion, but also to smother weeds and to take advantage of nitrogen fixation in the nodules of the legume roots and the accumulation of organic matter from the decay of fallen legume leaves.

The latex of *H. braziliensis* occurs in articulated latex vessels derived from the cambium in a series of cylinders which alternate with cylinders of phloem tissue. The latex vessels of each cyclinder form a much-branched, anastomosing system, but there is rarely any connection between adjacent cylinders of latex vessels. The number of such cylinders varies from two in the 'bark' of young seedling trees to as many as 50 in the thick bark of old trees. They are closest together near the cambium, and consequently this part of the bark yields the greatest quantity of latex if the tapping incision is made deep enough to cut them. It requires skill and experience to do so without damaging the cambium which lies so close beneath them. In *Hevea* the latex vessels spiral upwards to the right at an angle of about 30° from the vertical; to cut through the greatest number the tapping incision is made as a spiral at right angles to them, that is downwards, from left to

Fig. 13.3 *Hevea braziliensis*: Rubber. Tapping a rubber tree.

right, at an angle of about 30°. This spiral tapping cut is commonly extended about halfway round the trunk. Latex flows most freely in the early morning and runs down the channel made in the bark by tapping to be collected in a glass or earthenware cup at its base. Trees are first tapped when they are 4–7 years old. After the initial incision has been made in the bark of a tree the same cut is tapped repeatedly at daily or slightly longer intervals by removing a slice of bark about 1 mm thick from its lower edge, until eventually a 'panel' up to 2 m long has been removed from one side of the tree. Depending upon the frequency of tapping, the same panel may gradually extend downwards in use for up to 12 years. When its lower edge reaches ground level a new panel can be opened high up on the opposite side of the tree, so that the same tree remains in continuous production. Provided that the cambium has not been extensively damaged during tapping a smooth layer of new bark grows over the wound, and in time the same area can be tapped again. Tapping procedures vary in different countries and on different estates, but all are designed to ensure a steady supply of latex with sufficient time being allowed for the wound to heal before the same part of the tree is re-tapped.

The latex gathered by tapping is coagulated by mixing it in bulk with a small quantity of acid, acetic acid being commonly employed, but formic acid or sodium silicofluoride may also be used. The rubber particles coagulate as a white spongy mass which is then passed between rollers to squeeze out the watery latex, after which the strips of rubber are dried and smoked in a 'curing' house. For many of the modern uses of rubber, or for the sale of concentrated (centrifuged) latex, care must be taken that no natural coagulation takes place as the latex is being collected from the tree. When rubber is required for such special purposes a chemical anticoagulant is added to the latex in the collecting cup. It is always necessary to ensure scrupulous cleanliness when gathering and processing the latex, because the chance occurrence of pieces of bark or dirt in it spoil the appearance and decrease the quality of rubber. As well as dried, smoked sheet and concentrated latex, rubber is exported from producing countries as crêpe, which is manufactured by passing coagulated rubber through rollers which rotate at unequal speeds and so have a tearing action. Crêpe rubber dries quickly, but it is not smoked.

Further reading

Edgar, A. T. (1960). *Manual of Rubber Planting (Malaya)*, Kuala Lumpur: Incorp. Soc. of Planters.
Metcalfe, C. R. (1967). Distribution of latex in the plant kingdom, *Econ. Bot.*, 21, 115–25.
Polhamus, L. G. (1962). *Rubber*, London: Leonard Hill.
Schultes, R. E. (1956). The Amazonian Indian and the evolution of *Hevea* and related genera, *J. Arnold Arbor.*, 37, 123–47.
Taylor, K. W. (1951). Guayule – an American source of rubber, *Econ. Bot.*, 5, 255–73.

Appendix

Crop Production Data (Summarized from *Production Yearbook 1973*, Rome: FAO) (*Millions of metric tonnes*)

	1961–5 (annual mean)	1971	1973
1. Cereals (all crops)			
World	986.38	1312.53	1368.15
Africa	51.13	64.11	53.51
North/Central America	210.17	295.08	293.92
South America	40.19	54.13	56.38
Asia	392.08	492.65	508.65
Europe	158.58	216.82	223.93
Oceania	11.38	15.51	17.70
USSR	122.86	174.23	214.06
2. Rice (paddy)			
World	251.92	306.38	320.71
Africa	5.54	7.62	6.94
North/Central America	4.05	5.40	5.67
USA	3.08	3.89	4.21
South America	8.06	9.37	10.29
Brazil	6.12	6.59	7.45
Asia	232.20	280.43	293.79
Mainland China	86.04	109.03	111.52
India	52.73	64.60	67.60
Indonesia	12.39	18.66	20.32
Bangladesh	15.03	14.90	18.29
Japan	16.44	14.15	15.77
Thailand	11.26	13.74	14.65
Burma	7.79	8.17	8.56
Europe	1.52	1.81	1.92
USSR	0.39	1.43	1.76
3. Maize			
World	215.62	305.00	311.78
Africa	16.19	22.34	17.12
South Africa	5.23	8.60	4.20
Egypt	1.91	2.34	2.51
Kenya	1.11	1.50	1.30

Crop production data — *continued*

	1961–5 (annual mean)	1971	1973
3. Maize — *continued*			
North/Central America	105.82	157.86	157.90
USA	95.56	143.29	143.34
Mexico	7.37	9.34	9.50
South America	17.81	27.51	27.47
Brazil	10.11	14.13	14.60
Asia	36.77	47.05	48.90
Mainland China	22.76	30.06	30.30
India	4.59	5.10	6.80
Indonesia	2.80	2.61	2.50
Philippines	1.30	2.02	2.34
Europe	25.71	41.42	46.64
USSR	13.12	8.60	13.44
4. Sorghum			
World	35.70	50.71	51.77
Africa	8.88	9.43	8.08
Nigeria	4.20	3.14	3.00
Sudan	1.26	2.15	1.50
Ethiopia	0.83	1.07	1.10
North/Central America	14.61	24.90	26.57
USA	13.91	22.24	23.79
Mexico	0.45	2.33	2.47
South America	1.41	5.10	5.79
Argentina	1.36	4.78	5.16
Asia	10.39	9.48	9.78
India	8.85	7.72	8.00
Yemen	0.95	0.98	1.00
5. The millets (pearl, finger, proso, foxtail and barnyard millets)			
World	38.23	45.43	45.37
Africa	8.99	9.36	7.66
Nigeria	2.61	2.69	2.15
North/Central America	0.18	0.21	0.22
South America	0.19	0.19	0.24
Asia	26.12	33.53	32.77
Mainland China	17.13	23.00	23.00
India	7.23	9.20	8.50
Europe	0.06	0.04	0.03
USSR	2.64	2.04	4.40
6. Wheat			
World	254.39	353.88	377.02
USSR	—	—	109.68
USA	—	—	46.58
Mainland China	—	—	35.30
France	—	—	17.79
Canada	—	—	17.11
Argentina	—	—	6.50

Crop production data – *continued*

		1963	1971	1973
7.	**Roots and tubers** (all crops)			
	World	492.98	551.19	581.28
	Africa	60.28	71.88	74.01
	North/Central America	17.15	19.62	18.68
	South America	34.81	48.08	49.69
	Asia	156.67	196.08	199.82
	Europe	150.49	120.85	129.41
	Oceania	1.75	2.03	2.02
	USSR	71.83	92.65	107.65
8.	**Sweet potatoes**			
	World	107.03	133.64	133.37
	Africa	4.62	5.97	6.31
	Burundi	0.53	1.20	1.20
	Uganda	0.60	0.71	0.72
	North/Central America	1.32	1.22	1.26
	South America	2.43	3.29	3.42
	Brazil	1.55	2.21	2.35
	Asia	98.32	122.88	122.09
	Mainland China	83.15	111.39	111.00
	Indonesia	3.07	2.15	2.15
	Japan	6.73	2.15	2.00
	India	0.85	1.83	1.83
9.	**Cassava**			
	World	76.13	96.74	106.42
	Africa	31.57	39.85	42.49
	Zaire	6.07	8.43	10.00
	Nigeria	7.80	9.17	9.60
	North/Central America	0.53	0.68	0.74
	South America	25.42	35.08	37.41
	Brazil	22.25	30.26	33.00
	Asia	18.14	20.60	25.23
	Indonesia	11.68	10.42	10.10
	Thailand	2.11	3.11	6.40
	India	1.76	5.13	6.31
10.	**Taro** (cocoyam, dasheen and eddoe)			
	World	3.43	3.96	4.12
	Africa	2.63	3.19	3.32
	Nigeria	1.55	1.65	1.78
	Ghana	0.77	1.08	1.12
	Asia	0.71	0.67	0.70
11.	**Yams**			
	World	19.14	19.76	18.36
	Africa	18.86	19.41	17.99
	Nigeria	15.88	15.36	14.30
	Ivory Coast	1.23	1.55	1.55
	North/Central America	0.12	0.20	0.21
	South America	0.05	0.05	0.06
	Asia	0.04	0.02	0.03
	Oceania	0.06	0.08	0.07

Crop production data — *continued*

	1963	1971	1973
12. Grain legumes (all crops)			
World	42.82	43.85	43.84
Africa	4.01	4.96	5.07
North/Central America	2.41	2.54	2.70
South America	2.52	3.15	3.06
Asia	22.21	23.12	21.73
India	11.53	11.44	9.69
Mainland China	7.26	8.23	8.38
Europe	3.58	3.03	2.67
Oceania	0.07	0.10	0.12
USSR	8.03	6.95	8.44
13. Dry beans (*Phaseolus vulgaris, P. lunatus, P. radiatus, P. angularis, Vigna radiata, V. mungo*)			
World	9.97	11.22	11.22
Africa	0.89	1.21	1.31
North/Central America	1.90	1.94	2.06
South America	2.21	2.84	2.75
Brazil	1.94	2.50	2.40
Asia	3.95	4.34	4.22
Mainland China	1.26	1.43	1.45
India	1.83	2.03	1.80
Europe	0.95	0.82	0.79
14. Chick peas			
World	6.92	6.59	6.25
Africa	0.27	0.24	0.30
North/Central America	0.01	0.16	0.36
South America	0.01	0.01	0.01
Asia	6.31	6.04	5.44
India	5.36	5.20	4.47
Pakistan	0.68	0.49	0.55
Europe	0.23	0.14	0.14
15. Pigeon peas			
World	1.71	2.02	1.89
Africa	0.06	0.07	0.07
North/Central America	0.03	0.03	0.04
South America	0.01	0.01	0.01
Asia	1.61	1.91	1.77
India	1.58	1.88	1.75
16. Lentils			
World	1.05	1.08	1.09
Africa	0.16	0.18	0.20
North/Central America	0.03	0.05	0.04
South America	0.03	0.02	0.02
Asia	0.65	0.68	0.67
India	0.35	0.37	0.36

Crop production data – *continued*

	1963	1971	1973
17. Cowpeas			
World	0.88	1.04	0.96
Africa	0.80	0.99	0.91
Nigeria	0.61	0.80	0.75
North/Central America	0.03	0.02	0.02
South America	0.01	0.01	0.01
Asia	0.03	0.02	0.03

More recent estimates of cowpea production suggest that these FAO figures must be amended to include a production of at least 2.5 m. tonnes annually by Brazil, and smaller but still large amounts from other South American countries.

	1961–5 (annual mean)	1971	1973
18. Soyabeans			
World	32.47	48.66	62.88
Africa	0.07	0.07	0.08
North/Central America	19.79	32.52	43.54
USA	19.56	32.01	42.63
South America	0.41	2.44	5.54
Brazil	0.35	2.22	5.03
Asia	11.79	12.90	12.99
Mainland China	10.68	11.74	11.76
Europe	0.01	0.19	0.26
USSR	0.39	0.53	0.42
19. Groundnuts in shell			
World	15.88	19.04	17.33
Africa	5.36	5.62	4.45
Senegal	1.01	0.96	0.76
Nigeria	1.86	1.55	0.70
Sudan	0.33	0.29	0.63
North/Central America	1.05	1.53	1.74
USA	0.89	1.36	1.58
South America	1.01	1.34	1.04
Brazil	0.61	0.89	0.55
Argentina	0.36	0.39	0.44
Asia	8.41	10.48	10.03
India	5.12	6.18	6.00
Mainland China	2.07	2.68	2.70
20. Castor beans			
World	0.68	0.78	0.90
Africa	0.06	0.06	0.06
North/Central America	0.03	0.01	0.01
South America	0.32	0.34	0.44
Brazil	0.27	0.30	0.39
Asia	0.22	0.29	0.29
21. Sunflower 'seed'			
World	7.34	9.72	12.21
Africa	0.11	0.16	0.28

350 *Appendix*

Crop production data – *continued*

	1961–5 (annual mean)	1971	1973
21. Sunflower 'seed' – *continued*			
North/Central America	0.03	0.26	0.39
South America	0.73	0.90	0.96
Asia	0.18	0.59	0.69
Europe	1.21	2.09	2.44
USSR	5.07	5.66	7.34
22. Safflower 'seed'			
World	0.44	0.88	0.64
Africa	0.03	0.04	0.04
Ethiopia	0.03	0.04	0.04
North/Central America	0.33	0.66	0.48
Mexico	0.05	0.42	0.25
USA	0.28	0.24	0.23
Asia	0.08	0.16	0.08
India	0.07	0.15	0.08
23. Sesame seed			
World	1.69	2.05	1.94
Africa	0.37	0.53	0.50
Sudan	0.18	0.28	0.23
Ethiopia	0.03	0.09	0.10
North/Central America	0.17	0.20	0.14
Mexico	0.16	0.18	0.12
South America	0.08	0.14	0.09
Asia	1.07	1.17	1.19
India	0.44	0.45	0.48
Mainland China	0.35	0.37	0.35
24. Linseed			
World	3.44	2.80	2.54
Africa	0.07	0.08	0.09
Ethiopia	0.05	0.06	0.06
North/Central America	1.27	1.07	0.92
Canada	0.51	0.57	0.49
USA	0.74	0.46	0.42
South America	0.88	0.36	0.34
Argentina	0.76	0.32	0.30
Asia	0.53	0.57	0.54
India	0.43	0.47	0.44
Europe	0.22	0.22	0.18
USSR	0.44	0.46	0.45
25. Cotton seed (delinted)			
World	20.22	22.73	24.06
Africa	1.91	2.51	2.55
Egypt	0.85	0.90	0.91
Sudan	0.28	0.46	0.37
Nigeria	0.23	0.20	0.19
Uganda	0.14	0.16	0.19
Tanzania	0.09	0.11	0.15
North/Central America	6.78	4.82	5.54
USA	5.56	3.85	4.49

Crop production data – *continued*

		1961–5 (annual mean)	1971	1973
25.	**Cotton seed** (delinted) – *continued*			
	South America	1.56	1.55	2.01
	Brazil	0.91	0.95	1.22
	Asia	6.27	8.83	8.51
	Mainland China	2.30	3.30	3.05
	India	2.06	2.52	2.48
	Pakistan	0.76	1.42	1.30
	USSR	3.29	4.61	4.97
26.	**Copra**			
	World	3.54	3.96	3.94
	Africa	0.13	0.14	0.14
	North/Central America	0.23	0.21	0.20
	South America	0.03	0.03	0.03
	Asia	2.85	3.27	3.30
	Philippines	1.40	1.70	1.80
	Indonesia	0.64	0.73	0.80
	India	0.29	0.35	0.36
	Oceania	0.29	0.31	0.26
27.	**Palm oil**			
	World	1.44	2.27	2.54
	Africa	1.10	1.31	1.25
	Nigeria	0.67	0.66	0.65
	North/Central America	0.02	0.03	0.03
	South America	0.01	0.05	0.07
	Asia	0.31	0.88	1.18
	Malaysia	0.12	0.55	0.76
	Indonesia	0.15	0.25	0.30
28.	**Palm kernels**			
	World	1.09	1.34	1.36
	Africa	0.82	0.89	0.85
	Nigeria	0.44	0.43	0.42
	North/Central America	0.03	0.03	0.04
	South America	0.16	0.22	0.23
	Brazil	0.14	0.18	0.19
	Asia	0.07	0.19	0.24
	Malaysia	0.03	0.12	0.15
29.	**Tung oil**			
	World	0.12	0.12	0.12
30.	**Cotton lint**			
	World	10.93	12.18	13.08
	Africa	0.99	1.36	1.36
	Egypt	0.45	0.51	0.52
	Sudan	0.15	0.24	0.19
	Nigeria	0.12	0.10	0.09
	North/Central America	3.95	2.85	3.44
	USA	3.25	2.28	2.82
	Mexico	0.50	0.37	0.34

Crop production data – *continued*

	1961–5 (annual mean)	1971	1973
30. Cotton lint – *continued*			
South America	0.84	0.83	1.07
Brazil	0.48	0.50	0.64
Colombia	0.07	0.13	0.15
Asia	3.22	4.58	4.41
Mainland China	1.15	1.65	1.52
India	1.03	1.26	1.24
Pakistan	0.38	0.71	0.65
Turkey	0.27	0.52	0.48
Iran	0.12	0.15	0.20
Europe	0.21	0.19	0.22
USSR	1.70	2.35	2.55
31. Jute			
World	2.40	2.23	2.71
Brazil	0.05	0.05	0.05
Asia	2.34	2.17	2.65
India	1.02	1.02	1.13
Bangladesh	1.12	0.76	1.04
32. Kenaf, roselle and aramina			
World	0.96	1.07	1.25
Africa	0.02	0.02	0.02
North/Central America	0.01	0.01	0.01
South America	0.01	0.02	0.02
Asia	0.88	0.97	1.15
Thailand	0.28	0.34	0.49
Mainland China	0.23	0.37	0.38
India	0.30	0.21	0.22
33. Sisal			
World	0.67	0.63	0.68
Africa	0.41	0.35	0.34
Tanzania	0.22	0.18	0.15
North/Central America	0.03	0.02	0.02
South America	0.21	0.26	0.31
Brazil	0.18	0.21	0.26
34. Abacá			
World	0.12	0.06	0.06
Asia	0.11	0.05	0.06
Philippines	0.11	0.05	0.05
35. Made tea			
World	1.12	1.40	1.53
Africa	0.06	0.12	0.15
Kenya	0.02	0.04	0.06
Malawi	0.01	0.02	0.02
South America	0.02	0.03	0.04
Asia	0.99	1.18	1.27
India	0.36	0.43	0.46
Mainland China	0.18	0.26	0.31
Sri Lanka	0.22	0.22	0.21

Appendix 353

Crop production data — *continued*

	1961—5 (annual mean)	1971	1973

36. Coffee (*arabica* + *liberica* + *canephora*. Weights at 50 per cent moisture)

	1961—5	1971	1973
World	4.36	4.70	4.19
Africa	1.01	1.27	1.26
Ivory Coast	0.20	0.27	0.26
Ethiopia	0.14	0.17	0.18
Uganda	0.14	0.19	0.17
North/Central America	0.64	0.77	0.76
Mexico	0.15	0.19	0.19
South America	2.45	2.23	1.76
Brazil	1.82	1.55	1.03
Colombia	0.47	0.47	0.52
Asia	0.24	0.39	0.38
Indonesia	0.12	0.19	0.20

37. Cocoa 'beans' (fermented and dried)

	1961—5	1971	1973
World	1.28	1.61	1.35
Africa	0.93	1.16	0.94
Ghana	0.45	0.47	0.35
Nigeria	0.22	0.26	0.22
Ivory Coast	0.11	0.22	0.19
North/Central America	0.09	0.08	0.08
South America	0.23	0.33	0.30
Brazil	0.15	0.21	0.18
Asia	0.01	0.01	0.01

38. Rubber

	1961—5	1971	1973
World	2.22	3.03	3.45
Africa	0.16	0.22	0.25
South America	0.03	0.03	0.03
Asia	2.03	2.77	3.16
Malaysia	0.77	1.27	1.49
Indonesia	0.70	0.81	0.85
Thailand	0.20	0.32	0.38
Sri Lanka	0.11	0.14	0.15

39. Tobacco leaf

	1961—5	1971	1973
World	4.35	4.52	4.83
Africa	0.22	0.21	0.20
North/Central America	1.23	1.01	1.06
USA	0.98	0.77	0.79
South America	0.33	0.40	0.35
Brazil	0.20	0.25	0.18
Asia	1.81	2.05	2.24
Mainland China	0.70	0.80	0.97
India	0.34	0.36	0.36
Japan	0.17	0.15	0.15
Turkey	0.13	0.17	0.13
Europe	0.56	0.56	0.64
USSR	0.17	0.26	0.31

Appendix

Crop production data – *continued*

	1961–5 (annual mean)	1971	1973
40. Sugar cane			
World	472.38	598.38	634.39
Africa	30.68	49.66	53.16
South Africa	9.45	16.75	15.80
Egypt	4.75	7.49	8.00
Mauritius	5.13	5.26	6.80
Mozambique	1.49	2.90	3.50
North/Central America	126.22	151.42	156.81
Cuba	42.38	52.88	55.00
Mexico	25.08	36.33	36.50
USA	20.09	21.93	24.99
Dominican Republic	7.32	10.20	10.60
South America	114.03	136.23	165.03
Brazil	65.58	79.59	102.00
Colombia	13.85	16.79	17.30
Argentina	11.43	10.26	16.00
Ecuador	6.67	8.40	8.80
Peru	7.68	8.78	8.74
Venezuela	3.47	5.15	5.60
Asia	186.11	238.67	236.92
India	106.56	126.37	123.97
Mainland China	24.70	38.63	39.40
Philippines	13.16	17.81	21.25
Pakistan	15.85	23.17	19.95
Indonesia	10.46	11.23	12.35
Oceania	14.93	21.94	22.06
Australia	12.94	19.39	19.37
41. Banana (excluding plantains)			
World	24.61	33.84	34.98
Africa	2.95	3.94	4.12
Burundi	1.14	1.44	1.52
North/Central America	4.53	6.92	7.10
Honduras	0.62	1.50	1.60
Costa Rica	0.46	1.25	1.30
Mexico	0.86	1.12	1.11
Panama	0.56	1.01	0.96
South America	8.62	12.87	13.26
Brazil	4.09	6.81	7.30
Ecuador	2.83	3.51	3.00
Venezuela	0.75	1.00	1.01
Asia	7.93	9.44	9.82
India	2.65	2.85	2.90
Indonesia	1.38	1.79	1.89
Thailand	0.80	1.20	1.25
42. Mangoes			
World	9.97	11.75	11.88
Africa	0.39	0.45	0.44
North/Central America	0.47	0.60	0.60

Crop production data – *continued*

	1961–5 (annual mean)	1971	1973
42. Mangoes – *continued*			
South America	0.85	0.94	0.96
Brazil	0.58	0.66	0.67
Asia	8.26	9.76	9.87
India	7.26	8.45	8.55
Pakistan	0.31	0.65	0.67
Bangladesh	0.51	0.42	0.42
43. Pineapples			
World	3.27	4.16	4.41
Africa	0.29	0.50	0.58
North/Central America	1.21	1.29	1.26
USA	0.81	0.82	0.79
South America	0.48	0.65	0.73
Brazil	0.28	0.38	0.41
Asia	1.20	1.56	1.68
Philippines	0.15	0.28	0.38
Bangladesh	0.20	0.36	0.33
Malaysia	0.25	0.33	0.29
44. Oranges			
World	17.43	26.81	29.34
Africa	1.77	2.59	2.89
Egypt	0.29	0.71	0.77
Morocco	0.46	0.67	0.73
South Africa	0.47	0.42	0.54
North/Central America	6.18	10.26	11.57
USA	4.47	7.66	9.04
Mexico	1.23	2.00	1.90
South America	3.23	5.12	5.68
Brazil	2.01	3.15	4.00
Asia	3.03	4.68	4.74
Israel	0.56	1.11	1.15
India	0.71	0.90	0.92
Mainland China	0.53	0.74	0.76
Europe	2.97	3.82	4.09
Spain	1.62	1.87	1.93
Italy	0.89	1.46	1.57
45. Tangerines			
World	2.45	4.92	6.13
Africa	0.23	0.41	0.39
North/Central America	0.21	0.35	0.37
USA	0.20	0.32	0.35
South America	0.39	0.58	0.67
Brazil	0.17	0.25	0.35
Asia	1.28	2.84	3.72
Japan	1.06	2.49	3.33
Europe	0.62	0.72	0.95
Spain	0.47	0.37	0.56

Crop production data – *continued*

	1961–5 (annual mean)	1971	1973
46. Dates			
World	1.85	2.01	2.03
Africa	0.77	0.78	0.83
Egypt	0.41	0.34	0.35
Algeria	0.11	0.13	0.17
North/Central America	0.03	0.02	0.02
Asia	1.04	1.19	1.17
Iraq	0.34	0.45	0.40
Iran	0.30	0.28	0.30
Saudi Arabia	0.24	0.25	0.25
47. Onions (dry)			
World	11.62	14.88	15.12
Africa	0.94	0.99	0.89
Egypt	0.64	0.60	0.52
North/Central America	1.32	1.52	1.48
USA	1.17	1.35	1.33
South America	0.74	0.89	0.96
Asia	5.13	6.86	7.06
Mainland China	1.57	2.07	2.13
India	1.26	1.49	1.51
Japan	0.77	1.04	1.10
Europe	3.05	3.83	3.78
48. Tomatoes			
World	23.20	31.38	33.56
Africa	2.02	3.08	3.07
Egypt	1.07	1.64	1.58
North/Central America	6.14	7.37	7.84
USA	5.08	5.81	6.26
South America	1.18	1.64	1.70
Asia	3.14	5.01	5.46
Turkey	1.20	1.90	2.10
Europe	8.21	11.19	11.73
Italy	2.87	3.42	3.28
Spain	1.30	1.85	1.93
USSR	2.32	2.85	3.50

Index of Scientific Names

Acacia mearnsii, 72
Acacia senegal, 73
Agavaceae, 280−4, 286
Agave cantala, 280
Agave fourcroydes, 280
Agave lecheguilla, 280
Agave sisalana, 280−3
Albizzia, 107
Aleurites fordii, 294−7
Aleurites montana, 294−7
Alliaceae, 148
Allium ampeloprasum var. *porrum*, 151
Allium cepa, 148−51
    var. *aggregatum*, 151
    var. *cepa*, 148−51
    var. *proliferum*, 150
Allium sativum, 151
Alysicarpus vaginalis, 107
Aloe, 227
Aloe barbadensis, 227
Aloe ferox, 227
Aloe perryi, 227
Amaryllidaceae, 148
Anacardiaceae, 174, 192
Anacardium occidentale, 174, 192, 193, 302
Ananas ananassoides, 178
Ananas bracteatus, 178
Ananas comosus, 178−81
Ananas erectifolius, 178
Andropogon, 25
Andropogoneae, 25, 68, 334
Anomala, 255
Anonaceae, 328
Apium graveolens var. *dulce*, 249
Apocynaceae, 227, 336, 338
Aquifoliaceae, 214
Araceae, 123
Arachis hypogaea, 80−5
Areca catechu, 217
Artocarpus altilis, 188, 189
Artocarpus heterophyllus, 189

Arundinoideae, 23
Asclepiadaceae, 338
Aspergillus flavus, 84
Attalea funifera, 287
Avena fatua, 23
Avena sativa, 23
Aveneae, 23

Bambusoideae, 23
Bauhinia, 72
Beta, 1
Beta vulgaris subsp. *vulgaris*, 64
Boehmeria nivea
    var. *nivea*, 278, 279
    var. *tenacissima*, 279
Brachystegia, 72
Brassica, 145, 298, 299
Brassica campestris, 299
Brassica juncea, 145, 299
Brassica napus, 299
Brassica oleracea
    var. *botrytis*, 145
    var. *capitata*, 145
Brassica rapa, 145
Bromeliaceae, 178
Butryospermum paradoxum, 320, 321

Caesalpinioideae, 71−2, 75
Cajanus cajan, 99−101, 107
Calathea allouia, 127
Calopogonum mucunoides, 107
Camellia japonica, 197, 202
Camellia sasangua, 202
Camellia sinensis, 196−202
    var. *assamica*, 197
    var. *sinensis*, 197
Cananga latifolia, 328
Cananga odorata, 328, 329
Canavalia ensiformis, 102−4
Canavalia gladiata, 102−4
Canna edulis, 127
Cannabidaceae, 278

Index of Scientific Names

Cannabis sativa, 222, 278
Cannaceae, 127
Capsicum, 142, 245–8
Capsicum annuum, 245–8
  var. *annuum*, 245
  var. *minimum*, 245
Capsicum baccatum, 245
  var. *baccatum*, 245
  var. *pendulum*, 245
Capsicum chinense, 245, 246, 248
Capsicum frutescens, 245, 246, 248
Capsicum pubescens, 245
Carica papaya, 181–5
Caricaceae, 181
Carthamus lunatus, 294
Carthamus oxyacantha, 294
Carthamus tinctorius, 291–4
Carum carvi, 249
Caryocar, 302
Cassia angustifolia, 72
Cassia auriculata, 72
Cassia senna, 72
Castilloa elastica, 337
Ceiba pentandra var. *caribaea*, 252
Centrosema plumieri, 107
Centrosema pubescens, 107
Ceratopogonideae, 211
Cercospora, 81
Chenopodiaceae, 1, 64
Chlorideae, 26
Chloris gayana, 26
Chrysanthemum cinerariifolium, 291
Cicer arietinum, 98
Cinchona, 202, 222–5
Cinchona calisaya, 223
Cinchona ledgeriana, 223, 224
Cinchona officinalis, 224
Cinchona succirubra, 224
Cinchonoideae, 223
Cinnamomum camphora, 236, 326–8
Cinnamomum zeylanicum, 235, 236
Citrullus lanatus, 139–42, 299
Citrus, 159–68, 299, 325
Citrus aurantifolia, 166–7, 325
Citrus aurantium, 168
  subsp. *amara*, 325
  subsp. *bergamia*, 325
Citrus grandis, 167, 168
Citrus limon, 166, 325
Citrus medica, 167
Citrus paradisi, 164
Citrus reticulata, 167
Citrus sinensis, 164–6, 325
Clitoria ternata, 107
Clostridium, 269
Cocoideae, 313

Cocos coronata, 302
Cocos nucifera, 306–12
  var. *nana*, 307
  var. *typica*, 307
Coffea, 202–7
Coffea arabica, 202–7
  var. *arabica*, 202
  var. *bourbon*, 203
Coffea canephora, 202, 204, 207
Coffea liberica, 202, 204, 207
Coffeoideae, 223
Cola, 208, 214–17
Cola acuminata, 215
Cola nitida, 215
Colocasia esculenta, 123–5
  var. *antiquorum*, 124
  var. *esculenta*, 124
Colocynthis, 133
Compositae, 291–4, 336, 337
Convolvulaceae, 112
Corchorus aestuans, 146
Corchorus capsularis, 146, 269–72
Corchorus olitorius, 146, 269–72
Corchorus trilocularis, 146
Corchorus tridens, 146
Coriandrum sativum, 249
Crinipellis perniciosus, 208
Crocus sativus, 294
Crotalaria juncea, 107, 276–8
Cruciferae, 1, 145
Cryptostegia, 338
Cucumis melo, 138–9, 299
Cucumis sativus, 138
Cucurbita, 133, 134–7
Cucurbita ficifolia, 134
Cucurbita maxima, 134–6
Cucurbita mixta, 134, 137, 138
Cucurbita moschata, 134, 136, 137
Cucurbita pepo, 134, 136
Cucurbitaceae, 133–42
Cuminum cyminum, 249
Curcuma amada, 233
Curcuma angustifolia, 233
Curcuma aromatica, 233
Curcuma domestica, 232
Curcuma zeodaria, 233
Cyamopsis tetragonolobus, 104, 105
Cymbopogon, 331–4
Cymbopogon citratus, 331
Cymbopogon flexuosus, 331
Cymbopogon martinii
  var. *motia*, 334
  var. *sofia*, 334
Cymbopogon nardus, 333
  var. *lenabatu*, 334
  var. *mahepengiri*, 334

Index of Scientific Names 359

Cynodon dactylon, 26
Cyphomandra cetacea, 142

Daucus carota, 249
Delonix regia, 72
Derris elliptica, 108, 109
Desmodium intortum, 107
Desmodium uncinatum, 107
Digitaria decumbens, 60
Digitaria exilis, 24, 60–1
Dioscorea, 119–23
Dioscorea alata, 121–3
Dioscorea bulbifera, 121
Dioscorea cayenensis, 121–3
Dioscorea esculenta, 121, 123
Dioscorea opposita, 120, 121
Dioscorea rotundata, 121–3
Dioscorea trifida, 121
Dioscoreaceae, 120
Dolichos lablab, 96
Doryphora sassafras, 328

Echinochloa colona, 50, 56
Echinochloa crus-galli, 50, 56
Echinochloa frumentacea, 24, 50, 56–7
Echinochloa pyramidalis, 56
Echinochloa stagnina, 56
Elaeis guineensis, 312–20
Elettaria cardamomum, 233–5
    var. *cardamomum*, 234
    var. *major*, 234
Eleusine africana, 57, 58
Eleusine coracana, 26, 50, 57–60
Eleusine indica, 57
Enantiophyllum, 121
Eragrosteae, 26
Eragrostis tef, 26, 61
Eragrostoideae, 25
Eruca sativa, 145
Eryoxyla, 255
Erysiphe cichoracearum, 134
Erythrina, 107
Erythroxylon coca, 226
Erythroxylon novogranatense, 226
Euchlaena mexicana, 25, 33
Eucitrus, 159
Eugenia caryophyllus, 240–3
Euphorbiaceae, 117, 295, 302, 336, 337

Festucoideae, 20–3
Ficus elastica, 338
Foeniculum vulgare, 249
Funtumia elastica, 338
Furcraea gigantea, 286
Fusarium oxysporum, 134, 155

Garcinia hanburyi, 187
Garcinia mangostana, 187, 188
Geraniaceae, 326
Geranium, 326
Glycine gracilis, 85
Glycine javanica, 107
Glycine max, 85–7, 107, 297–8
Glycine ussuriensis, 85
Glyricidia sepium, 107
Gossypium, 252, 253–68
Gossypium arboreum, 255–8, 267
Gossypium barbadense, 256–8
Gossypium herbaceum, 255, 256, 258
    race *africanum*, 255
Gossypium hirsutum, 256–68
    var. *latifolium*, 256
    var. *punctatum*, 257
Gossypium raimondii, 256
Gossypium thurberi, 256
Gossypium tomentosum, 256
Gramineae, 16–70, 331
Guizotia abyssinica, 291, 294–5
Guttiferae, 187

Helianthus annuus, 291, 298
Hemilia vastatrix, 203
Hevia braziliensis, 336, 338–44
Hibiscus cannabinus, 146, 273–6
Hibiscus esculentus, 146–8
Hibiscus sabdariffa, 146
    var. *altissima*, 276
    var. *sabdariffa*, 276
Hordeum vulgare, 23
Hyparrhenia, 25

Ilex paraguariensis, 214
Indigofera, 107
Ipomoea batatas, 111–16
Ipomoea tiliacea, 112
Ipomoea trifida, 112

Kerstingiella geocarpa, 102
Klotzschiana, 155

Labiatae, 327, 329, 330
Lablab, 88
Lablab niger, 96
Lablab purpureus, 96–8
Lablab vulgaris, 96
Lactuca sativa, 291
Lageneria siceraria, 133
Landolphia, 338
Leguminosae, 71
Lens esculenta, 98–9
Liliaceae, 120, 148, 227
Linaceae, 297

Linum usitatissimum, 253, 297
Loganaceae, 227
Lonchocarpus, 108
Luffa cylindrica, 133
Lycopersicon esculentum, 142–4, 299
    var. *cerasiforme*, 143

Malvaceae, 147, 208, 253, 279
Mangifera indica, 174–8
Manihot esculenta, 116–19
Manihot glaziovii, 118, 337
Maranta arundinacea, 126, 127
Marantaceae, 126
Marasmius perniciosa, 208
Maydeae, 25
Medicago sativa, 107
Metroxylon rumphii, 128, 129
Metroxylon sagus, 128, 129
Microcyclus ulei, 342
Mimosoideae, 71–2, 75
Moraceae, 188, 336–8
Musa, 153–9, 248–86
Musa acuminata, 154, 159
Musa balbisiana, 154, 156
Musa paradisiaca, 154
Musa sapientum, 154
Musa textilis, 284–6
Musaceae, 284
Myristica fragrans, 244–5
Myristicaceae, 244
Myrtaceae, 186, 240–4, 329

Nicotiana, 142, 218–21
Nicotiana otophora, 219
Nicotiana rustica, 218
Nicotiana sylvestris, 219
Nicotiana tabacum, 218–21

Ocimum americanum, 330
Ocimum basilicum, 330
Ocimum kilimandscharicum, 327
Ocimum sanctum, 330
Ocimum viride, 330
Ocotea cymbarum, 328
Ocotea pretiosa, 328
Olea europeae, 290, 302
Oleaceae, 302
Orchidaceae, 238
Oryza barthii, 27
Oryza glaberrima, 25, 27
Oryza sativa, 25, 26–33
    subsp. *indica*, 27–8
    subsp. *japonica*, 28
    subsp. *javanica*, 28
Oryzeae, 25
Oryzoideae, 21, 25

Pachyrrhizus erosus, 108
Pachyrrhizus tuberosus, 108
Palmae, 128, 217, 306, 313
Paniceae, 24
Panicoideae, 20–5, 331
Panicum maximum, 24
Panicum miliaceum, 24, 50, 55–6
Panicum miliare, 24, 56
Panicum sumatrense, 24, 56
Panicum turgidum, 56
Papaver setigerum, 225
Papaver somniferum, 225–6
    var. *hortense*, 226
    var. *somniferum*, 226
Papaveraceae, 225
Papilionoideae, 71–108
Parkia clappertoniana, 73
Parkia filicoidea, 73
Parthenium argentatum, 337
Pastinaca sativa, 249
Pedaliaceae, 299
Pelargonium capitatum, 326
Pelargonium graveolens, 326
Pelargonium odoratissimum, 326
Pelargonium roseum, 326
Pennisetum americanum, 24, 50–4
Pennisetum clandestinum, 24
Pennisetum purpureum, 24
Pennisetum typhoides, 24
Persea americana, 190–2, 302
    var. *drymifolia*, 192
Petroselinum crispum, 249
Phaseoleae, 88
Phaseolus, 7, 10, 12, 79, 87–91, 348
Phaseolus aconitifolius, 88, 91
Phaseolus acutifolius, see *Vigna acutifolius*
Phaseolus angularis, see *Vigna angularis*
Phaseolus aureus, 88
Phaseolus calcaratus, 88
Phaseolus coccineus, 88
Phaseolus lunatus, 88, 89–90
Phaseolus mungo, 88
Phaseolus vulgaris, 88–9
Phoenix dactylifera, 168–74
Phormium tenax, 286
Physalis peruviana, 142
Phytophthora, 168
Phytophthora palmivora, 208
Pimenta dioica, 243, 244
Pimenta racemosa, 243, 329, 330
Pimpinella anisum, 249
Piper betle, 217
Piper nigrum, 217, 236–8
Piperaceae, 217, 236
Pistacia vera, 174

Pogostemon cablin, 329
Pooideae, 21
Portulaca oleracea, 145
Pseudoperonospora cubensis, 134
Psidium guajava, 185–7
Psidium littorale, 186
Psophocarpus palustris, 105
Psophocarpus tetragonolobus, 105
Pueraria phaseoloides, 106, 107

Quelea, 54

Raphanus sativus, 145
Raphia, 287
Rauvolfia serpentina, 227
Rhizobium, 76–9, 82, 86, 105
Rhizobium japonicum, 86
Rhus toxicodendron, 174
Ricinus communis, 302–6
Roystonea oleracea, 129
Rubiaceae, 202, 223
Rutaceae, 160, 325

Saccharum, 25
Saccharum barberi, 66
Saccharum edule, 66
Saccharum officinarum, 64–70
Saccharum robustum, 66
Saccharum sinense, 66
Saccharum spontaneum, 66, 67
Sansevieria, 286
Sassafras albidium, 328
Scrophulariaceae, 299
Secale cereale, 23
Sesamum indicum, 299–301
Setaria italica, 50, 54–5
Soja, 85
Solanaceae, 1, 142–5, 218, 245
Solanum guiotense, 142
Solanum melongena, 142, 143
Solanum muricatum, 142
Solanum tuberosum, 1, 111, 142
Sorghum arundinaceum, 45
Sorghum bicolor, 25, 43–9
Sorghum propinquum, 45
Sterculiaceae, 208, 214
Stizolobium atterimum, 107
Stizolobium deeringianum, 107

Stocksiana, 255
Strychnos nux-vomica, 227
Sturtiana, 255
Stylosanthes gracilis, 106, 107
Syagrus coronata, 302

Tacca leontopetaloides, 128
Taccaceae, 128
Tamarindus indica, 72
Taraxacum kok-saghyz, 337
Tephrosia candida, 107
Theaceae, 197, 208
Theobroma cacao, 207–14, 320
Thurberana, 255
Tiliaceae, 146, 270
Trifolium alexandrinum, 107
Trifolium semipilosum, 107
Tripsacum dactyloides, 33
Triticeae, 23
Triticum aestivum, 17, 23
Triticum durum, 23

Umbelliferae, 248–9
Urena lobata, 279–80
Urticaceae, 278

Vanilla fragrans, 238–40
Vetiveria zizanioides, 334
Vigna, 88, 91–6, 107, 348
Vigna aconitifolia, 91
Vigna angularis, 91
Vigna mungo, 88, 91, 95, 348
Vigna radiata, 88, 91, 95–6, 348
Vigna unguiculata, 91–5, 107
    subsp. *cylindrica*, 93–4
    subsp. *dekindtiana*, 92
    subsp. *sesquipedalis*, 93–4
    subsp. *unguiculata*, 92–5
Vigna vexillata, 107
Voandzeia subterranea, 101–2

Xanthosoma sagittifolium, 125, 126

Zea mays, 25, 33–43
Zea mexicana, 25, 33
Zingiber officinale, 230–2
Zingiberaceae, 230–5

General Index

Abaca fibre, 284–6, 352
Abata kola, 215
Acid limes, 167
Adventitious embryony, 162
Adzuki bean, 11, 91
Aestivation, 71
Aflatoxin, 84
African rice, 25, 27
Albedo, 163
Aleurone, 19
Allspice, 186, 243–4
Aloes, 227
Alyce clover, 107
Amelonado cocoa, 210
Amylopectin, 29, 49
Amylose, 29, 49
Anise, 249
Antelope grass, 56
Arabian coffee, 202–7
Arabica coffee, 12, 13, 202–7
Arrack, 306
Aramina fibre, 279, 352
Arecoline, 195
Arrow, sugar cane, 68
Arrowroot, 126–7
Asparagus bean, 93
Assam tea, 197
Aubergine, 142–3
Avaram, 72
Avocado pear, 190–2, 235, 302

Backcrossing, 67
Bambarra groundnut, 101
Bamboo, 16
Banana, 12, 153–9, 354
    classification, 155–6
    morphology, 156–9
    production, 354
    origin, 12, 154
    wilt, 155
Barley, 12, 16, 23
Barnyard millet, 50, 56

Bast fibres, 253, 268–80, 352
    aramina, 279–80
    hemp, 278
    Italian hemp, 278
    jute, 269–72
    kenaf, 273–6
    ramie, 278–9
    roselle, 276
    sann hemp, 276–8
    sunn hemp, 276–8
Basil, 320
Bay, 243, 329–30
Bay rum, 320
Benniseed, 299–301, 350
    oil extraction, 301
    production, 350
Beet sugar, 64
Beriberi, 33
Betel nut, 195
Betel palm, 217
Betel pepper, 195, 217
Bermuda grass, 26
Berseem clover, 107
Bhang, 222
Bird pepper, 142
Black gram, 88, 91, 95
Black pod, cocoa, 208
Black tea, 201
Black wattle, 72
Black eye pea, 92
Bottle gourd, 133
Bourbon sugar cane, 66
Bow-string hemp, 286
Breadfruit, 188–9
Breadnut, 188
Bread wheat, 23
Broad bean, 12
Broccoli, 145
'Boot' stage, cereals, 47
Bromelain, 180
Brucine, 227
Budding, citrus, 168

Bulrush millet, *see* pearl millet
Butter bean, 89
Buttercup squash, 135
Butternut squash, 136

Cabbage, 12, 145
Cabbage palm, 129
Caffeine, 195, 201
Caffeol, 207
Calysaya bark, 223
Camellia, 197
Camphor, 235, 236, 323, 326—8
Cananga oil, 328—9
Canary banana, 155
Cane sugar, 64
Cantala fibre, 280
Cantaloupe melon, 138—9
Capitulum, 291
Cape aloe, 227
Cape gooseberry, 142
Capsaicin, 248
Caraway, 249
Cardamom, 233—5
Carthamin, 294
Carrot, 249
Caryopsis, 16, 17, 19
Cashew, 174, 192—3
Cashew apple, 192
Cashew seed oil, 193, 303
Cashew shell oil, 192
Cassava, 12, 116—19, 347
    cultivars, 119
    morphology, 117—18
    origin, 12, 116
    production, 116, 347
    propagation, 118—19
Catilloa rubber, 337
Castor, 301—2, 349
Catjang, 92, 93
Cauliflower, 145
Cauliflorous, 189, 211
Cavendish banana, 155
Cayenne pepper, 248
Cayenne pineapple, 180
Ceara rubber, 118, 337
Celery, 249
Cheribon sugar cane, 66
Cherry tomato, 143
Chick pea, 12, 98, 348
China grass, 279
China tea, 197
Chilli pepper, 12, 142, 245—8
Chocolate, 213, 321
Chupon, 209
Cinnamon, 186, 235—6
Cinnamon oil, 236

Citral, 161, 331, 333
Citron, 167
Citronella oil, 333
Citrus fruits, 159—68, 355
    grapefruit, 164
    lemon, 166
    lime, 166—7
    mandarin, 167
    pummelo, 167—8
    rough lemon, 168
    Seville orange, 168
    shadock, 167—8
    sour orange, 168
    sweet orange, 164—6
    tangerine, 167
Citrus oils, 325
Citrus scab, 168
Chinese yam, 120
Cleistogamy, 75, 83
Clove, 186, 240—3
Clove oil, 243
Cluster bean, 104
Cob, maize, 25, 34
Coca, 226
Cocain, 226
Cocoa, 4, 12, 195, 207—14, 353
    butter, 213, 320
    chocolate, 321
    classification, 209—10
    morphology, 210—13
    processing, 213—14
    production, 208, 353
    witches broom disease, 208
Coconut, 12, 287, 306—12, 351
    apple, 310
    coir, 310
    copra, 311
    dispersal, 307
    flowering, 309
    fruit, 309
    germination, 310—11
    milk, 312
    morphology, 308—10
    oil, 312
    origin, 12, 307
    production, 351
    water, 310
Cocoyam, 123—5
Codein, 225
Coffee, 12, 202—7, 353
    history, 202—3
    morphology, 203—4
    leaf rust, 202
    processing, 205—7
    production, 203, 353
    pruning, 203

Coir, 287
Common bean, 88
Common millet, 24, 50
Congo jute, 279
Copra, 311, 351
Cotton, 12, 252, 253—68, 350, 351
    Asian, 255, 256
    cultivated cottons, 258—68
    dispersal, 257
    diploid cottons, 254—6
    ginning, 266
    Ishan, 257
    lint, 264, 266—7
    morphology, 260—4
    production, 258, 350, 351
    Sea Island, 257, 258
    seed oil, 267, 298
    tetraploid cottons, 254—6
    tree cotton, 255
    Upland, 257
    wild, 254
Coriander, 249
Corn, *see* maize
Cover crops, 105—7, 342
Cowpea, 12, 91—5, 107, 349
Creole sugar cane, 66
Criollo cocoa, 209
Crown bark, 224
Cucumber, 138
Cucurbits, 133—42
Cumin, 249
Curare, 227
Cush-cush yam, 121
Cushites, 45
Cystein, 79
Cystine, 79

Dasheen, 123—5
Date, 168—74, 356
    dry, 173
    morphology, 169—72
    origin, 169
    production, 356
    soft, 173
Dent corn, 40
Deli oil palm, 312
Derris, 108
Dickenson pumpkin, 135
Dhal, 95, 98, 99
Doldrums, 4
Drugs, 218—27
    bhang, 222
    coca, 226
    ganja, 222
    hashish, 222
    marijuana, 222

    opium, 225—6
    quinine, 222—5
    tobacco, 218—22
Drug aloes, 227
Drying oils, 290, 291—8
    linseed, 297
    niger seed, 294
    safflower, 291—4
    soyabean, 297—8
    tung, 294—7
Dura oil palm, 318
Durra sorghum, 45

Eddoe, 123—5
Egg plant, 142—3
Egyptian clover, 107
Elephant grass, 24
Enfleurage, 324
Essential oils, 161, 163, 186, 195, 201, 230, 323—35
    basil, 330
    bay, 329—30
    camphor, 326—8
    cananga, 328
    cinnamon, 236
    citronella, 333, 334
    citrus, 325
    clove, 243
    definition, 323—4
    eugenol, 243, 330
    extraction, 324
    geranium, 325—6
    gingergrass, 334
    khuskhus, 334
    lemon grass, 331—3
    myristicin, 245
    palmarosa, 334
    patchouli, 329
    pimento berry, 243
    rosha, 334
    saffrole, 327
    sassafras, 327
    vetiver, 334
    ylang-ylang, 328
    zingiberine, 232
Evapotranspiration, 7

Farming systems, 6—9
    cereal, 6—9
    root and tuber, 6
Fats, 306—21
    cocoa butter, 320
    coconut oil, 312
    palm oil, 312—20
    shea butter, 320, 321
Fennel, 249

Fertile crescent, 10
Fibres, 252–88
    bast, 268–80
    structural, 280–7
    surface hairs, 253–68
Field mustard, 299
Finger millet, 12, 26, 50, 57
Fish poison, 108
Flag leaf, 47
Flavedo, 162, 325
Flax, 253
Flint corn, 40
Floating rice, 29
Floret, 18
Floury-2, 41
Fodder legumes, 75
Forage legumes, 75, 105–7
Forastero cocoa, 209
French bean, 88
Fruit crops, 153–92
    avocado, 190–2
    banana, 153–9, 354
    breadfruit, 188–90
    citrus, 159–68, 355
    date, 168–74, 356
    guava, 185–7
    mango, 174–8, 354
    mangosteen, 187–8
    pawpaw, 181–5
    pineapple, 178–81, 355
Fusarium wilt, 134
Fuz, cotton, 254, 264, 267

Gamboge, 187
Gamma grass, 33
Ganja, 222
Garlic, 151
Garri, 119
Gbanja kola, 215
Genome, 57, 154, 255
Geraniol, 326, 333
Geranium oil, 325
German millet, 54
Gherkin, 138
Ginger, 230–2
Gingergrass, 334
Ginning cotton, 266
Glume, 18
Gluten, 17
Glycerol, 320
Golden Delicious squash, 135
Goa bean, 105
Grapefruit, 164, 325
Grain legumes, 4, 7, 75, 79–105
    adzuki bean, 91
    Bambarra groundnut, 101–2
    black gram, 95
    chick pea, 98
    cluster bean, 104–5
    common bean, 88–9
    cowpea, 92–5
    green gram, 95–6
    groundnut, 80–5
    hyacinth bean, 96–8
    lentil, 98–9
    lima bean, 89–90
    mat bean, 91
    pigeon pea, 99–101
    soyabean, 85–7
    sword bean, 102–4
    tepary bean, 90–1
    winged bean, 105
Grain legume production, 348
Gram, 98
Greater yam, 121
Green gram, 88, 95–6
Green tea, 202
Gros Michel banana, 155
Groundnut, 4, 12, 80–5, 349
    branching habit, 80–2
    cultivars, 80–2
    morphology, 82–4
    production, 349
Growing season, 7
Guava, 185–7
Guayule rubber, 337
Guinea corn, 45
Guinea grass, 24
Gum arabic, 73
Gummosis, 168

Halophytic, 306
Hard fibres, *see* structural fibres
Haricot bean, 88
Hashish, 222
Hegari, 45
Heliotropin, 327
Henequen, 280, 283–4
Heroin, 225
Hesperidium, 160
Honeydew melon, 138–9
Horse latitudes, 4
Hubbard squash, 135
Hungarian millet, 54
Hungry rice, 24, 60
Hyacinth bean, 96–8, 107
Hypanthium, 83, 241

Inbred lines, 43, 138
Incompatibility
    gametophytic, 219
    sporophytic, 211

India rubber, 338
Indian arrowroot, 233
Indian mustard, 145, 299
Indica rice, 27
Indigo, 107
Ingera, 26, 61
Intertropical convergence zone, 5
Iodine number, 290
Ionine, 331
Istle, 280
Italian hemp, 278
Italian millet, 54
Ixtle, 280

Jack bean, 102
Jackfruit, 189
Jamaica pepper, 243
Japanese barnyard millet, 24, 50, 56
Japonica rice, 28
Jassids, 260
Jat, 197
Javanica rice, 28
Jew's mallow, 146
Julie mango, 177
Jute, 12, 269–72, 352
  production, 352
  quality, 271–2
  retting, 272
  tossa, 146, 270
  white, 270

Kafir sorghum, 45
Kapok, 252
Kaoliang, 45
Kenaf, 273–6, 352
Kenya white clover, 107
Kentucky Field pumpkin, 136
Key lime, 167
Khuskhus, 334
Kidney bean, 88
Kikuyu grass, 24
Kola, 195, 214–17
Kudzu, 107
Kwashiorkor, 111

Labellum, 232, 239
Lady's finger, 146
Lagos rubber, 338
Large Cheese pumpkin, 136
Latex tubes, 336, 342
Laticifers, 336
Lauric acid, 312
Leaf blight, rubber, 342
Leaf fibres, see structural fibres
Ledger bark, 224

Leek, 151
Leghaemoglobin, 78
Legumes, 71–109
Lemma, 18
Lemon, 163, 166
Lemon grasses, 331–4
Lemon oil, 325
Lentil, 12, 98, 348
Lesser yam, 123
Lettuce, 12, 291
Liberica coffee, 202
Ligule, 17
Lima bean, 89
Lime, 163, 166–7
Lime oil, 325
Linamarin, 119
Linen, 253
Linoleic acid, 290, 301
Linolenic acid, 290
Linseed, 297, 350
Lint, 258, 264, 266–7, 351
Linters, 267
Lipase, 320
Little millet, 24, 56
Local adaptation, 4, 8, 48, 92
Locust bean, 73
Lodicule, 18
Loofah, 133
Lycopersicin, 144

Macaroni wheat, 23
Mace, 244–5
Macrocarya oil palm, 318
Maguey fibre, 280
Maize, 7, 12, 16, 25, 33–43, 298, 345, 346
  caryopsis, 40
  improvement, 43
  morphology, 36–40
  oil, 36, 298
  origin, 12, 33–4
  production, 36, 345, 346
Malabar lemon grass, 331
Male sterility, 43, 49, 54, 151
Mandarin, 167
Mango, 174–8
  flowering, 176
  morphology, 174–5
  production, 354, 355
  propagation, 178
Mango ginger, 233
Mangosteen, 187–8
Manila hemp, 284–6
Manioc, see cassava
Marblehead squash, 135
Marihuana, 222

General Index 367

Marijuana, 222
Marrow, 134–7
Masticatories, 214–17
 betel, 217
 kola, 214–17
Mat bean, 91
Maté, 214
Mauritius hemp, 286
Melon, 138–42
Metaxenia, 172
Methionine, 42, 79
Mexican lime, 167
Millets, 16, 49–61
 bulrush, 50–4
 common, 55–6
 finger, 57–60
 foxtail, 54–5
 German, 54–5
 Hungarian, 54–5
 hungry rice, 60–1
 ingera, 61
 Italian, 54–5
 Japanese barnyard, 56–7
 little, 24
 pearl, 50–4
 proso, 55–6
 Siberian, 54–5
 tef, 61
Millet production, 346
Milo, 45
Mishrig date, 173
Monkey nut, 80
Monopodial, 144, 260
Morphine, 225
Moth bean, 91
Multiple fruit, 180
Mung, 95
Musk melon, 138–9
Mutation, 10
Mycorrhiza, 161, 169, 178, 197, 210
Myristic acid, 312
Myristicin, 245

Napier grass, 24
Naranjilla, 142
Nastic movement, 83
Navel orange, 163
New World cotton, 256
New Zealand hemp, 286
Niacin, 79
Nicotine, 218
Niger seed, 291, 294, 295
Nitrogen fixation, 76–9, 342
Nitrogenase, 76
Noble cane, 66
Nobilization, 66

Non-drying oils, 301–6
 cashew nut, 302
 castor, 302–6, 349
 groundnut, 80–5, 302, 349
 olive, 12, 290, 302
 ouricuri, 302
 sawarri, 302
 souari, 302
Nucellus, 19
Nutmeg, 244–5

Oat, 16, 23
Oil of Neroli, 161, 325
Oil palm, 312–20, 351
 cultivars, 318
 fruit, 317–18
 kernel oil, 320
 morphology, 313–17
 origin, 312
 palm oil, 318, 320
 production, 312, 313, 351
 propagation, 318
 sex expression, 315
Oils and fats, 289–322
 classification, 290–1
 extraction, 291
Okra, 146–7
Old World cotton, 255
Oleic acid, 290, 320
Olive, 12, 290
Olive oil, 290
Onion, 148–51, 356
Oolong tea, 202
Opaque-2, 41
Opium, 195, 225–6
Orange, 11, 164–6, 325, 355
Orange blossom oil, 161, 325
Orange oil, 325
Orthotropic, 203
Ouricuri oil, 302

Paddy, see rice
Palay rubber, 338
Palea, 18
Palm fibres, 286–7
Palm oil, 318, 320
Palmarosa oil, 334
Palmitic acid, 312, 320
Palms, 128, 168–74, 217, 286–7, 306–20
Panama disease, 155
Panama rubber, 337
Pangola grass, 60
Panicle, 17
Papain, 181, 185
Papaya, 181–5

Para rubber, 338–44, 353
  history, 340
  latex, 343–4
  morphology, 340–2
  production, 338, 353
  propagation, 342
  tapping, 342–3
Parboiling, 33
Parchment, coffee, 204
Parthenocarpy, 161, 188
Parsley, 249
Parsnip, 249
Pasture legumes, 75
Patchouli oil, 329
Pawpaw, 181–5
Peanut, see groundnut
Pearl millet, 4, 8, 12, 24, 50–4, 346
Pectin, 163
Pedicelled spikelet, 25, 48, 68–9
Peg, groundnut, 83
Penicillin, 195
Pepino, 142
Pepo, 133
Pepper, 12, 236–8
Pepsin, 185
Pericarp, 19
Persian lime, 167
Peruvian bark, 223
Petit grain, 325
Phloem fibres, see bast fibres
Piassava fibre, 287
Pigeon pea, 99–101, 107, 348
Pimento, 243
Pimento oil, 243
Pina cloth, 181
Pineapple, 12, 178–81, 355
  morphology, 179–80
  origin, 178–9
  production, 178, 355
  propagation, 180–1
Pisifera palms, 318
Pistachio, 174
Plagiotropic, 203
Plantain, 154–9
Plucking table, tea, 200
Pneumathode, 313
Pod corn, 34
Poi, 125
Poison ivy, 174
Polyembryony, 161
Polyisoprene, 336
Pomace, 306
Potato, 12, 142
Powdery mildew, 134
Prop root, 36, 46, 53, 68
Proso, 24, 50, 55

Protandry, 34, 150
Pseudostem, 156, 284
Pulp vesicle, 163
Pulses, see grain legumes
Pulvinus, 83
Pummelo, 164, 167
Pumpkin, 134–7
Purslane, 145, 146
Pyrethrum, 291

Queens pineapple, 180
Quelea, 54
Quinine, 195, 202, 222–5

Rachilla, 18
Radicle, 20
Radish, 145
Rag, citrus, 163
Rai, 299
Rainfall, 4
Ramie, 278–9
Rape, 298
Raphia, 287
Ray floret, 291
Reserpine, 227
Retting, 268–9, 272, 275
Rhea, 279
Rhodes grass, 26
Ricin, 306
Ricinoleic acid, 306
Rice, 4, 7, 12, 16, 25, 26–33, 345
  breeding and IRRI, 33
  classification, 27–9
  floating, 29
  morphology, 30–1
  origin, 27
  paddy, 28, 29
  production, 26, 345
  swamp, 28
  upland, 29
Rocket cress, 145
Robusta coffee, 202
Roots and tubers, 6, 111–28, 347
  arrowroot, 126–7
  cassava, 116–19
  cocoyam, 123–5
  dasheen, 123–5
  eddoe, 123–5
  manioc, 116–19
  production, 347
  sweet potato, 111–16
  tannia, 126
  yam, 119–23
Root stocks, 162, 342
Root nodules, 76–9
Rose oil, 326

Roselle, 276
Rosette disease, 85
Rosha oil, 334
Rostellum, 239
Rotenone, 108
Rough lemon, 168
Rubber, 12–13, 336–44, 353
    castilloa, 337
    ceara, 337
    guayule, 337
    India, 338
    Lagos, 338
    palay, 338
    Panama, 337
    para, 336, 338–44, 353
    production, 338, 353
Rye, 12, 16, 23

Sabaean lane, 12, 50, 58
Safflower, 291–4, 350
Safflower carmine, 294
Saffron, 294
Safrole, 327
Sago, 128–9
Sann hemp, 276–8
Sassafras, 327
Sawarri oil, 302
Scion, 168, 342
Scutellum, 19
Semi-drying oils, 290, 298–301
    benniseed, 299–301
    cotton seed, 298
    maize, 298
    rai, 299
    rape seed, 298–9
    sesame, 299–301
    simsim, 299–301
    sunflower, 298
    tomato seed, 144
Seminal root, 17, 20
Senna, 72
Sesame, *see* benniseed
Sessile spikelet, 48
Seville orange, 168, 325
Shaddock, 167
Shallot, 151
Shallu sorghum, 45
Siberian millet, 54
Sieva bean, 89
Silk route, 12
Silver skin, coffee, 204
Simsim, *see* benniseed
Sisal, 12, 280–3, 352
    fibres, 283
    morphology, 281
    production, 352
    propagation, 281–2
Snap bean, 88
Snake melon, 138
Socotrine aloe, 227
Soft cane, 65
Soft date, 173
Soft fibre, *see* bast fibre
Solar radiation, 2
Solo pawpaw, 181
Sorghum, 4, 7, 8, 16, 25, 43–9, 346
    caryopsis, 49
    cultivars, 45
    flowering, 48
    improvement, 49
    morphology, 45–8
    origin, 12, 45
    production, 45, 346
Soyabean, 11, 85–7, 107, 297–8, 349
Souari oil, 302
Sour orange, 168, 325
Spanish-Valencia groundnuts, 81
Spadix, 123, 126, 315
Spathe, 123, 126, 150, 172, 308, 315
Spice, 230–51
    allspice, 243–4
    anise, 249
    caraway, 249
    cardamom, 233–5
    chilli pepper, 245–8
    cinnamon, 235–6
    clove, 240–3
    coriander, 249
    cumin, 249
    fenel, 249
    ginger, 230–2
    Indian arrowroot, 233
    mace, 244–5
    mango ginger, 233
    nutmeg, 244–5
    pepper, 236–8
    pimento, 243–4
    turmeric, 232–3
    vanilla, 238–40
    zeodary, 233
Spice islands, 13
Spikelet, 17
Staminode, 125, 175, 211–12, 232, 235
Star grass, 26
Stearic acid, 320
Stipel, 73
Stipule, 73
Stock, 168, 342
Strawberry guava, 187
Streptomycin, 195
Structural fibre, 253, 280–7, 352
    abacá, 284–6, 352

Structural fibre – *continued*
   bow-string hemp, 286
   cantala, 280
   coir, 287
   henequen, 283–4
   istle, 280
   ixtle, 280
   maguey, 280
   manila hemp, 284–6, 352
   Mauritius hemp, 286
   New Zealand hemp, 286
   palm fibres, 286–7
   piassava, 287
   raphia, 287
   sisal, 280–3, 352
Strychnine, 227
Stylo, 107
Sucrose, 64
Sudd, 56
Sugar beet, 64
Sugar cane, 12, 13, 25, 64–70, 354
   morphology, 68–9
   nobilization, 66–7
   origin, 66
   production, 65, 354
Summer squash, 134–7
Sunflower, 291, 298, 349–50
Sunn hemp, 107, 276–8
Surface hairs, 252, 253–68
Sweet lime, 167
Sweet orange, 164–6, 325, 355
Sweet potato, 4, 12, 111–16, 347
   cultivars, 116
   morphology, 112–14
   origin, 12, 112
   production, 116, 347
   tubers, 114–16
Sword bean, 102
Symbiotic bacteria, 75
Sympodial, 144, 260
Syncarp, 180
Synthetic variety, 43, 54

Tamarind, 72
Tangerine, 167, 355
Tannia, 126
Tannin, 72, 188
Tapioca, 119
Taro, 123, 347
Tassel, maize, 25, 34
Tea, 12, 195–202, 352
   history, 196
   morphology, 197–8
   origin, 12, 197
   production, 196, 352
   pruning, 200

   quality, 201
   seed oil, 202
Tef, 26, 61
Temperature, 2–3, 87
Tehuacan Valley, 10
Tenera palm, 318
Teosinte, 33
Tepary bean, 90
Terpenes, 323
Testa, 19
Tetrahydrocanabinol, 222
Tex, 268
Theobromine, 195
Thiamine, 33, 79
Tick clovers, 107
Tillers, 17
Tobacco, 142, 218–21, 353
   history, 218
   husbandry, 220
   morphology, 219
   processing, 221
   production, 353
Tomato, 12, 142–4, 299, 356
   morphology, 144
   origin, 12, 143
   production, 356
Toopee tambu, 127
Tossa jute, 146, 270
Totaquina, 222
Tree cotton, 255
Tree tomato, 142
Tristeza, 168
Tropical environment, 1–9
Tryptophan, 79
Tripsin inhibitors, 79
Tuba root, 108
Tubers, *see* roots and tubers
Tung, 294–7, 351
Turnip, 145

Water melon, 139, 299
West Indian lime, 167
Wheat, 12, 16, 346
White guinea yam, 121–3
White jute, 270
Winged bean, 105
Winged yam, 121
Winter squash, 134–7
Wild oat, 23

Xanthophyll, 180
Xenia, 40

Yams, 12, 119–23, 347
   morphology, 120–1
   origin, 12, 121

production, 119–20, 347
species, 121–3
taxonomy, 121
Yam bean, 108
Yard-long bean, 92, 93
Yellow bark, 223

Yellow guinea yam, 123
Ylang-ylang, 328

Zanzibar aloe, 227
Zein, 40
Zeodary, 233